# Complexity Metrics in Engineering Design

Matthias Kreimeyer • Udo Lindemann

# Complexity Metrics in Engineering Design

## Managing the Structure of Design Processes

Springer

Matthias Kreimeyer
Engineering Architecture -
Architecture
Dachauer Straße 667
80995 München
Germany
matthias.kreimeyer@man.eu

Udo Lindemann
TU München
Lehrstuhl für
Produktentwicklung
Boltzmannstr. 15
85748 Garching
Germany
udo.lindemann@pe.mw.tum.de

ISBN 978-3-642-20962-8 e-ISBN 978-3-642-20963-5
DOI 10.1007/978-3-642-20963-5
Springer Heidelberg Dordrecht London New York

Library of Congress Control Number: 2011933716



*Cover design*: eStudio Calamar S.L.

Printed on acid-free paper

Springer is part of Springer Science+Business Media (www.springer.com)

To Monica

# FOREWORD BY THE AUTHORS

To manage and improve engineering design processes in a methodical and systematic manner, an important issue that needs tackling is their analysis, interpretation and goal-oriented improvement. Although approaches for managing complex processes exist, a systematical, method-based analysis and improvement is still highly difficult.

To support the systematic and holistic analysis and improvement of an engineering design process, this book presents a measurement system that makes use of complexity metrics to embody various patterns of the interplay of a process' entities (e.g. tasks, documents, organizational units, etc.). These metrics are used to draw inferences about the process' behavior (e.g. timeliness, need for communication, forming of opinions, etc.). This way, knowledge about a process can be extracted from existing process models, or new process models can be structured systematically by addressing desirable patterns. This supports management in reducing the risks in process planning through better understanding how the structure of a process impacts the behavior of a process. Generating such a means of process analysis and management provides a major contribution both for academia and industry, especially for the improvement of large and complex engineering design processes. The metrics embody the foundations of network theory and the management of structural complexity to generate a practice-oriented application.

The metrics are supported by a meta-model for process modeling. The meta-model uses multiple-domain matrices, integrating existing process models across common domains and relationship types. The modeling method is enhanced with additional constructs of modeling that act as a bridging between existing dependency models and established process models.

Furthermore, the analysis approach is operationalized by a framework to select the metrics in accordance with the goals of the process analysis. To this end, the metrics are classified and allocated to the common goals of process analysis with regard to the structure of a process, producing eight different guidelines. To enable a flexible application, a modular set-up consisting of three steps is chosen: As a starting point, the strategic level is addressed using common goals of process analysis. Then, these goals are concretized by typical questions that can be posed in their context. Finally, these questions are answered using the metrics and parts of the meta-model.

The overall approach is detailed using three case studies from automotive development; on the one hand, the modeling and goal-oriented analysis of the body-in-white design of a premium class mid-size sedan is shown and, on the other hand, the detailed analysis and extraction of possible weak spots within the concept design, programming, and testing of electronic control units for an SUV is regarded. A third case study on general automotive design is used to illustrate all individual metrics. Results from the case studies point e.g. to particularly robust parts of the process, to critical structural bottle-necks, to the core drivers for iterations or rework, and, more generally, to potential weak spots in the overall structure of a process.

The book is based on a rigorous scientific approach to illustrate the origin of the presented results as well as the limits of their applicability. At the same time, much attention was put to illustrating all details in their industrial relevance to bridge the scientific approach and its industrial application.

Therefore, the book provides both academia and industry with new insights, above all a comprehensive collection of complexity metrics and their interpretation towards common problems in process management. It expands literature in structural complexity management into this field without limitation to its significance to other areas of application, as e.g. the design and management of complex product architectures.

At the same time, the research in this book was motivated to come "full circle", i.e. it was created in a way that both the modeling scheme, the analysis approach and the overall guidance about how both modeling and analysis work together were integrated in a more general framework. This endeavor thus guides the overall outline of the book. Nevertheless, none of these constituents to the solution are designed to be exclusive, so that, for example, the complexity metrics can also be based on models other than the multiple-domain matrices that are used here.

Munich, March 2011

Dr.-Ing. Matthias Kreimeyer

Prof. Dr.-Ing. Udo Lindemann

# THE RELEVANCE OF COMPLEXITY METRICS

Industry and scientific research require methods to support management of complex engineering development processes in a way that recognises and exploits the characteristics of their structural complexity. In particular, there is a pressing need to find ways to exploit the structural knowledge represented in process models in support of process management.

This research addresses this need through development of a systematic and scientifically rigorous yet practical approach to modelling and analysing processes. The approach is clearly demonstrated by application to different case studies of automotive design. It thereby presents a significant contribution to practitioners wishing to understand and improve their complex processes. It also fills a major gap in the scientific literature by further developing and systematising the emerging area of structural complexity management in engineering design.

The empirical background of this research highlights the complexity of engineering design and clearly outlines the problem that, even when models of the activities, information flows, resources etc. are available, such models are sufficiently complex that problem areas cannot be identified by inspection. The concept of structural analysis serves here as a promising means to address this by identifying potential 'problem areas' within a complex process.

The main body of this research considers a comprehensive state of the art drawn from the fields of system theory, graph-theory, matrix-based methods for structural complexity management, network theory, process management and software engineering. Contributions from these disciplines are combined, using an established approach of system analysis, enhanced with a clear goal-orientation. The solution is therefore based on three constituents:

An enhanced method of process modelling is first introduced that encompasses a means of combining existing process models. This modelling scheme is, above all, constructed in a way that it serves as a means of making the use of complexity metrics compatible with existing models that, similarly, represent dependencies in a system.

Based thereon, 52 complexity metrics are explained to analyze a process. The metrics address the clear and pressing need for a rigorous approach to formalise and prepare the large volumes of data required for process analysis in many practical situations, as it is often the case with complex systems. At the same time, the abstract approach is illustrated with extensive tables to support the interpretation of any findings. Above all, however, the substantial set of 52 metrics should form a major resource for further research in structural complexity management for engineering design.

Third, both modeling and analysis approach are combined offering a goal-oriented conduction of process analysis. This completes the description of the new approach and convincingly supports the technical discussion of previous chapters by showing how the approach can be linked to the real challenges faced in industry.

Supported by three studies, the book clearly illustrates how the method of earlier chapters can be applied. The practical application of structural analysis to understand and improve complex processes is clearly demonstrated and critically reflected upon.

Cambridge, December 2010 — Professor P. John Clarkson

Engineering Design Centre, University of Cambridge

# BACKGROUND OF THIS RESEARCH

This work results from a series of research projects on the management of structural complexity at the Institute of Product Development at the Technische Universität München. Based on a rigorous research approach as a basis to the systematic obtainment of the results presented in this research, the authors' involvement in numerous research projects provides an experiential basis to design a methodology that fulfils all requirements.

The authors were, among other activities, involved in a major study to identify and conceptualize a possible reorganization of the development departments involved in the design and simulation of the body-in-white of a large German automotive manufacturer[1]. In fact, this project provided the motivation for the research presented in this research, as the initial study at the company showed that almost all problems were interconnected, and the systematic determination of improvement potential, while only "reorganizing" the existing structure, appeared as an almost insurmountable problem.

The authors were also involved in various projects to improve process management in engineering design. At a strategic level, a management framework based on common management models was developed in cooperation with a management consulting firm to better guide the development of automotive safety features [KREIMEYER et al. 2006d]. At the operational level, a project to set up guidelines to access the various committees inside a large automotive manufacturer was run to improve decision making; to do so, the overall structure of the various decision processes was analyzed to obtain specific routes through the various decision-making bodies. Another project was carried out to research the potential and implementation of architectural standards across all models of a premium class automotive manufacturer; here, the goal was to establish all necessary processes to implement the definition of sustainable architectural standards, derive individual models, maintain and update them, and integrate future technologies in a cost-efficient manner.

Part of the research presented in this research was done in collaboration with another large German automotive manufacturer [KÖNIG et al. 2008] [KREIMEYER et al. 2008d]. In combination with the data available from the reorganization project described above, these two comparable projects provided ample empirical data and relevant access to industry to guarantee an approach both pragmatic and relevant. A third set of empirical data was available publicly [Braha & Bar-Yam 2004].

---

[1] For an overview see [DEUBZER et al. 2007]; a problem description is given in [KREIMEYER et al. 2005] and [KREIMEYER et al. 2007b]; the core concept is detailed in [HERFELD et al. 2006] and [KREIMEYER et al. 2006a] and completed in [KREIMEYER et al. 2007a].

At the same time, two large studies of engineering design were conducted. One study focused on the collaboration patterns in the "digital factory planning" in automotive companies and their ties to the engineering design and supply industry [KEIJZER et al. 2007]. A second study was carried out as a benchmark comparing the engineering divisions of three firms producing diesel engines of various sizes (400 to 100.000 horsepower). Both studies generated a broad picture of the necessities and particularities of engineering design.

Furthermore, the authors have been active for a long time in research on structural complexity management. As co-founders of the research group "Systems Engineering" at the Institute of Product Development[2], the authors repeatedly organized the International Dependency and Structure Modeling (DSM) Conference[3], which serves as an international platform for practitioners and researchers on structural complexity management. The authors were also co-foundera of the Special Interest Group "Managing Structural Complexity" within the Design Society, and still act as chairs of this Special Interest Group[4]. The authors also re-launched the web-portal www.DSMweb.org[5] to provide the international research community on structural complexity as well as interested practitioners with a comprehensive set of material, publications, and tutorials to facilitate the application of methods to manage structural complexity. At the same time, the authors were directly involved in re-launching the Special Interest Group on "Modeling and Management of Engineering Processes (MMEP)" within the Design Society[6].

Ultimately, much of this experience resulted in the successful application of the Collaborative Research Center SFB 768 on "Managing cycles in innovation processes–integrated development of product service systems based on technical products". Within this research center, the authors led the research group on "Development of models and processes" and supported both project A2 "Modellierung und Analyse disziplinen-übergreifender Zusammenhänge" ("Modeling and Analysis of trans-disciplinary Relationships") and B1 "Prozessplanung für die zyklengerechte Lösungsentwicklung" ("Process Planning for the Efficient Development of Product Service Systems").

[2] see http://www.pe.mw.tum.de, viewed on 20 February 2009

[3] see http://www.dsm-conference.org, viewed on 20 February 2009

[4] see http://www.designsociety.org/index.php?menu=35&action=21, viewed on 20 February 2009

[5] see http://www.DSMweb.org, viewed on 20 February 2009

[6] see http://www-edc.eng.cam.ac.uk/mmep, viewed on 20 February 2009

## Acknowledgements

This book would not have been possible without the help of numerous colleagues and friends. My gratitude extends to all of them. Above all, the continuous support of Udo Lindemann and John Clarkson brought this book to its final state, and their guidance and input provided the most important foundations.

An important tribute goes to my colleagues at the university, many of whom have become close friends through our close collaboration. In particular the support of my friends Frank Deubzer and Ulrich Herfeld need to be mentioned. I have gained many valuable ideas and much energy during our ongoing discussions and our fruitful collaboration in different contexts. Maik Maurer, Thomas Braun, and Wieland Biedermann have helped me develop the ideas of complexity management much further, both at the institute and at Teseon GmbH. Finally, I want to thank those colleagues who have helped me with both proofreading and providing valuable feedback. This, in particular, helped finalizing all the little details, especially Stefan Langer, Christoph Baumberger, Ralf Stetter, Udo Pulm, and Willem Keijzer. Of course, the many helping hands deserve my gratitude, among them Nikolas Bradford, Matthias Gürtler, and Caspar Sunder-Plassmann. Similarly, the support of my colleagues at MAN Nutzfahrzeuge, especially Wilhelm Heintze, enabled me to make this book possible.

In addition, special thanks goes to the international research community, especially Tyson Browning, Ed Crawley, Mike Danilovic, Steven Eppinger, Georges Fadel, Andrew Kusiak, Anja Maier, Don Steward, and David Wynn. Their input and ideas and their continuous provision of new references helped me to complete many of my concepts, whose full development would otherwise not have been possible.

Lastly and mostly, I am deeply grateful to my parents, to Monica, and to my friends for supporting me all the while. Their patience and comprehension eave me the strength necessary for the finishing touch.

Munich, December 2010 Matthias Kreimeyer

## CONTENTS

## LIST OF ABBREVIATIONS

| | |
|---|---|
| ARIS | Architecture of Integrated Information Systems |
| BSC | Balanced Scorecard |
| BPMN | Business Process Modeling Notation |
| CAD | Computer Aided Design |
| CAE | Computer Aided Engineering |
| CFC | Control-Flow Complexity |
| DMM | Domain Mapping Matrix |
| DSM | Design Structure Matrix |
| EBB | Elementary Building Block |
| eEPC | Extended Event-driven Process Chain |
| EPC | Event-driven Process Chain |
| GQM | Goal-Question-Metric |
| HOQ | House of Quality |
| IDEF | Integrated Definition Method |
| IT | Information Technology |
| IUM | Business-Process Modeling (Integrierte Unternehmensmodellierung) |
| MCC | McCabe Complexity |
| MDM | Multiple-Domain Matrix |
| NVH | Noise Vibration Harshness: domain of simulation |
| oEPC | Object-oriented Event-driven Process Chain |
| OMEGA | Objektorientierte Methode für die Geschäftsprozessmodellierung und –analyse |
| PERT | Program Evaluation and Review Technique |
| PMM | Process Module Methodology |
| QFD | Quality Function Deployment |
| SADT | Structured Analysis and Design Technique |
| S-GQM | Structural Goal Question Metric |
| SMS | Structural Measurement System |
| SPA | Structural Process Architecture |
| UML | Unified Modeling Language |
| YAWL | Yet Another Workflow Language |

## List of Software Tools

| | |
|---|---|
| ABAQUS | Solver for non-linear calculus by ABAQUS Inc. |
| ANIMATOR | Postprocessor by Dassault Systèmes |
| ANSA | Preprocessor by BETA CAE Systems S.A. |
| CATIA | Commercial CAD System by Dassault Systèmes |
| COVISE | COllaborative VIsualization and Simulation Environment: postprocessor for computational flow dynamics simulations by VirCinity GmbH |
| ENSIGHT | Computational flow dynamics postprocessor tool by CEI Corp. |
| EVA | EVAluator: Postprocessor software |
| FALANCS | Stress and strain simulation tool by LMS International |
| FEMFAT | Finite Element Method - FATigue: Fatigue simulation tool by Engineering Center Steyr GmbH & Co KG |
| GS Mesher | Surface and Solid Mesher by MacNeal-Schwendler Corporation |
| HYPERMESH | Preprocessor by Altair Engineering GmbH |
| ICEM | Parametric CAD surface modeler by ICEM Ltd. |
| LOOMEO | Software to support Multiple-Domain Matrices by Teseon GmbH |
| MEDINA | Pre- and postprocessor by T-Systems Enterprise Services GmbH |
| MS OFFICE | Office Product Range by Microsoft Corp. |
| NASTRAN | NAsa STRuctural Analysis: Solver by MacNeal-Schwendler Corporation |
| PAM CRASH / PAM VIEW | Solver and viewer for crash simulation by ESI Group |
| PATRAN | Pre- and postprocessor by MacNeal-Schwendler Corporation |
| PERMAS | Solver for linear calculus by Intes GmbH |
| POWERFLOW | Computational flow dynamics simulation tool by Exa Corp. |
| SFE Concept | CAD program with integrated preprocessor and mesher by SFE GmbH |
| STAR CD | Simulation tool for fluid flow, heat transfer and stress by CD-adapco |

# 1. Complex processes in engineering design

## 1.1 Preface

As globalization increases, the time to market continues to decrease and customers can choose among a variety of suppliers and demand better prices, better quality, and more and more customized products [COOPER & EDGETT 2005].

Companies have to cope with this trend, especially in their engineering departments [SPATH et al. 2001]. To do so, several strategies have become available, which continue to evolve over time. The claim to operate in a "lean" manner, for example, has recently found its way from the factory floor to the engineering department [GRAEBSCH et al. 2007]. One of the constants to raising efficiency for many decades now has been process management [SMITH 1996].

The aim of process management is a better definition and control of the processes[7] with respect to the "three sacred cows": "time, quality and budget" [KNEUPER 2007] [PMI 2003]. Process management works under the assumption that a better definition and control of a process enables a manager to know more about its price, duration and possible risks [DINSMORE & CABANIS-BREWIN 2006]. These goals, however, demand an in-depth knowledge of the processes that govern a company.

Process management now includes many facets, for example, scheduling, communication, resource management, and others [BECKER et al. 2005], and has developed many different models, methods, and tools. This is not only the case in general business process management[8], but also in the management of engineering design processes[9] [HALES & GOOCH 2004] [CLARKSON & ECKERT 2005]. In fact, many approaches from the management of business processes remain valid in engineering design, yet their application is complicated by the fact that creativity, moving targets, the management of uncertainties and the limited ability to plan any generation of knowledge during the process have to be considered [HATCHUEL & WEIL 2003] [VAJNA 2005, p. 371]. These specific facts — together with the need for a detailed division of labor — have made it necessary to incorporate many points of synchronization in any engineering design process, thus causing all entities in a process to be tightly interwoven [COATES et al. 2000]. This is especially the case for products that are of an interdisciplinary character, e.g., mechatronic devices or product-service systems.

---

[7] The term process, being central to this research, is defined in section 2.2.1; in this research, a process involves the processing of tasks, including their inputs and outputs, as well as the necessary organizational aspects, such as the company organization, resources, and milestones.

[8] Compare Section 2.2.1 for a closer review of business process management

[9] An overview of current practices and strategies is given, e.g., by LINDEMANN, comparing different strategies and problem-solving models [LINDEMANN 2007, pp. 33-35]; a rather formal approach is given by BALDWIN, visualizing engineering design as a transformation of input parameters into output parameters [BALDWIN & CLARK 2000].

As a commonly practiced strategy to raise efficiency and shorten process lead times through parallelization in engineering design, concurrent engineering [BULLINGER & WARSCHAT 1996] has brought with it an intensified need for a networked engineering design process. In concurrent engineering processes, more than ever before, tasks are not simply put into sequence, with one task waiting for the preceding tasks to finish, but they are processed in parallel, and interlinked to be synchronized on the go to reduce the cycle time while the individual artifacts within the process are gradually concretized. This has created an even greater need for densely networked processes, as currently even partial results have to be checked for their mutual dependencies [KREIMEYER et al. 2008c].

A deeper look into engineering design processes reveals that such networks of a process exist on many levels. Not only are the tasks interlinked [EPPINGER 2001], but also the documents [ILIE et al. 2008], the IT Systems [BURR et al. 2003], and, above all, the protagonists of the process, i.e., engineers and management who communicate[10] with each other [SCHÖN 1983, p. 76]. In fact, HERFELD concludes that the management of an engineering design process necessitates a balanced improvement and the mutual calibration of all involved perspectives that are relevant [HERFELD 2007, p. 100][11].

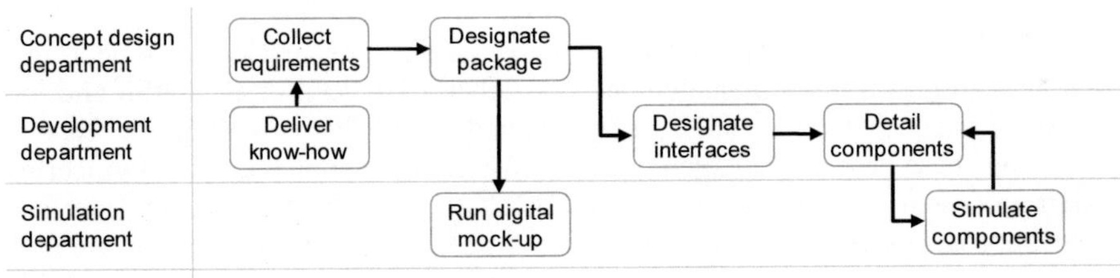


*Figure 1-1: Example of a process model*

The methods used in process management make these different networks explicit as process models or process maps. In fact, "business process models are an important knowledge source for managerial decision making" [DALAL et al. 2004]. Figure 1-1 shows a process model in swimlane notation: Organizational units are represented as lanes, where those tasks that a unit is responsible for are shown as a flow chart. Figure 1-2 shows a complex model of a "real-world" automotive design process (see also section 1.2), equally organized as a swimlane model. Such models depict a common time-based plot of the processes taking

[10] An overview of the role of human communication in engineering design can be found in [MAIER 2007, p. 28].

[11] HERFELD concludes that the basic views necessary for a well-balanced process improvement are product architecture (requirements, functions, and components), human actors (in his case simulation and embodiment design engineers in an automotive company), information (geometry models, simulation models, communication in general), tools (3D-design tools, finite element simulation, data management), and the process per se (synchronization points, milestones, availabilities of resources).

place in a company. Usually, the models represent tasks, actors, data objects and supporting resources, and they thus represent a partial view of the many network perspectives there are to a process. These networks are commonly plotted out as "boxes and arrows", i.e., qualitative models whose elements are interdependent or associated. Often, detailed information on process behavior, e.g., runtimes or the probability of a decision, is not available.

So far, there has been little work to tap the full extent of knowledge embedded in these process maps [ZAKARIAN & KUSIAK 2000] [MENDLING 2008, p. 103]. While some quantitative methods are available, e.g., the Critical Path Method (see appendix 10.1.12), little work has been done regarding the possible meaning of patterns that arise in the structure of a process. Most of the research available so far concentrates on the role of iterations in engineering design (e.g., [BADKE-SCHAUB & GEHRLICHER 2003] [WYNN et al. 2007]).

The research results presented here bridge this gap by showing how the structure of a process relates to its behavior, as suggested, for example, in [MALIK 2003, p. 93]. They present an approach that is based on analyzing the interplay within the network of entities of a process. From this network and its characteristics, inferences about the behavior of the process are drawn — for instance which actors are "central" (i.e., a characteristic of the structure of the process organization), and hence with whom effective communication might be critical to process performance or predictability (i.e., the behavior of the process). To do so, complexity measures are applied to obtain a condensed view of the characteristics of processes comprised of many entities. Overall, the goal is, therefore, to extract the knowledge about the behavior embedded in process models to deduce implications for their improvement.

In general, processes have become increasingly complex due to rising product complexity and reduced process lead times. This increased complexity needs to be managed, and understanding the specific aspects of complexity can reduce risks and better define and control a process. In this research, complexity metrics are, therefore, developed that support the structured analysis of a process.

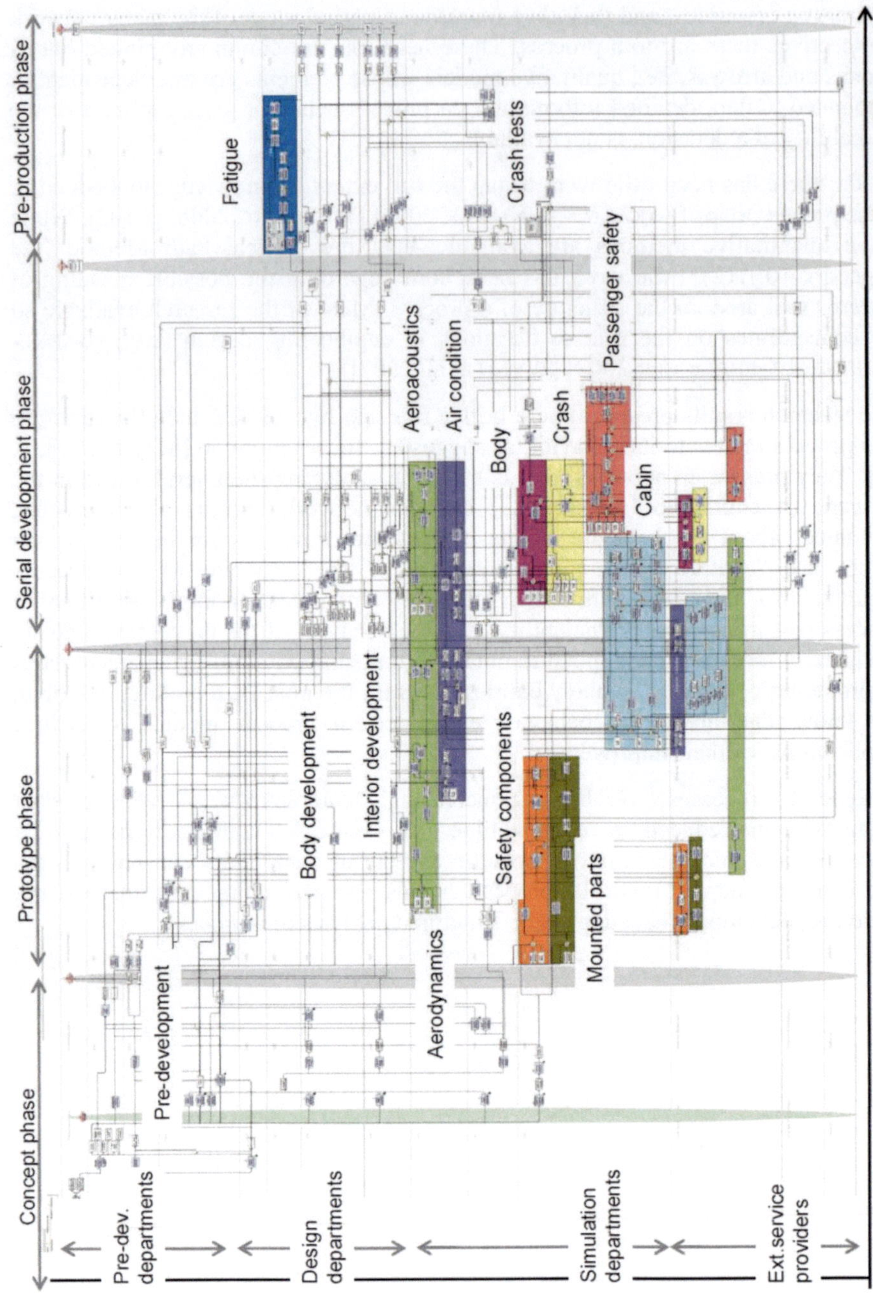


*Figure 1-2: Detailed overview of the process of developing the body-in-white of a premium class sedan*

## 1.2 A practical application: A design process at Audi AG

Before introducing the research approach and the solution design, an example of a process analysis in automotive body design is explored. This example highlights the different aspects of process analysis as commonly encountered in process management. From it, a problem description, the relevant hypotheses, and the overall goal of the research are drawn to illustrate the background of this research.

The process represents a real development process as practiced at Audi AG, Germany. AUDI is a brand of the Volkswagen Group, manufacturing passenger cars and SUVs for the premium segments of nearly all international markets.

First, the context of the actual process is described. Then, the model representing the process is shown. As this research focuses on how the resulting process chart can be systematically analyzed, typical analyses are run to show how the complexity of the process chart can productively and systematically be analyzed.

Figure 1-2 provides an example of a process in automotive body design. The process chart was set up as part of the initial work in a process reengineering project. Its scope is the interaction between embodiment design and simulation departments, comprising 134 different business objects that are processed concurrently by 160 different tasks. The process involves four major phases, 14 organizational units (three of them external service providers), and 27 different IT systems. There are 54 major decisions modeled as OR-decision (for reasons of simplicity, management agreed not to differentiate XOR and OR in the model). The process is structured as a swimlane model for each organizational unit along an implicit left-to-right time axis that is not to scale. Nine sub-processes take place, and they are colored according to their specific focus (different simulations are run); most of them take place concurrently in two organizational units (internally and at an external service provider).

### 1.2.1 Description of the process

The process focuses on the interaction of all embodiment design engineers and simulation engineers, both internally and externally, who are involved in developing the body-in-white (Figure 1-3) for serial production. The body-in-white includes the car body as well as doors, hoods, lids without further components (axes, motor) and trim (windshields, seats, upholstery, electronics, etc.). Simulation only considers, in this example, vibration, deformation (such as crash), and air flow load cases for the body-in-white. Overall, the body-in-white is a highly complex product. Here, approximately 400 components and 130 different load cases are used.

Developing such a complex product demands a complex process organization. The process starts with the official launch of the project[12]. It ends with the start of

[12] It is hard to differentiate in such large development projects between whether the process that takes place for approximately five years is a project of repetitive character (for every new model and its derivates) or a process. Here, the term "process" is chosen to emphasize the fact that the repetition of the process taking place can lead to corporate learning about the process to raise its efficiency.

detailed production preparation after testing and pre-series are finished, i.e., when design and simulation are no longer involved to a major extent. Thus, the process starts with the business object "customer need", delivered by the marketing department; it finishes after all components have reached the release level "ready for purchasing".

*Figure 1-3: Body-in-white of a premium class sedan*[13] *[VOLKSWAGEN AG 2007] (p. 143 & p. 147)*

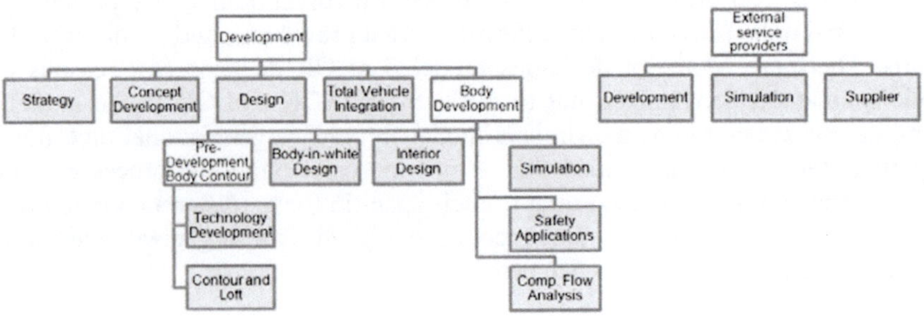


*Figure 1-4: Organizational structure of the organizational units involved in the process (shaded)*

As Figure 1-4 shows, 14 organizational units are involved (shaded). The strategy department delivers the first input; the concept department and the design department prepare the initial concepts (two to four) with the support of the body development department, including the technology design, the body-in-white department, the simulation departments, and others. After a decision has been made about which concept is detailed for serial development (generally, elements of all different concept designs are combined into a final design), the latter three departments complete the overall design, which is progressively tested and refined until the "Final Design Freeze" milestone.

All of this is supported by various external services in simulation and development, and by suppliers who not only deliver the final components but also

[13] For nondisclosure reasons, the real nature of the case study is withheld; the case study focuses on the design process of a mid-size sedan derivative at Audi AG, Germany.

support development as an integrated partner. In the process chart, these service providers are modeled only as archetypical organizational units.

The process is supported by a complex organizational structure. The development is done jointly by the pre-development departments, the body design department, and the interior development department. During the concept phase, a concept for the body and for the interior is prepared; it involves a technology model that collects all chains of dimensioning and all technical aspects of the car (e.g., the envelope of the wheels moving under steering and suspension movement), and the contour of the car (in collaboration with the designers). The technology model is initially set up and constantly kept up-to-date to serve as a point of reference for all other activities. During this early phase, the other departments only act as support to feed know-how on serial production into the concept phase. At all time the body design department tries to work ahead of the interior design to ensure a good fit. As the level of detail rises, the development task is transferred to the respective development departments (body-in-white, interior, safety applications) and their development teams (about 800 engineers). This transfer of responsibility takes place at the end of the concept phase.

While limited simulations and estimations take place in the early phases, detailed simulation only occurs after the concept is released at the "project decision" milestone. At first, a scaled model of a predecessor is used, and little by little information is transferred from the development departments to the simulation departments with growing concretization. Each of these iterations takes about four months. The more detailed the models are, the better simulations that can be run; thus, in the beginning only worst-case scenarios are simulated for core load cases, whereas variant models of all components and their functionalities can be reviewed later. In the process model, the growing degree of integration of the simulation departments into the development process can easily be spotted in the lower half of Figure 1-2 (the different types of load cases are marked as colored sub-processes), where there are more and more tasks shown over the progression of time. After initial structural simulations, the air flow is optimized, including air conditioning. At the same time, using the same basic simulation model, vibration (i.e., eigenfrequencies, noise, and harshness) and deformation (crash, passenger, and pedestrian protection) are simulated. Most geometry and simulated models are, in fact, prepared by external partners (so-called "extended workbench"), coordinated by the internal development engineers. When the first components are available about halfway through the prototype phase, these are tested to validate the simulation results and to test load cases that cannot be simulated with sufficient quality (e.g., fatigue). The prototype phase concludes when a series of full prototypes has been validated and a final concept is ready for serial development at the "vehicle concept decision" milestone.

After the final vehicle is decided on, all components are prepared for serial production during the serial development phase. Again, many simulations and tests are run to ascertain the properties of all components and assemblies. At the end of that phase, the design is "frozen", i.e., no more changes to the geometry are permissible, as the production tools are about to be ordered, which represent a large investment and, therefore, should not be further altered.

During the final (pre-production) phase of the development process, typically, endurance-related issues are solved, and the final release is prepared to enable purchasing production equipment. Figure 1-5 provides an overview of the relevant phases and the milestones that start and end a phase; in fact, the concept phase consists of two phases, with an intermediate review. In the process these phases are further detailed to work packages of two to three weeks' size.

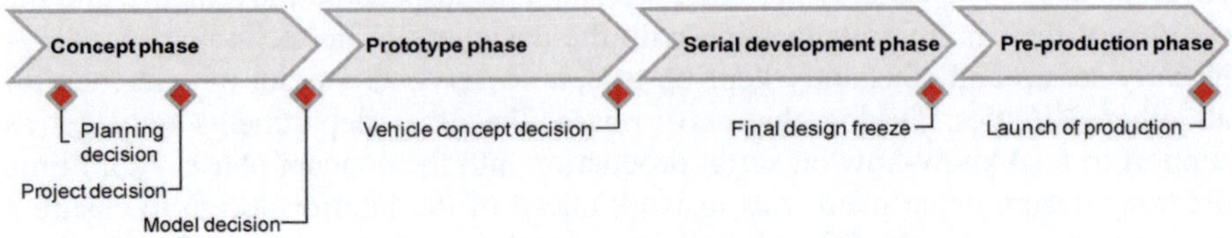


*Figure 1-5: Overview of major milestones and phases of process in focus*

The tasks of the engineers are executed using complex software tools. Most tasks are supported by specific tools, which are strongly interrelated both via mutual interfaces between them and across the exchange of information throughout the process. For each, different models and other information objects are required as input, and most tools are linked to one another to some extent by generating output data that is further processed. Figure 1-6 lists the occurrence of different systems in the process, showing the variety of different systems in use.

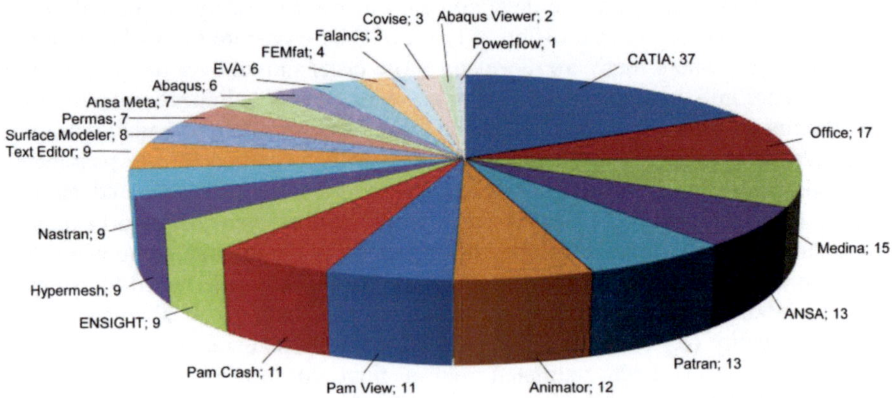


*Figure 1-6: Employment of different IT systems in the process (name of system & occurrence)*

## 1.2.2 Modeling the process as an EPC process chart

The model is set up as an EPC model (see appendix 10.1.1), following common rules of modeling, as a network of alternating tasks and business objects[14]. For all tasks, the supporting IT systems are added as attributes, and the responsible organizational units are allocated using horizontal swimlanes. Milestones appear as columns in the model, collecting necessary business objects.

The model was designed using the ARIS Toolset. A printed map with font size 7 reaches a printout size of A0 (841mm x 1189mm); it is thus barely readable. At the given level of detail ("control level," see Figure 1-7), it is impossible to recognize all 1089 relations (between any two entities) that are modeled. The model is set up at a medium level of detail according to Audi AG's specification of the "control" level.

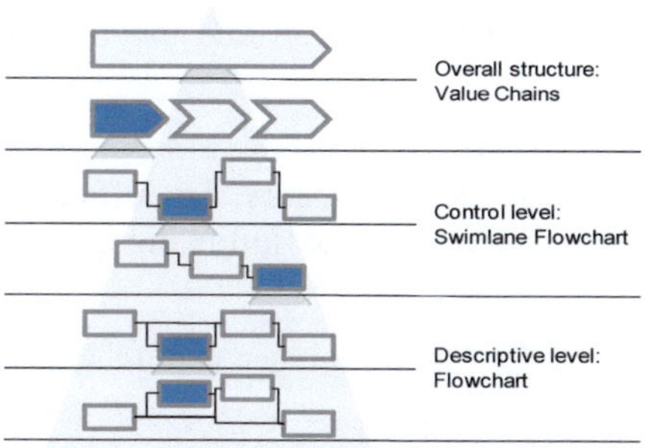


*Figure 1-7: Levels of detail of process charts*

## 1.2.3 Deficits when analyzing the process chart using existing methods

A process chart is almost always the starting point to analyze and improve a process. The primary reason to build such a process model is to gain an overview of entities that are relevant to the improvement project. In the context of the process chart from Figure 1-2, the intent was to improve collaboration between the different departments, i.e., their information exchange and their interfaces. To do so, an overview of all business objects and their purpose was needed. More generally, a process chart satisfies different demands. Above all, it helps gain an overview of a process, i.e., what entities are involved to what extent. As such, process charts are frequently used to document the organization, to reorganize a company by focusing on its processes, to facilitate continuous process management, to support process control, to certify a company according to, for example, ISO 9000 standards, to allow benchmarking with other divisions or companies, to manage knowledge embedded in the process, to select and customize enterprise resource planning tools, to introduce new software and

---

[14] EPC prescribes an alternation of functions and events, with business objects as a third kind of entity; however, the company used a simplified modeling scheme.

workflow management, or to simulate processes to better anticipate risks [BECKER et al. 2005, p. 45].

However, addressing these issues directly by looking at a large process model is almost impossible, as a quick glance at Figure 1-2 proves. In fact, many of the goals above are only served implicitly by a process chart, which is often further analyzed to access the knowledge embedded in the process.

In practice, the problems are less abstract: In the case of the process study presented here, a short survey[15] among the engineers participating in the process showed the following core aspects necessary for an efficient process:

- Transparent definition of responsibilities
- Accessibility of tools and methods throughout the process
- Fast transfer of information
- Constant harmonization of business objects among design and simulation engineers

A study on the quality of communication was also carried out for the same process [MAIER et al. 2008]. It pointed to additional key elements for efficient communication in the process in question. In particular, the following points turned out to be most important:

- Understanding the mutual information needs: Each engineer should know about what information is available where.
- Orientation and transparency: Each engineer should be able to locate his or her own task in the process and understand its relevance for the overall process.
- Reflection about interaction: Each engineer should be able to consider how he or she can improve collaboration by communicating in the network.

Unfortunately, a lack of available support for the management of complex processes prevented the identification of key entities and their mutual relations to access the root causes and possible drivers of the process. While the process chart shown (Figure 1-2) was available, no inferences could be drawn from this structure to detail the above goals with the knowledge represented in the process model. It was simply "too complex". At the same time, investing more modeling effort, for example to simulate the process, served little purpose, as no systematic basis existed detailing the model [KREIMEYER ET AL. 2010].

Available process management tools offer limited support of a systematic analysis. Media breaks, for example, can be detected, and lists of elements, such as the direct input that is necessary to reach a milestone, can be provided. More complex analyses, for example, the determination of a critical path to finish the process in time, were not possible, because the extensive data on durations of all tasks was not available. Equally, it might be intuitive to count those entities that

[15] For detailed results, see [KREIMEYER et al. 2005] and [KREIMEYER et al. 2006b].

are modeled[16] the most often (body structure: 11 times; cockpit: 8 times; mounted parts: 7 times). However, this still does not indicate how they are embedded in the process. As such, only a few first impressions could be collected, and it is hard to determine the core business objects of the information exchange.

A major driver of resource consumption is iteration. However, no direct indication of where these iterations would typically show up could be detected. While there are very few short iterations (i.e., direct rework or improvement among a few entities), those of medium length, involving 15 to 25 tasks (and thus business objects) play a major role. This is partly in the nature of the process chart, as only the interaction between the different tasks is modeled, but not, for example, the rework that needs to be done to each task. However, in order to improve the collaboration across departmental interfaces, such "longer" iterations are of interest, as they dominate the collaboration. Table 1-1 and Table 1-2 show what business objects and what links between tasks and business objects appear the most often. Whereas the technology model is among the top four objects, it is much less involved in controlling the iterations than the crash results, even though the technology model is designed as a central means of coordinating all design efforts by collecting all relevant core measures and information.

*Table 1-1: Occurrence of business objects in iterations*

| Business object | Occurrence in iterations |
|---|---|
| Crash simulation results | 85371 |
| Body simulation results | 70004 |
| Passenger safety simulation results | 66600 |
| Technology model | 62208 |

*Table 1-2: Occurrence of control influences of tasks on business objects within iterations*

| Control influence | Occurrence in iterations |
|---|---|
| Coordinate crash simulation → body simulation results | 69839 |
| Set up model for crash simulation → simulation model for crash | 53904 |
| Coordinate body simulation → body simulation results | 50070 |
| Coordinate passenger safety simulation → passenger safety simulation results | 46916 |
| Design concept of cockpit → concept of cockpit | 33952 |

[16] EPC uses an object-oriented modeling concept. It is thus possible (and quite common) that an object, such as a task or a business object, is instantiated several times across the process model, e.g., to represent iterations.

In the case of the crash simulation results, this could mean that a core team that manages the information on this data has a strong influence on how smoothly the process runs. If the team were able to influence the crash results in a way that all partners agreed on, an early exchange of (possibly immature) information, and thus unwanted iterations, might be prevented.

Equally, tasks that lead to the important business objects appear quite often, although they are a small percentage of all the number of iterations; the results of the tasks listed in the tables are the most commonly reworked ones across department frontiers. Again, the controlling influence of the crash simulation is easily visible. However, the setup of the technology model is not among the top five elements, because that technology model is drawn from many sources and, thus, its formal set-up is not as relevant as other tasks. At the same time, the concept design of the cockpit also drives iterations.

To further estimate the impact of the crash simulation results, this object was turned into the root node of a hierarchy representing the "avalanche" of subsequent tasks and business objects that depend on it to better visualize its impact. During the project, this was done manually for all tasks, as no algorithmic support was available. Figure 1-8 shows the 101 entities that can be reached directly or across other intermediate entities from the root node at the top. As the figure shows, the subsequent tasks and business objects are reached via different levels; yet, all subsequent entities are dependent on this initial entity.

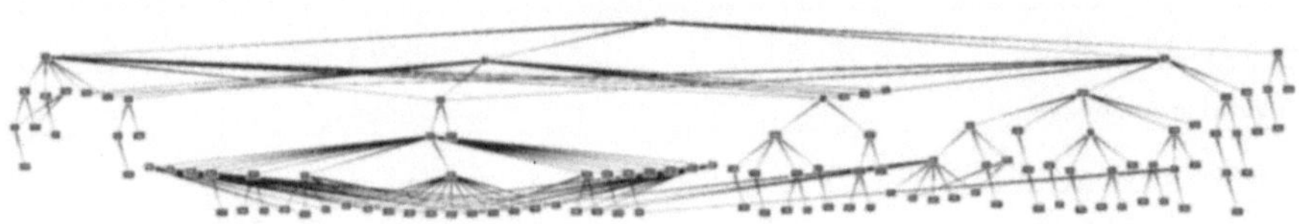

*Figure 1-8: Hierarchy with crash simulation results as root node (generated with Loomeo)*

## 1.2.4 Conclusion: Systematic analysis of a process chart

The complexity[17] of a design process spawns not simply from the tasks the process consists of, but other entities that are relevant to the process and contribute to the fact that an engineering design process is perceived as complex. In fact, a complex product demands a complex process and a complex enterprise organization; yet, these different views are not independent of each other [EPPINGER 2001]. This complexity is mostly due to the different types of entities, their large number, and the relationships among them that often create knock-on changes. Yet, there is no methodical support to systematically analyze the relationships among the different entities in a complex process at such a high level of process management.

[17] The term "complexity" is reviewed in detail in section 2.1.1.

As the short case study shows, a systematic in-depth analysis of a process is, thus, a complex undertaking. Yet, there are several characteristics of the structure of the process, such as tasks or business objects, which are central to information exchange and govern the overall timeliness of the process or necessitate a great amount of communication to ensure coherent results. For effective process improvement, a comprehensive overview of the characteristics of a process is first necessary which prioritizes an in-depth analysis of possible improvement measures. At the same time, understanding the governing structural patterns helps reduce risks in process management.

In fact, most results of the project that followed the case study shown above were based on the few results that were just outlined. Whereas these are based on the experience of the engineers who participated in setting up the model, the results are erratic, nevertheless, as they were not obtained in a systematic manner.

## 1.3 The need for systematic analysis in practice

As shown, no systematic support of analyzing a process in terms of its structure is available yet, which allows a high-level analysis to determine its possible weak spots and to prioritize further investigative efforts. To represent processes[18] at a given level, process models are used. These usually take shape as process charts, i.e., large maps that represent the process in a flow-oriented manner. These maps are commonly found in any company, and using them efficiently to improve processes by using the knowledge represented in these charts is still very difficult.

Processes serve, in the context of this research, as a form of dependency modeling with specific semantics. Being very common in industry, they therefore are an adequate means of illustrating what shape dependency models can take in industry. The complexity metrics shown later are, above all, tailored to generically work with such dependency models.

### 1.3.1 The problem: Systematic analysis of a process chart

As the example showed, extracting inferences about the process behavior from a process map is a difficult issue that has not been methodically supported so far. Here, the complexity of the process model represents the actual barrier (Figure 1-9) to gaining an in-depth understanding of the process: Resolving the problem at an even more abstract level can possibly aid understanding the chart better, and it will later be shown that structural characteristics and complexity metrics, for example, can support the aggregated characterization of a structure. However, it is equally necessary to find the way back from abstraction to a level of application to make the methodology suitable for use in industrial practice. Therefore, a complete solution to the problem requires not only analysis at an abstract level, i.e., the upward path of the model shown in Figure 1-9, but the interpretation also needs to be methodically supported, i.e., the downward path.

[18] The terminology specific to this research is only introduced here; details are given in section 2.1.1.

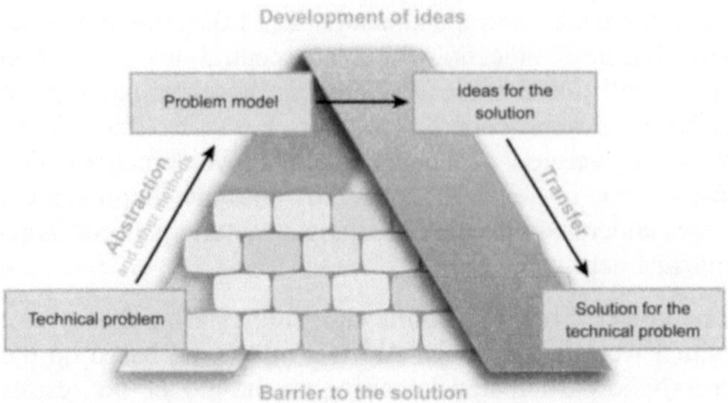


*Figure 1-9: Solving technical problems via abstraction, based on [LINDEMANN 2007, p. 29]*

The knowledge of the process is, to a large extent, found in the interplay of tasks, business objects, organizational units, and other entities. This interplay forms a network-like structure of all entities that are involved in the process; in the introductory case study, not only were tasks and business objects tightly coupled, but also, for example, the different departments and the supporting IT systems. This leads to the basic assumption that the intentional design of this network-like system of a process governs its behavior, and that if the process' entities were coupled differently, the process would exhibit a different behavior.

This interplay takes shape — at a finer level of detail — as certain structural patterns that are referred to as "structural characteristics"; these patterns are the basic constellations of a few entities and their relations with other entities. A process is, therefore, assembled from many of these patterns, and the literature (e.g., [BAR-YAM 1997]) shows that these patterns embody small units of behavior that point to the behavior of the overall process. Making the patterns accessible, therefore, means making the knowledge embedded in the process accessible.

Describing this structure[19], however, is not simple. There are many perspectives to a process, such as tasks, business objects, people, or IT systems. These perspectives are not independent of each other and thus contribute to the patterns. At the same time, the process behavior is only generated by the interplay of all perspectives. To gain a complete overview of a process, therefore, it is necessary to review all entities and their involvement in every possible pattern. However, improving a process is most useful in those places that drive the overall process or that are, at least, of high impact. Therefore, identifying those patterns that stand out from the rest of the process is a good approach to find the relevant patterns.

[19] The term "structure" is critical for this research. Defined in section 2.1.1, it addresses, at its core, the pattern that is generated from a constellation of objects and their mutual relations. For example, an engineer and his communication with his partners in the process characterize the importance of the engineer for the process from the standpoint of the structure of communication in the process.

The problem is further aggravated by the fact that process models are often inconsistent or incomplete. While such low quality of the model is problematic, it is often the most valid, when the process modeled is actually controversial; in other cases, only parts of a process are modeled explicitly, while their environment is intentionally neglected. Nevertheless, often the part of the process the process chart focuses on is still meaningful and can be analyzed. In general, however, the quality of the analysis will only be as good as the input to it.

Finally, many process models only provide qualitative information for a process (i.e., "boxes and arrows"), and thus make it impossible to use simulation approaches or more sophisticated methodologies to analyze the process. Yet, the patterns (and thus the knowledge) are already embodied in these "boxes and arrows". The challenge is to relate the structure to the behavior of the process [KAUFFMAN 1993] [HOLLAND 1996] [BAR-YAM 1997]. Of course, there is no absolute truth about this inference, as different companies (and thus different processes) have different cultures, which then lead to different foci of process organization; while in one company, the concept of having single employees as center coordinators of design knowledge might be desirable, a different company might prefer to store knowledge in a database, for example, and not depend on single employees that much. Yet, there are certain patterns that may appear in a process, and these patterns are linked to one or more common kinds of behavior in a process. Figure 1-10 shows an example of engineers in a process, communicating via channels (e.g. team meetings). Whereas in the pattern on the left no designated coordinator of communication is discernable, in the pattern on the right all major flows of information go through the person at the center.

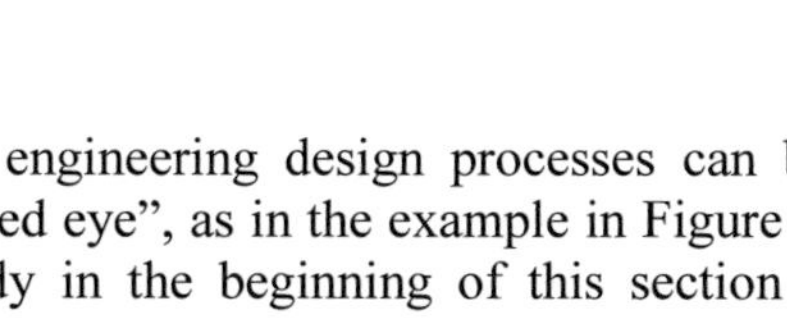

*Figure 1-10: Example of two possible communication patterns in a process*

As engineering design processes can be very large, the identification "by the naked eye", as in the example in Figure 1-10, is typically not possible, as the case study in the beginning of this section was able to prove. Therefore, a formal approach is necessary to handle large systems that are densely crosslinked. It may also need to be supported by a computer-based tool or by algorithms that can be computed in realistic run-times.

### 1.3.2 Basic hypotheses and research questions

With the focus on extracting knowledge about a process's behavior from the constellation of its elements and relations, this research was based on three fundamental hypotheses. These are introduced here, as they delineate the approach developed in this research, and their knowledge permits the solution that is presented to be better understood.

The first hypothesis represents the basic understanding of a system (i.e., a process); commonly, processes are seen as a time-oriented flow of tasks and documents. Here, however, all supporting entities as well as their coupling will also be considered. An engineering design process is thus not simply a set of tasks that can be put into interaction, but rather forms a **network of multiple layers consisting of different classes of entities**[20] (such as tasks, organizational units, milestones, resources) [GAUSEMEIER et al. 2006, p. 223] [ZACHMAN 1987]. This is generally due to the high degree of integration seen in almost any kind of product today; in turn, it both causes and necessitates many different stakeholders in a process, who need to collaborate [SPATH et al. 2001]. Therefore, when improving a process, it is necessary to gain a detailed understanding of the process and all involved domains that enable the process [HERFELD 2007, pp. 92-93]. An engineering design process[21] is a complex socio-economic construct that is unique within each company and for each product. Typically, the holistic analysis of a process should involve different views of the process, represented by the available domains, as well as its relation to the product architecture [SOSA et al. 2004b].

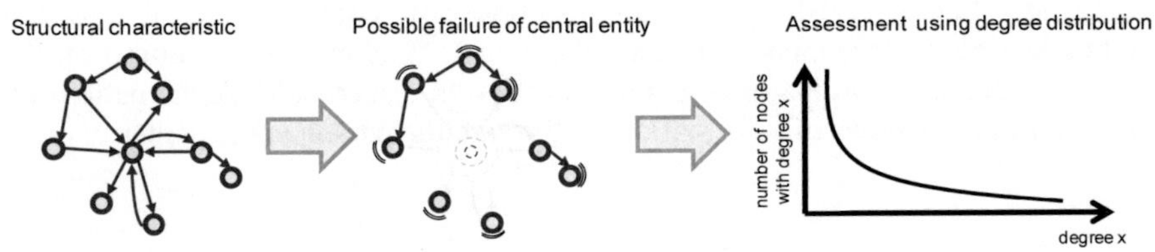


*Figure 1-11: Example of a structural characteristic (a "hub"), the possible related behavior, and the assessment*

The second hypothesis points to the assumption that it is possible to identify certain **patterns of entities**[22] **in a process that drive the behavior of the process**. This phenomenon is referred to as *inference* [KAUFFMAN 1993] [HOLLAND 1996] [BAR-YAM 1997] [CANTAMESSA et al. 2006]. In fact, being a network of multiple layers, a process forms a complex system. This system "process" only emerges because the goal-oriented and purposeful configuration of its entities provides value over the pure sum of all the entities [BOARDMAN & SAUSER 2006]. Thus, methods of understanding, modeling, and managing systems can be applied [MILLER et al. 2006]. Figure 1-11 shows an example of the degree distribution of a network, which allows the assessment of the homogeneity of a

[20] These are later addressed as domains (i.e., a class of one type of entity of a process) and relationship types (i.e., one class of relations between entities). See section 2.1.1

[21] While we speak of processes, in fact, many processes actually exist as projects [LINDEMANN 2007, p. 16]. See page 69 for more details.

[22] This constellation of entities will later be referred to as structural characteristics of a network. Possible structural characteristics are explained in section 2.1.6.

process (details are found in appendix 10.5.12): For a network that is structured around a central hub, there is a high risk of failure of the overall process associated with the failure of the central entity connecting the overall process. This can be identified using a (schematically represented) degree distribution.

The third hypothesis proposes that the **identification of structural outliers** is an appropriate means for the high-level analysis of the behavior of a process. As outliers, such instances are identified that particularly stand out with regard to their involvement in a pattern of entities. While, of course, a process has a limited number of entities that is often too small to obtain statistically significant results, the concept of the outliers essentially embodies the Pareto principle[23] [REED 2001] by highlighting the core entities of a system.

In fact, the identification of outliers makes it possible to pinpoint entities that are of extremely high or low impact, thus significantly driving a pattern of entities [HAWKINS 1980]. Outliers are, therefore, those results that are "numerically distant" from the main population of results, and they commonly show up in histograms or other distributions [BARNETT & LEWIS 1998, p. 16].

The approach presented here is not meant to rate the outliers in terms of their possible negative or positive contribution, even though every structural characteristic present in the process will inevitably contribute to the process quality. However, as the implications of an outlier vary for different companies, the neutral term "outlier" was chosen to indicate that an outlier is only meant to point to a possible problem without judging if there actually is a weak spot in the process.

Based on these hypotheses, the research focuses on a main research question:

> How can a process be systematically analyzed (I) in terms of the structure of the relations of its entities (II) in a goal-oriented manner (III) to point a user to possible weak spots (IV) and their meaning (V)?

Figure 1-12 visualizes the idea behind this research question: The initial goal is to analyze the process in a comprehensive and systematic manner. To do so, process models are to be used to access knowledge about the behavior using the network-like structure of the process. This analysis is to be given in a goal-oriented manner, i.e., by providing target-oriented analyses in a compact form that point to possible weak spots in the overall process structure. Lastly, the interpretation of these findings is to be supported to better draw inferences about the actual impact and behavior of the identified weak spots.

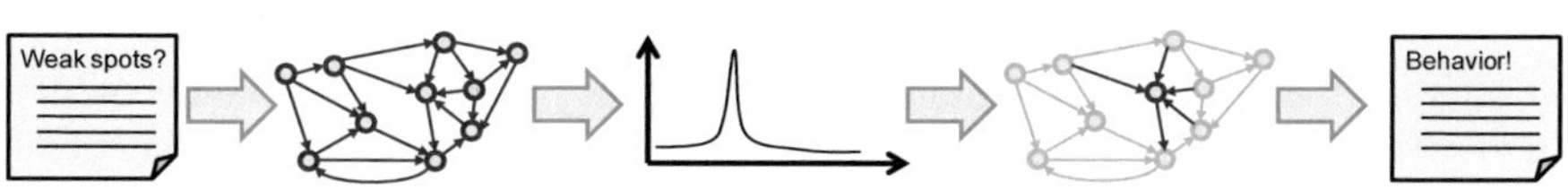


*Figure 1-12: Focus of this research*

[23] Also called the 80/20 rule.

The intent of this research, therefore, is in developing a methodical support to analyze process charts in a systematic manner to obtain a complete overview of the behavior of the process. The behavioral description needs to be as correct as possible with the available input data, i.e., in the first step, it should not be supported by additional simulations or other methods but enable a compact picture of the overall process and its improvement potential. The results of the analysis need to be consistent to allow for a comparison of the process' entities to point to possible weak spots. They need to be intuitive and supported by clear guidelines that allow transferring the results back to an operational level.

### 1.3.3 The approach used in this research

The solution presented was developed from a strategy similar to that of Agile Development. In a repetitive pattern, small entities (i.e., the metrics and individual aspects of matrix-based process modeling) were developed (based on requirements from various companies), tested, and improved. Each time, a second step consisted of integrating the partial solutions into the overall context, again involving testing, improving, and completing. As such, the research approach used is in line with common iterative research approaches in technical sciences [MINNEMANN 1991, p. 16]. It is also consistent with research approaches in economics and management, in general, where empirical (requirements, test, and adaptation) and conceptual work collude [KORNMEIER 2007, p. 43]. This research is, in fact, very similar to action research [ARGYRIS et al. 1985, pp. 36-40].

This research can be classified as a development of the application of principles (as commonly done in economics and management science [KORNMEIER 2007, pp. 23]) at a high level of abstraction. At the same time, the generalization of the approach is intended to explain why processes behave in a certain manner due to their structure. However, the approach does not present a means of prognosis per se, but only elements of prognosis [KREIMEYER et al. 2006c] [HEYMANN 2005, pp. 513].

## 1.4 Context of developing complexity metrics

As the research question shows, this research is intended to generate a methodology that facilitates the systematic analysis of a process by regarding and evaluating its structure. Therefore, the following goals, requirements, and limits to the solution need to be considered.

### 1.4.1 Goals of this research

To address the problems encountered in the analysis of a large and complex engineering design process, this research was initiated to meet the following goals:

- Establishment of a structural process modeling method:
  - Show that a process consists of multiple layers of a network
  - Develop a pragmatic process model that allows this structure to be accessed integrating common process modeling methodologies (to

generate a solution that is suitable for different kinds of process models) as a common basis for the design of structural metrics

- Development of structural metrics tailored to engineering design processes:
  - Develop a coherent set of metrics that can describe the structure of a process network
- Setup of a selection framework to guide a process analysis project:
  - Set up a framework to select appropriate metrics in relation to the goals of process analysis
  - Show the possible significance for each metric to extend the applicability of the framework
  - Show how a framework can be applied in process improvement

The process model creates the foundation to answer the research question by providing a common basis and a consistent structure to develop structural metrics. It uses essentially the first hypothesis by modeling the process as a network instead of a purely time-oriented flow. At the same time, it addresses the fact that different process models might serve as a starting point to analyze an existing process chart. The second goal addresses the main research question directly, collecting goals, possible metrics, and an approach to identify the main drivers for the behavior, combining hypotheses two and three. Last, the overall framework combines the modeling and the analysis method into a measurement system that enables a goal-oriented application.

### 1.4.2 Basic requirements of the solution

Generally, this research is intended to deliver a rigorous contribution to design research, extending the existing body of knowledge and understanding of engineering design processes, while, at the same time, delivering a practicable methodology to industry. To this end, this research is based on the common aspects of research methodology.

At the solution level, there are different requirements of developing a "good" method. This entails a solution that is complete, correct, consistent, and clear. **Complete** refers to several aspects at the same time. On the one hand, the approach to be developed needs to be complete in regard to its setup, i.e., it should support the planning of an analysis, the necessary modeling and the analysis itself. On the other hand, the approach needs to consider, in each of these aspects of the solution, all possible scenarios, i.e., the planning of the analysis should contain all relevant elements of an analysis; the modeling needs to embody all possible modeling constructs, and the analysis should provide a means of analysis for the possible behaviors found in the engineering design processes. A **correct** analysis approach is one that is homomorphous with the process it focuses on. As such, the elements of the approach to be designed should represent the empirical object as closely as possible. The analysis approach to be developed needs to be **consistent** in itself and with the existing use of similar methods; this relates especially to metrics that are in use in engineering and software design. Finally, the developed approach needs to be as **clear** and as self-explanatory as possible, necessitating detailed guidelines about how to apply the method and interpret the results obtained. As the solution presented here consists of three parts, the requirements are grouped accordingly.

The **process modeling framework** needs to enable the *complete representation* of the structure of a process. It should do so *in a formalized way*, making an automated assessment possible. It should be *in line with common approaches to the management of structural complexity*, thus not developing a new methodology but extending existing methods where necessary. It should, as such, *integrate the structural aspects of existing process models* to be compliant with the state of the art in process management. It should, furthermore, *be able to represent large systems in a manageable fashion*, allowing, for example, the automated integration of partial models into an overall model that can be recombined, reused, or parsed from different sources. It should, thus, provide an adapter to structural process analysis.

The **metrics** need to be, above all, *relevant* to process management. As such, they should be collected from sources that have *empirically validated* their usefulness and applicability. They also need to respect the necessary measurement foundation[24], which describes the quality of "good" measures. Ultimately, the metrics should allow an *intuitive understanding* of the structure to the extent that is possible for such an abstract entity.

Lastly, the **metrics selection framework** should enable a straightforward *navigation of all necessary aspects* of process modeling and metrics-based assessment *based on relevant concepts, goals, and interests* of process management. It should, therefore, classify the metrics comprehensively and show their mutual dependencies. Ultimately, it should be created in such a way that it can later be extended.

### 1.4.3 Targeted audience

With the author's experience and the three case studies originating from automotive design, the focus is on processes similar to automotive design. These are characterized by a multitude of requirements from various sources that result in a highly integrated product that is concurrently designed by many engineers from different backgrounds. The high degree of division of labor corresponds to a great number of specific artifacts in the process (files, prototypes, etc.) that are processed by highly specialized resources. The processes are commonly coordinated by development engineers [HERFELD 2007, pp. 18-20] [SAPUAN et al. 2006]. Other similar contexts could be, for example, aircraft design, the development of production machinery, or mechatronic household appliances.

The approach is, therefore, tailored for managers, consultants and project engineers who continually have to plan, improve, and control processes. At the same time, the solution is meant to improve scientific understanding of

---

[24] Measurement foundation is detailed in section 2.3.1. It describes how measures for a system can be set up in a systematical and error-free manner. There are various models that explain "good" complexity metrics, especially provided by the set of Weyuker's Criteria [WEYUKER 1988].

engineering design processes by providing formalized access to the structure of engineering design processes.

### 1.4.4 What this book is not about

As there are many aspects of process management, and because this research regroups and recombines many different streams of research from different disciplines, a brief overview of possible misconceptions of this research is given to ensure that it is understood correctly.

With the focus on structure, the approach developed is tailored to analyze qualitative models. It does not provide any quantitative model or a more complete description and analysis of a process outside its structure, e.g., in terms of cost, run-time of the process, or the amount of manpower needed to design a product. It intentionally only considers the need to systematically analyze qualitative models (i.e., "boxes and arrows") to discover possible weak spots of a process, which will be referred to as outliers.

Thus, the methodology will not provide any rating as to whether a process is good or bad. Rather, the approach is meant to methodically access possible outliers that can turn out as problematic in a given context, depending on the company culture. Thus, the measures in this process are referred to as metrics and not as performance indicators.

As such, the approach presented is meant to complement existing views on process analysis and potentially to review the structure and, in the long run, interlink this structural analysis with other approaches in order to better understand engineering design processes. Thus, this approach is not meant to replace any existing paradigms, such as business process reengineering or continuous improvement within a company, but it complements these paradigms by providing means to access the structure of a process.

Therefore, the results of this research are not planning methods; they only contribute to better plan processes using existing methodologies by generating knowledge about risks in process management. This is done through gaining insight into how the behavior of a process relates to the structure that is set up during process planning and execution, i.e., when actors, tasks, and resources are combined into an overall architecture.

In this context, it needs to be stressed that the focus is not on how such models can be generated to depict a real life process, but the focus is on how a model needs to be structured. Thus, information collection via interviews or similar methods is omitted, and it is assumed that (partial) models are available, and that they are of reasonable quality.

### 1.4.5 Related fields of science

Figure 1-13 regroups the related major fields of science that are most important to this research. In general, the focus of this research is on engineering design **processes**; however, a number of inputs are taken from management sciences, especially topics that are related to process management, its application, and the modeling of processes.

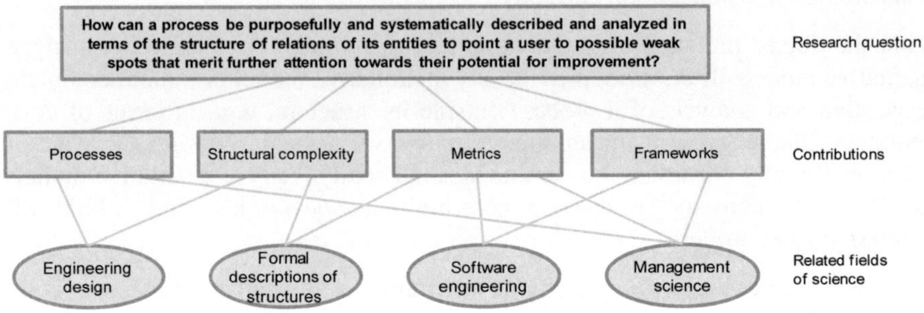


*Figure 1-13: Overview of related fields of science*

The access to **structural complexity** is not available as a dedicated field of science, but rather it avails itself of different approaches in engineering sciences, especially matrix-based methods, and in applied mathematics, mostly graph theory and network science, which is another interdisciplinary field of science bordering almost any other field of research. The different forms of system sciences contribute equally to this research.

Software Engineering provides numerous means of **measuring** structures; in fact, processes are, in many ways, similar to software programs, being based on a number of interrelated resources and representing a flow of information [CARDOSO 2006]. Many approaches and the foundations available can, therefore, be adapted almost directly.

Ultimately, all three fields of science contribute to their systematization in a **framework**. In fact, frameworks from all fields of science are relevant; however, those frameworks most closely related to managing metrics are available in software engineering, which is why the section on frameworks is linked to this science in Figure 1-13.

## 1.5 Structure of this book

As shown in Figure 1-14, following the introduction given by industrial need, as provided in the initial case study, the research question is explored throughout the state-of-the-art shown in chapter 2. Here, the foundations of systems and their structures are presented by collecting possible models and methods for the analysis and meaning of structures. Engineering design processes are then analyzed for their structural content. The structure inherent to common process models is collected to later constitute a meta-model suitable for structural process models, and typical goals of process improvement are collected that focus on structural analysis in a framework designed later in this research. A literature review on different kinds of metrics that connect to structures is also presented to show current models and approaches and to discuss their capabilities and transferability to process management. These will later serve as a basis to assemble a comprehensive set of structural metrics. Lastly, different frameworks are reviewed to enable a goal-oriented analysis of the structure of a process.

Chapter 3 establishes an overview of the three subsequent solution chapters and places them in a common context. A basic procedural model is used as a broad framework.

Chapter 4 introduces a matrix-based modeling scheme predicated on Multiple-Domain Matrices. To incorporate all needs of process modeling as reviewed as part of the state-of–the-art, this modeling scheme is extended to incorporate different modeling needs, such as Boolean operators or product attributes. The modeling scheme itself consists of a meta-model that serves as a reference for the establishment of a structural process model using commonly accepted semantics. It is also used as a basis for the formulation of metrics and their interpretation.

Chapter 5 regroups the structural characteristics that govern different systems to form a generic procedure to design structural metrics. In the second step, the metrics that were developed as part of this research are presented, and their suitability and adherence to the foundation of measurement and the requirements of process management is reviewed. Lastly, the metrics are classified to generate a complete picture of their applicability.

Chapter 6 makes use of this classification to generate a framework connecting the metrics to the different goals of process improvement to allow a goal-oriented application of the structural metrics. To do so, first a set of goals for structural process analysis is established which, in the second step, is completed by different questions that operationalize each goal. Each question is then completed with a set of metrics, domains, and relationship types that can be used to provide answers to the question. Connecting these three constituents back to the metrics, the goal-oriented interpretation using each metric's structural significance is laid out.

In chapter 7 two examples of process analysis from automotive design are presented. The first case study regards the use of metrics to characterize another engineering design process in general. The second case study takes up the initial case study from chapter 1, using the framework to present and review the applicability of a goal-oriented process analysis.

The conclusion in chapter 8 summarizes all chapters to determine if the research question was answered and to record the strength and weaknesses of the approach shown.

To complete this research, a comprehensive appendix details all aspects that can only be briefly addressed in the main body of this research.

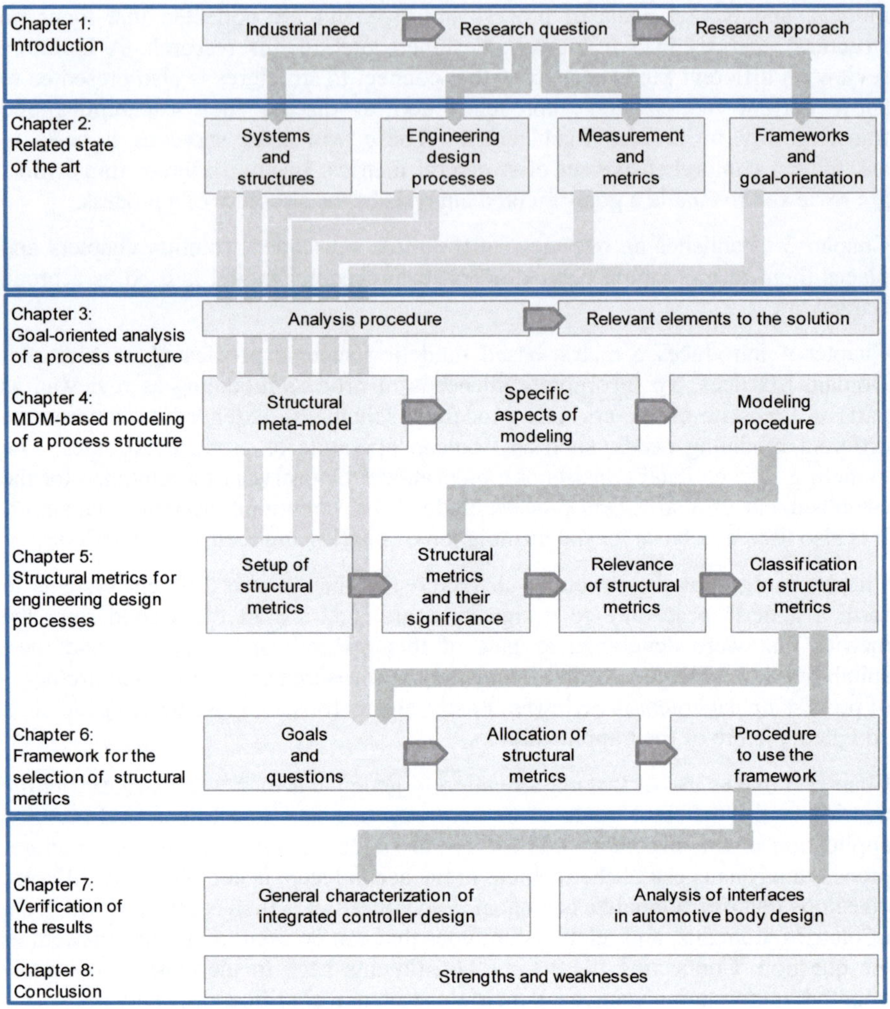


*Figure 1-14: Structure of this book*

# 2. The foundations of complexity metrics

To analyze the structure of engineering design processes using metrics, foundations from the different areas of relevance shown in Figure 2-1 are used. These were identified using the DRM (Design Research Methodology) approach [BLESSING & CHAKRABARTI 2009, p. 63].

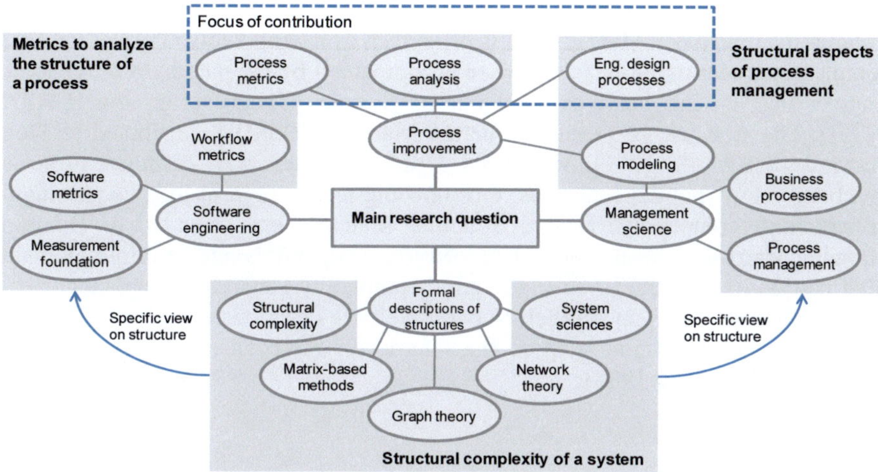


*Figure 2-1: Main areas of relevance to this research*

In this chapter, each area is explained in detail and related to the needs of this research. First, the foundations of systems in general and the structure are explained. Different means of dependency modeling for systems are explained, including relevant analysis methods. These foundations are used to explore what structural content is relevant for process management regarding both the goals of an analysis and the actual process models. Last, existing metrics are reviewed that are able to assess structures in general and, more specifically, in processes.

## 2.1 Structural complexity of a system

Complexity is present in many disciplines. Commonly, complexity means "consisting of parts or entities not simply coordinated, but some of them involved in various degrees of subordination; complicated, involved, intricate; not easily analyzed or disentangled" [SIMPSON et al. 1989]. Indeed, complexity has many facets. Computational complexity refers to the computability of an algorithm [PAPADIMITRIOU 1994]; information processing understands complexity as the total number of properties transmitted [NEWELL 1990]; and physics sees it as the probability of reaching a certain state vector [HEISENBERG 1999, HEISENBERG 2007]. In engineering, complexity generally addresses the high coupling of the entities of a technical system [MAURER 2007], and software science focuses on assessing program code for its complexity, and thereby the risk of introducing errors into the code.

Complexity science originated from Cybernetics, founded by [WIENER 1948], and Systems Theory, founded for the most part by Ludwig von Bertalanffy [VON BERTALANFFY 1950]. It was also influenced by Dynamic System Theory, which belongs to the field of applied mathematics for the description of dynamic systems [PADULO & ARBIB 1974].

Structural Complexity Management is often seen as having evolved out of the first complex engineering projects that were accompanied by the paradigm of Systems Engineering, having itself evolved out of Systems Theory (e.g., the [NASA 1995]). The first use of Design Structure Matrices (DSM) is attributed to Don Steward [STEWARD 1981], who used DSM to better plan complex projects involving many interdependencies, thus opening up the field of today's paradigm of structural complexity. Design Structure Matrices later evolved to Domain Mapping Matrices [DANILOVIC & BROWNING 2004], and then to Multiple-Domain Matrices [MAURER 2007]. While structural complexity generally regards technical (i.e., planned) systems, in parallel, Network Science describes complex systems of random or natural origin, such as the internet or molecules [BARABÁSI, 2003] [WATTS & STROGATZ 1998].

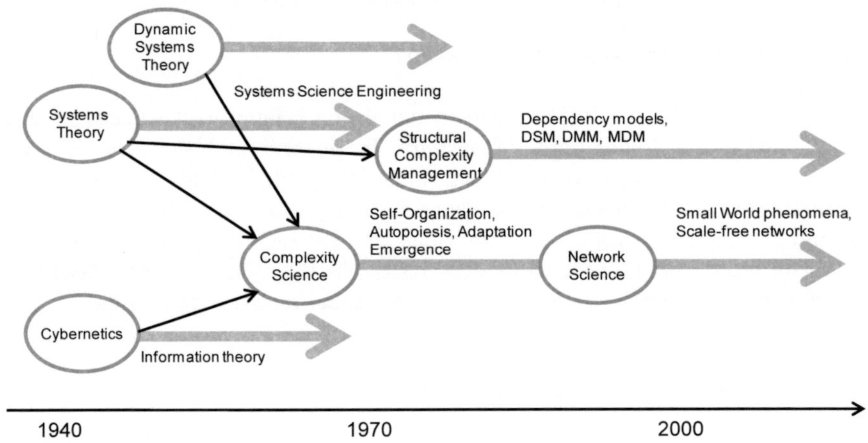


*Figure 2-2: Evolution of sciences related to structural complexity (after [ART AND SCIENCE FACTORY 2009])*

Figure 2-2 shows the development of each field of science from a historical point of view. The originating methods are discussed as follows. First, a review of systems and how they are constituted to derive a specific view on what the structure of a system is provided. For this purpose, the terms *system*, *complexity*, *structure*, *domain* and *relationship type* are explained. Using these basic notions, different kinds of dependency modeling are then introduced. Here, graph theory serves as a formal basis for all other dependency models and introduces a number of concepts that keep appearing in other dependency models (e.g., cliques as tightly coupled clusters). Then, models and methods for matrix-based dependency models and for large networks are reviewed in detail to collect a complete set of analysis methods and the requirements to model the dependencies within a system.

### 2.1.1 General notions of managing structural complexity

Understanding processes as a complex system that has a network-like character and a certain structure necessitates certain basic notions and terminology, which are explained in this section.

#### System

The concept of a "system" is essential to the analysis of processes, as a process[25] represents a special form of a system [WASSON 2006]. A "system" is a set of entities that have relations among each other, either by interacting with or being independent of each other, and the system is delimited from the environment by a system border and connected to the environment by inputs and outputs [HALL 1963] [LINDEMANN 2007, p. 337].

WASSON defines a system as an "integrated set of interoperable entities, each with explicitly specified and bounded capabilities, working synergistically to perform value-added processing to enable a user to satisfy mission-oriented operational needs in a prescribed operating environment with a specified outcome and probability of success" [WASSON 2006, p. 18]. The "integrated set" refers to the fact that a system consists "of hierarchical levels of physical entities, entities, or components". The entities are interoperable, as they must be compatible with each other to allow greater value of the system [BOARDMAN & SAUSER 2006]. This is also addressed by the "explicitly specified and bounded capabilities", meaning that every entity "should work to accomplish some higher level goal". To work "synergistically" underlines the additional value [RECHTIN 1991, p. 29] and the leverage factor of a network. Ultimately, the "value-added processing" alludes to the classical process term, which is based on the assumption that a process provides some additional value to the customer. This aspect of customer orientation is further emphasized by the expressions "specified outcome" and "success" in the definition [PATZAK 1982].

In a more general manner, MAURER provides a definition of a complex system that is, in many respects, similar to that of WASSON: "A system is created by compatible and interrelated parts that form a system structure, possess individual properties, and contribute to fulfill the system's purpose. Systems are delimited by a system border and connected to their surroundings by inputs and outputs. Changes to parts of a system can be characterized by dynamic effects and result in a specific system behavior". MAURER thus adopts the aspect of the relationships ("compatible and interrelated") and boundaries ("system border") but also provides a new aspect, that is, a system response that is not necessarily linear ("dynamic…specific system behavior") to the definition [AUE & DUSCHL 1982] [MAURER 2007, p. 31].

[25] The term process will be defined later in section 2.2.1.

Keeping the focus of this research in mind, the following definition of a system is used:

> A system is a set of entities of (possibly) different types that are related to each other via various kinds of relations. The system is delimited by a system border, across which inputs and outputs of the system are possible as an interaction with the environment. The system fulfills a purpose, which guides the meaningful arrangement of entities and relations. The behavior of the system is, in turn, due to the arrangement of the system's elements.

To a certain extent understanding a system denies traditional cause-and-effect chains and the certainty of determinism that, as practice shows, are not applicable to complex systems and complex processes anyway. This, however, implies that a reductionist approach which only centers on certain objects is too limited when regarding complex systems [BANATHY 1997].

### Modeling complex systems

To manage a complex system, modeling is used to better understand it [BROWNING 2002]. In comparison to the object being modeled, a model[26] represents a target-oriented, simplified formation analogous to the original, which permits drawing conclusions based on the original [LINDEMANN 2007]. These conclusions normally comprise "making predictions and testing hypotheses about the effects of certain actions" [BROWNING 2002]. In this way, systems are made more transparent to improve understanding.

A model thus represents an abstraction of a real system, serving a certain purpose the model was made for; at the same time, a model usually involves pragmatics for those points that appear to be of little relevance to the purpose during the generation of the model [RECKER 2007]. This implies that the modeler influences the model in its expressiveness even before it is analyzed [MENDLING 2008, p. 7]. The development of any model-based analysis framework must thus incorporate possibly "unclean" models.

### Nodes and edges, entities and relationships

In regards to a system's entities and relations, the management of structural complexity can be understood as an application of Graph Theory[27]. As different terms prevail in the different disciplines, Table 2-1 provides an overview of how the entities and relations within a system are commonly denominated.

---

[26] A good model acknowledges the "Guidelines of Modeling" to achieve inter-subjectivity of the models: correctness (i.e., the model is syntactically correct), relevance (only the interesting parts of the object being modeled are represented in the model), economic efficiency (the trade-off between the expense to build the model and making it as complete as possible), clarity (to ensure that a reader is able to understand the model), comparability (the consistent use of naming conventions and modeling schemes), and systematic design (the clear separation of different views) [BECKER et al. 1995].

[27] Graph theory is outlined in more detail in section 2.1.2.

*Table 2-1: Different terminologies to describe the parts of a system*

| Term in Systems Theory | Entity | Relation |
|---|---|---|
| Term in Graph Theory | Node, vertex | Edge, arc |
| Other terms | Element: especially in matrix-based methodologies, the term "element" is used to refer to an entity that is entered into a row or a column | Relation: refers to any kind of association between two entities (directed or not)<br>Dependency: often implies a direction |

## Domains

From a structural point of view, a system can be disentangled into a network-like model of entities and their relations. These entities can be of different kinds, e.g., documents, organizational units, and work packages. However, if many such kinds are mixed, a productive analysis is impossible. Each kind of entity represents a specific view, called a "domain"[28]. The purpose of a domain is to create "homogeneous networks" that allow elements to be compared during analysis [MAURER 2007, pp. 71-72]. The term "domain" can, therefore, be defined as a specific view of a complex system, comprising one type of entity.

Three principles apply to the use of domains to model the structure of a complex system that involves specific views (visualized in):

- **Instantiation**: An entity is an instantiation of a domain. Thus, all entities that belong to one domain are of the same type, e.g., "information object".
- **Decomposition**: A domain can be refined by creating sub-domains that are congruent with a part of the super-domain to which they belong, e.g., the domain "document" and the domain "prototype" can both be "information objects" in a process.
- **Recombination**: A system can be assembled by combining relevant domains and relationship types (see next page). By assembling different domains and relationship types, different views of the system can be generated.

## Relationship types

Like the domains (see above), the relationships within a domain (or between two domains) need to be uniform to allow a systematic modeling and a purposeful analysis. While a domain contains entities of a kind, one relationship type refers to one class of relations that are similar [MAURER 2007, pp. 71-72]. The relationship type therefore defines the kind of (directed or undirected) relation between two entities. It is described as "(entity of domain A) is related to (entity of domain B)". The term "related" can be replaced to specify the type of relationship; of course,

[28] A domain is comparable to a "class" of objects in object-oriented programming paradigms.

reflexive relationship types ("intra-domain") are possible as well as mappings between two domains ("inter-domain").

Again, three principles apply:

- **Instantiation**: A relation is an instantiation of a relationship type.
- **Differentiation**: A relationship type can be refined by detailing the description in a coherent manner with the superior relationship type. However, such a refined relationship type applies to fewer entities and is harder to model coherently [MAURER 2007, pp. 71].
- **Recombination**: A system can be assembled by combining relationship types.

Figure 2-3 visualizes these three principles for both domains and relationship types. At the highest domain it uses domains and relationship types to set up the meta-model of a system. Here, the basic class of an entity is defined. If a pre-defined meta-model is used, it actually provides the domains and relationships. If necessary, both domains and relationships can be refined and broken down into more specific sub-domains or differentiated relationship types. This is necessary, for example, when two domains are linked by two different relationship types simultaneously such that each transport a specific meaning, as shown in the example on page 131. To build a model, the chosen domains and relationship types are instantiated.

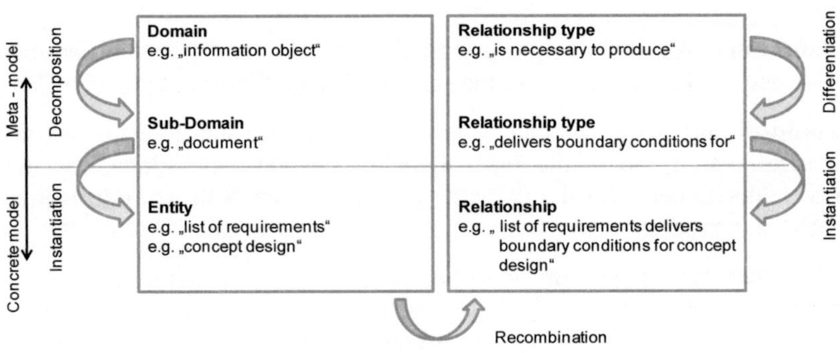


*Figure 2-3: Principles of using domains and relationship types to describe the structure of a complete system*

Figure 2-4 embodies these principles into the basic meta-model that is used for structural modeling, using only domains and relationship types and their decomposition. It represents a common dependency model that is only able to represent entities and relations.

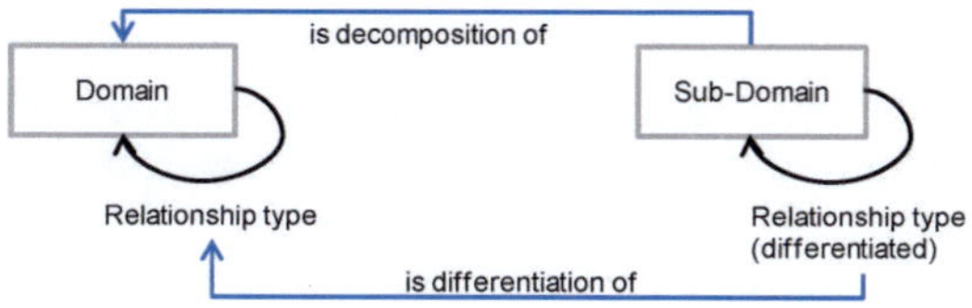


*Figure 2-4: Meta-model for structural modeling*

## Views to a system: Native and aggregated networks

Complex systems can consist of more than one domain. These domains can be coupled within themselves or among themselves. In turn, it is possible that two domains are not directly related but only via a third domain. However, for many analyses it is necessary to use intra-domain networks, i.e., networks that consist of only one domain and of one relationship type that connects only the entities of the one domain among themselves. Therefore, aggregate views are needed at times to reduce the relationships via an intermediate domain to using just one single domain of reference.

In fact, according to the availability of models and data, a network can be *native*, i.e., the information in the model is a direct result of the data acquisition [MAURER 2007, p. 78] , or it can be *aggregated*, i.e., it is computed based on other networks that are available. Figure 2-5 visualizes the difference. In common Multiple-Domain Modeling (compare section 2.1.3), this aggregation is used to condense relations across one or more domains into single intra-domain networks to facilitate analysis. Section 2.1.3 illustrates in greater depth how this is done.

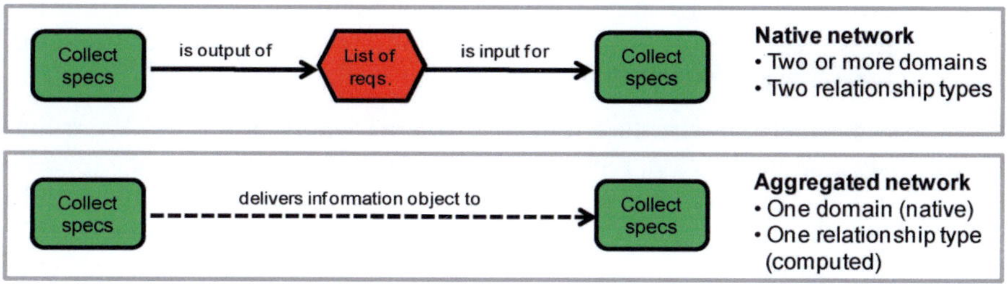


*Figure 2-5: Native and aggregated networks*

## Structure

Although the term *structure* (Latin: *structura*: orderly combination, derived from the verb *struere*: combine, stack, stratify) appears to be one of the most common words in engineering terminology, there seem to be very few definitions available. Commonly, the term structure is used in engineering for organizational structures and product structures, i.e., the purposeful organization of an enterprise into manageable entities or the setup of a product's components, functions, etc., into a meaningful architecture.

In mathematics (algebra and model theory), the term *structure* refers to a set of distinct entities, including the functions that transform these objects and the relations among the objects and functions. Systems Theory narrows this definition down to the patterns of interaction of the elements towards a behavior of the system [OLIVER et al. 1997, p. 21], which possibly serves a purpose [BOARDMAN & SAUSER 2006]. Structure is also closely connected to the phenomenon of emergence (see section 2.1.4) and sophistication, which is sometimes used as a measure for complexity and its logic decomposition into compact parts [KOPPEL 1987]. Both concepts indicate that structure involves the regrouping of entities of a system according to the logics (and the relationships between the entities) of a specific perspective. This is in line with the common understanding of structure as "the aggregate of entities in their relationships to each other" [MERRIAM-WEBSTER ONLINE DICTIONARY 2009]. A structure can be a hierarchy (a cascade of one-to-many relationships) or a network featuring many-to-many relationships [PULLAN & BHADESHIA 2000]. MALIK and BEER both refer to structure as the degree of organization which helps to bring order to a complex system and maintain it in order to enable the management of the system [MALIK 2003, p. 211] [BEER 1972, p. 65]. [GOLDENFELD & KADANOFF 1999] link complexity to structure (i.e., a basic pattern) by pointing out that everything has a basic structure according to how it is built; yet all systems come in variations that deviate from this basic structure, causing complexity. Structure is, therefore, defined as follows:

> Structure regards the pattern in the network formed by dependencies (edges) between a system's entities (nodes). It reflects the semantics of this network; the structure of a system therefore always contributes – in its constellation – to the purpose of the system. A structure takes shape in structural characteristics, i.e. a particular constellation of nodes and edges. The characteristic gains its meaning by the way the pattern is related to the actual system it is part of, i.e. it must serve a special purpose in the context of the overall system. A structural characteristic only possesses significance in the context of the system it is describing.

In process management, structure refers to patterns within processes. Workflow patterns[29], as a particular case, represent the basic decision structures of a process [VAN DER AALST, et al. 2003]. These patterns, however, relate to possible constellations of splitting and joining the control-flow of a process using AND, OR, or XOR operators. Yet, the patterns relate to the different perspectives of a process, as also outlined by [CARDOSO, 2005a]: control-flow, data, and resources. This idea can be extended to detect possible errors in a process, i.e., they search process models for typical structures that point to an error in the model [VAN DONGEN et al. 2006] [MENDLING 2008, pp. 59]. MENDLING does so, in fact, using metrics to formalize the patterns.

[29] A comprehensive overview is available at http://www.workflowpatterns.com (accessed 10.05.2009).

## Complexity

Complexity often involves the difficulty of handling a system, as it is hard to estimate the outcome of an action (therefore, popular lore goes "never change a running system") [LEE 2003]. Apart from these notions, complexity is sometimes defined as a degree of disorder [SHANNON & WEAVER 1998].

Like the approach by [CARDOSO 2006b], complexity is thus characterized by:

- **Structure**: A complex system is a potentially highly structured system which indicates a structure with variations [GOLDENFELD & KADANOFF 1999] [OLIVER et al. 1997, p. 29].
- **Configuration**: Complex systems have a large number of possible arrangements of their parts [KAUFFMAN 1993] [HOLLAND 1996] [BAR-YAM 1997].
- **Interaction**: A complex system is one in which there are multiple interactions between many different parts [RIND 1999].
- **Inference**: System structure and behavior cannot be inferred from the structure and behavior of its parts [KAUFFMAN 1993] [HOLLAND 1996] [BAR-YAM 1997].
- **Response**: Parts can adjust in response to changes in adjacent parts [KAUFFMAN 1993] [HOLLAND 1996] [BAR-YAM 1997].
- **Understandability**: A complex system is one that by design or function, or both, is difficult to understand and verify [WENG et al. 1999] [IEEE 1991].

The completeness of a solution (or rather the lack of determining the completeness of a complex solution), as well as uncertainty about a complex system's state and coupling are not part of the definition, although they, too, are commonly related to complexity [DANILOVIC & SANDKULL 2002].

## Process Complexity

In particular, the contextuality of a system (i.e., its strong relation to the environment) and radical openness (i.e., its possible unforeseeable interaction with entities not originally suspected to be relevant) are seen as the core drivers of complexity [CHU et al. 2003]. This implies that a complex system has many entities which have many different relationships, as stated for systems[30].

Processes are systems in themselves; CARDOSO points out that a process exhibits multiple facets of process complexity [CARDOSO 2005a]. He bases his conclusions on the similarities of executing a process to running a software, as it has structure (software is a highly structured system, cf. above), arrangements (a program has many possible arrangements), interacts with different parts, inference (the behavior of parts may be different than the sum of the parts), response (program gives response to input) and understandability (a software program is difficult to understand and verify). He thus transfers and adapts the work from software

[30] Further definitions of complexity can be found in [SUH 1999]. [WEBER 2005] discusses different ways of looking at complexity in engineering design.

engineering to describe process complexity as the number of tasks in the process (task complexity), the degree of cross-linking in the arrangement of tasks, and the related decision points such as AND, OR, or XOR (control-flow complexity), the degree of dependency of the information objects and their mapping to tasks, resources etc. (data-flow complexity), and the degree of interdependency of resources and their attribution to the tasks in the process (resource complexity).

Cardoso thus deduces that the definition provided by the IEEE Glossary is appropriate to describe process complexity [CARDOSO 2005b]. It is adapted here to include not simply the sequence of tasks, but all necessary supportive domains and their relationships.

> Process complexity is "*the degree to which a process is difficult to analyze, understand or explain. It may be characterized by the number and intricacy of activity interfaces, transitions, conditional and parallel branches, the existence of loops, roles, activity categories, the types of data structures, and other process characteristics*" [IEEE 1991].

Focusing on the network of entities and relationships of the system "engineering design process", in the following, different models, methods, and tools that are commonly used to understand, represent, and analyze such systems are presented. First, approaches to manage complex network structures are reviewed; then, the concept of structure is extended to engineering design processes in order to collect metrics that embody structure in a process.

In summary, processes can be understood as a special class of systems that are constructed from entities and their relations among each other. Within this network-like structure, certain patterns can be identified that serve as structural characteristics. To productively manage such a complex system, a detailed classification of entities and relations into appropriate domains and relationship types provides a systematic basis to model a system and to compare the entities among each other. The actual entities and relations can be taken from any process model, as will be shown in section 2.2.3.

There is a consensus that the structure of a system, i.e., the purposeful constellation of entities and relationships, drives its behavior. The identification of the patterns that repetitively occur within such a structure can serve as a basis to draw inferences about the behavior of a system. It is, therefore, necessary to collect these patterns and relate them to their structural significance to better analyze a system and understand its behavior. The structural characteristics that are available so far will thus be explained in the following sections.

Lastly, complex systems tend to have many domains that are not independent of each other. The aggregation of different views onto a system provides a compact means of generating manageable models that do not reduce the structure of a system but condense it to its individual impact on a domain of reference. It is, therefore, necessary to further explore the creation of such aggregate views to better identify structural patterns within large processes.

### 2.1.2 Graph Theory

Graph Theory[31] provides the mathematical foundations to study any kind of network, called a graph. A graph is an ordered pair $G = (N, E)$, where $N$ is a set of nodes (also called vertices) and $E$ is a set of edges (also called arcs); being a 2-element subset of $N$, a graph is thus a formal description for "boxes and arrows" when drawing a network on plain paper. It is commonly understood to date back to the works of Euler and the "Seven Bridges of Königsberg", i.e., the mathematical solution to the question whether a walk through the city of Königsberg could be routed in a way that all bridges would be crossed only once.

Graph Theory describes networks in a generic way, attributing to them the following basic properties:

- They can be directed ("digraph") or undirected, or both ("mixed graph").
- They can have a weight associated to nodes or edges ("weighted graph").
- They can have loops ("simple graph") or not.
- An edge can connect a node to itself ("loop").
- They can have multiple edges between two nodes ("multigraph"), one, or none, or one edge connecting one node to many others ("hyperedge").
- They can have edges not associated with any node ("half-edges", "loose edges").

Graphs have basic characteristics used to compute more complex analyses:

- Two edges of a graph are called "adjacent" if they share a common node; equally, two nodes are called "adjacent" if they are connected by an edge.
- An edge is called an "incident" (in a directed graph), if it is directed towards a node; the opposite direction is called an "outgoing" edge.
- The number of edges that connect to a node is called a "degree" of the node.
- Nodes are discernable ("node-labeled graph"), edges as well ("edge-labeled"). In particular cases, nodes or edges can also be treated as not distinguishable.

Graphs also contain certain basic structures that can be used to describe them:

- Elements in a graph can be "disconnected", i.e., a node has no edge to any other node.
- A graph is "complete" if every pair of nodes is connected by an edge, i.e., if the graph contains all possible edges. Such a graph, in which every node is connected to every other node, is also called a "clique".
- If a graph is "strongly connected", it does not necessarily have any cliques in it, but every node can be reached from every other node.

---

[31] There are many textbooks available on graph theory that need not be included here; for a better understanding, refer to [DIESTEL 2006]. [GROSS & YELLEN 2005] also is used as a common, very detailed, textbook; a good introduction in German is available by [TITTMANN 2003].

- A graph is "bipartite" if the set of nodes can be grouped into two sets *U* and V in a way that each edge has one node in *U* and one in *V*. This implies that the nodes in *U* are disconnected; equally no node in *V* is connected to another node in *V*.
- A "path" is a set of adjacent edges listed in a specific order; the path can be attributed by its length. A "walk" more specifically addresses an alternating sequence of nodes and edges that form an open or closed (="cycle") walk. The shortest path between two nodes is also called a "geodesic".
- A "cycle" is a path that starts and ends with the same node.
- If a graph has no cycles, it is called a "forest". If it is a connected graph that has no cycles, it is called a "tree".
- A "spanning" tree is the minimal graph necessary to connect all edges in a graph.
- A "planar" graph is a graph whose edges do not cross each other.
- A "subgraph" is a graph *S* contained within a graph *G*: *G* is the "supergraph" of *S*.
- "Graph labeling" is used to assign integer labels to nodes and edges; this can be used for the "coloring" of a graph, assigning a color to each node with no tuple of neighboring nodes being of the same color.
- The "genus" of a graph refers to the fact that the graph can be drawn as a planar graph on a surface of genus n (e.g., a sphere, a torus, etc.).

Graphs are commonly modeled as "boxes and edges". There are many methods to draw a graph, serving different purposes. A common algorithm is a force-based layout that arranges the nodes in a way such that nodes which are closely connected arrange as neighbors, repelling less connected nodes [FRUCHTERMAN & REINGOLD 1991].

Mathematically, graphs can be modeled as an Adjacency Matrix, which is similar to a DSM (see below). Adjacency Lists are also a possibility, listing which node is connected to what other node. MAURER ET AL. used this to indicate the usefulness of graphs to represent DSMs, while drawing equally on the extensive means of systematic analysis available in Graph Theory to extend analysis tools suitable for matrix-based methodologies [MAURER et al. 2004]. Other models, e.g., the Distance Matrix, are not considered here.

Commonly, the models and methods of Graph Theory provide the basis for analyzing structures, as shown in the next section (see Table 2-2 on page 50). Graph Theory also provides the basic means of describing large networks in Network Theory.

### 2.1.3 Matrix-based methodologies to manage structures

Research on matrix-based complexity management[32] has come a long way. Originating from a focus on process management using the **Design Structure Matrix (DSM)** [STEWARD 1981], a whole community has developed around this research. DSM is able to model and analyze dependencies of one single type within one single domain. For example, for a product, the domain "components" can be considered. Using the relationship type "change of component 1 causes change of component 2", an assembly can be analyzed with regard to the overall change impacts in order to model possible change propagations [KELLER et al. 2005]. [BROWNING 2001a] classifies four types of DSMs to model different types of problems: Component-, team-, activity-, and parameter-based DSMs. [JARRATT 2004] also proposes eight basic types of a DSM related to product architectures. However, all of these classifications present special cases of DSM application and should not understood as the only possibilities of modeling, but as suggestions for commonly accepted and useful models.

| | Task 1 | Task 2 | Task 3 | Task 4 | Task 5 | Task 6 |
|---|---|---|---|---|---|---|
| Task 1 | | X | | | X | |
| Task 2 | | | X | X | | |
| Task 3 | X | | | | | X |
| Task 4 | | | | | | X |
| Task 5 | | | | X | | |
| Task 6 | | | | | | |

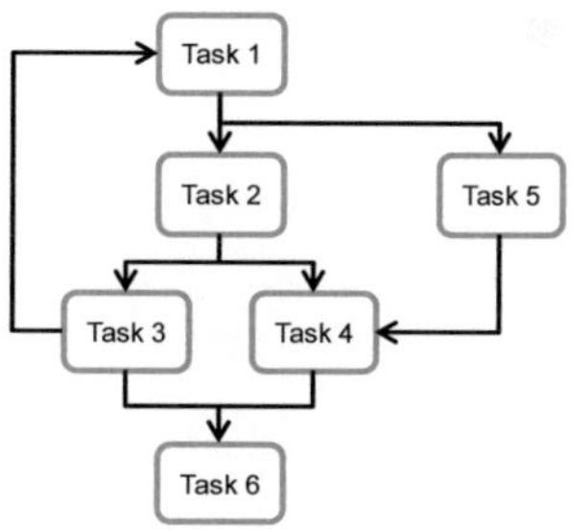


*Figure 2-6: Binary Design Structure Matrix of a simple process*

Binary DSMs only represent the existence of a relation, whereas numerical DSMs [BROWNING & EPPINGER 2002] implement a weighted graph to represent the strength of a relation. DSMs can either be directed (as shown in Figure 2-6), or non-directed; in the latter case, a DSM can either be written as a triangular matrix, i.e., only the upper or lower triangular matrix is used, or as a symmetrical DSM, i.e., the upper and lower triangular matrices are symmetric to the diagonal. Elements in DSMs are never reflexive, i.e., a relation from an element to itself is not permissible.

[32] A recent compendium on different types of matrix-based methods is available as [LINDEMANN et al. 2009]. Similarly, [BONJOUR 2008] provides a complete overview of recent developments (in French). Also, the web portal www.DSMweb.org (accessed 10.7.2009) provides a reference data base related to matrix-based dependency modeling.

| | Task 1 | Task 2 | Task 3 | Task 4 | Task 5 | Task 6 |
|---|---|---|---|---|---|---|
| Person 1 | X | | | | X | |
| Person 2 | | X | | | | |
| Person 3 | | | X | | | X |
| Person 4 | | | | X | | |

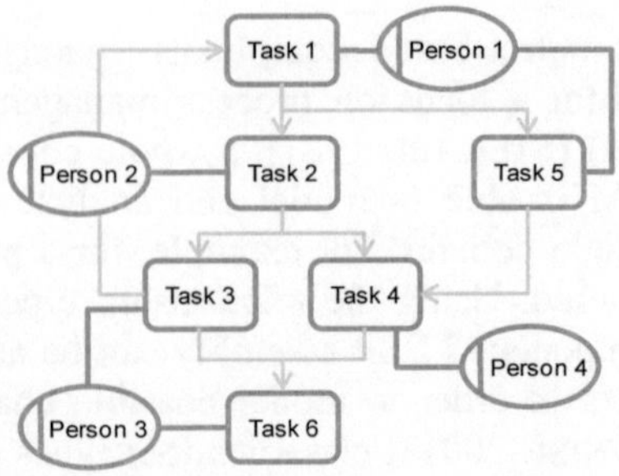


*Figure 2-7: Binary Domain Mapping Matrix for the process shown in the previous figure*

DSM was later extended to **Domain Mapping Matrices (DMM)** [DANILOVIC & BROWNING 2007]. The goal was to enable matrix methodology to include not just one domain at a time but to allow mapping between two domains [YASSINE, A. et al. 2003]. DMMs are thus rectangular, and again they can be binary or numerical.

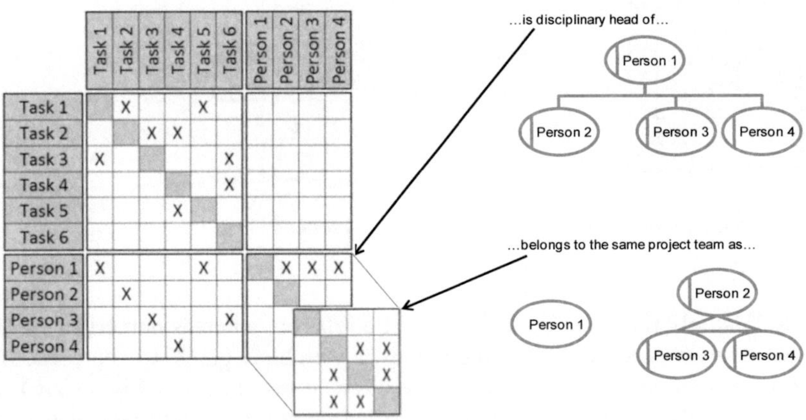


*Figure 2-8: Multiple Domain Matrix combining the DSM and DMM from the two previous figures and introducing two additional DSMs for one domain, using two different relationship types*

MAURER has taken this approach further to model whole systems consisting of multiple domains, each having multiple elements, connected by various relationship types [MAURER 2007]: the **Multiple Domain Matrix (MDM)**. Figure 2-8 illustrates this concept. It shows how a MDM basically is a DSM with more detailed DSMs along its diagonal and DMMs outside the diagonal. It also depicts how multiple relationship types create several representations of a specific submatrix of the overall MDM.

The MDM allows a system's structure to be analyzed across multiple domains, condensing each single analysis into one aggregated DSM that represents multiple domains at one time. That way, the MDM is able to apply algorithms for DSM analysis meaningfully across several domains [WALDMAN & SANGAL 2007] [CRAWLEY & COLSON 2007].

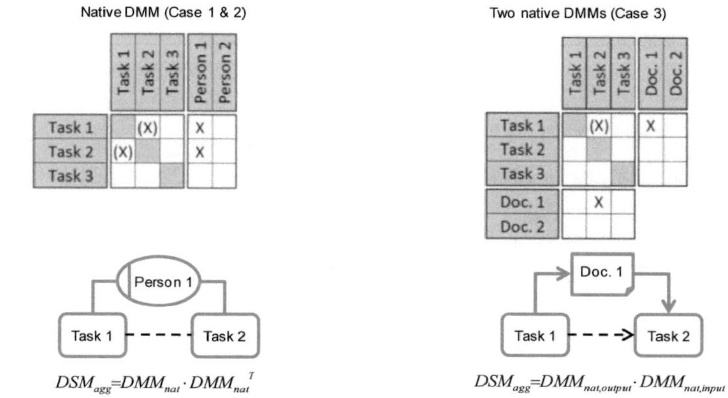


*Figure 2-9: Possible cases 1 to 3 for computing an aggregated DSM within a MDM [MAURER 2007, pp. 113-116]*

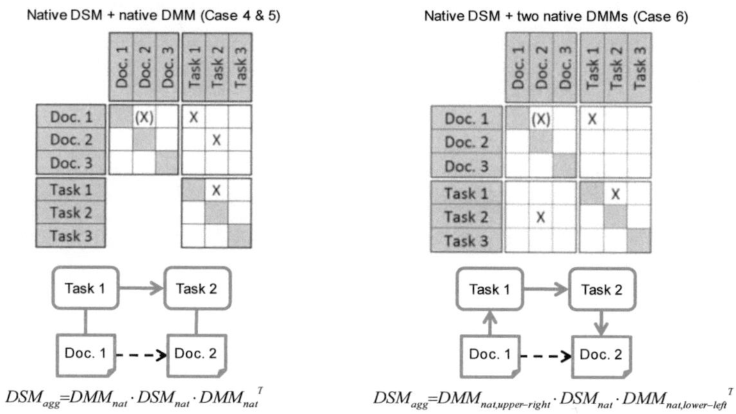


*Figure 2-10: Possible cases 4 to 6 for computing an aggregated DSM within a MDM [MAURER 2007, pp. 113-116]*

The main advantage of a MDM is that relations across more than one domain can be aggregated into a DSM using a domain mapping logic, using one native DMM (differentiating the direction of the DMM mapping as cases 1 and 2), two DMMs (case 3), one DSM and one DMM (differentiating the direction of the DMM mapping as cases 4 and 5), and ultimately two DMMs and one DSM (case 6). Figure 2-9 and Figure 2-10 illustrate these cases.

As it is very important "to determine which aspects are to be considered in order to answer a specific question" [Maurer 2007, p. 18], these cases can be used to collect and compile the existing information on the structure in accordance to the goal of the analysis. Maurer proposes two ways to do so: the opportunistic application ("see what you can get") and the requirements-driven application ("define what you need") [Maurer 2007, p. 93]. Each way is, at the same time, equivalent to the strategy of acquiring the data for the model. An MDM can thus be used either to gather native data or to combine data in a way suitable for an analysis. [Biedermann & Lindemann 2008] extend the proposition and suggest a way of calculating cycles across domains of an MDM instead of elements of a DSM to calculate single DSMs out of a series of matrices in an MDM; however, this strategy remains largely unexplored.

There are several **strategies to analyze** the DSMs generated. Classically, a DSM is used for sequencing, tearing, banding and clustering. In sequencing, the rows and columns of a flow-oriented DSM are rearranged in a way that as few relations as possible remain below the diagonal, thus reducing the number of active feedbacks, leading to an ideal sequence. However, such an ideal sequence cannot always be found[33]. Tearing consists of choosing the set of feedback marks that obstruct sequencing the DSM. The relations that need to be removed are called "tears". Banding rearranges the rows and columns in a way that blocks of parallel entities remain, which, for example, in a process can be executed independently of each other. Thus, a "band" represents a group of elements being active in parallel. Clustering is executed to find those clusters of entities that are mutually related. Figure 2-11 provides an overview. Maurer explains all techniques in detail [Maurer 2007, pp. 225-240]. Similarly, there are algorithms for the systematic analysis of DMMs, mostly focused on clustering [Danilovic & Browning 2007] [McCormick et al. 1972].

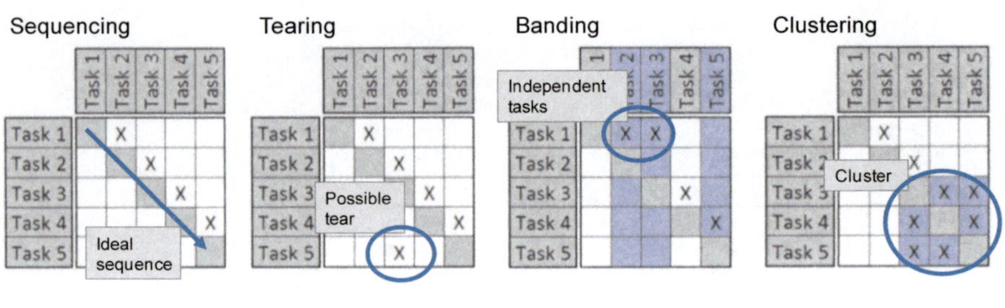


*Figure 2-11: Classic DSM analysis techniques*

Furthermore, DSMs can be analyzed for **structural characteristics**. The analysis is based on identifying patterns within a structure, i.e., typical constellations of entities and relations. A structural characteristic is defined as follows:

---

33 See page 164 for the limits of triangularization.

A structural characteristic is a particular constellation of entities and relations, i.e., it is formed by a particular pattern formed from nodes and edges in the graph [CARDOSO, J. 2006a] [MAURER 2007, p. 123]. The characteristic gains its meaning by the way the pattern is related to the actual system it is part of, i.e., it must serve a special purpose in the context of the overall system [BOARDMAN & SAUSER 2006]. A structural characteristic only possesses significance in the context of the system it is describing.

Figure 2-12 categorizes the common basic structural characteristics that are currently available in different works. The structural characteristics shown form the basic building blocks to analyzing a structure. An overview is available in [MAURER 2007, pp. 225-240], therefore the classification is not detailed further[34]. Eppinger similarly proposes that there are patterns across several domains [EPPINGER 2001]; however, he does not formalize them.

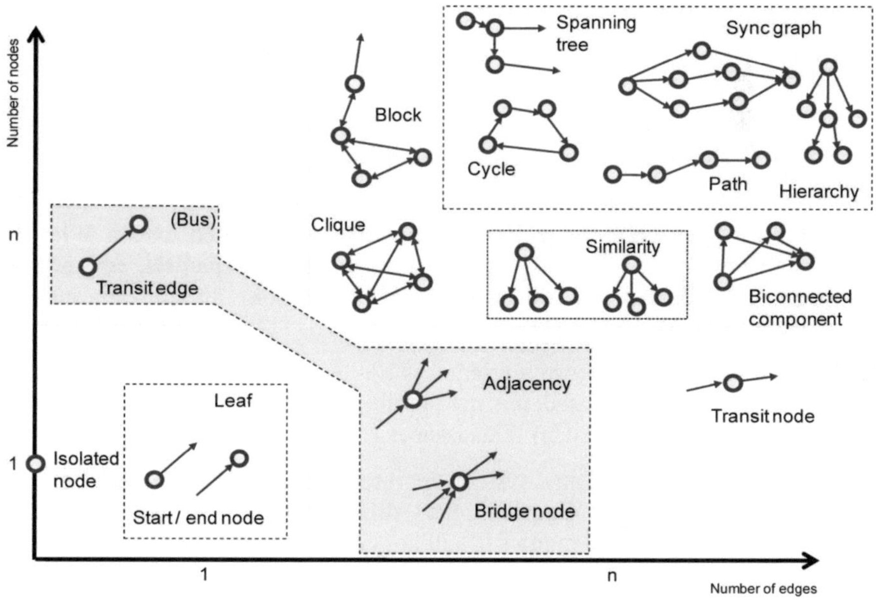


*Figure 2-12: Overview of basic structural characteristics*

Table 2-2 illustrates the available basic phenomena in graph theory [GROSS & YELLEN 2005] and shows which structural characteristics (Figure 2-12) they relate to. Although there is no complete one-on-one relationship between phenomena and structural criteria, the table regroups what phenomenon a structural criterion focuses on. For each, the table shows whether the mathematical phenomenon has an application in engineering design or not. As can be seen, there are a number of phenomena that have no application (yet).

[34] Structural characteristics will later be used to create structural metrics. They are detailed in section 5.1.

*Table 2-2: Phenomena in graph theory and adoption as structural criteria (based on [GROSS & YELLEN 2005])*

| Graph Theory | Structural Criteria available | Structural Criteria still missing |
|---|---|---|
| Cliques, subgraph | Strongly connected components | |
| Walks | Cycles, paths, distance | |
| Trees | Leafs, roots, spanning trees, knots | |
| Adjacency and degree | Neighborhood, bridges, degree / activity, independence, connectivity /attainability | |
| Genus | | Planarity, thickness |
| Weighted graphs and networks | Weighted nodes, weighted edges, minimum spanning tree, shortest path | |
| Coloring | | Chromatic number, k-coloring, color-classes |
| Multipartite graphs | N-partite graphs | Disjunctive sets |
| Eigenvalues | | Eigenspectra |

There are still **shortcomings to matrix-based methodologies** [KREIMEYER et al. 2008a].

Generally, attributes to an edge are only possible to a limited extent. Whereas matrix-based models are mostly designed as qualitative models and not as quantitative models, this fact still hinders the transfer from other models such as Petri-nets into a DSM.

Linking elements of a matrix to a relation is not possible in a matrix. While two nodes of a system can be related, it is not possible to link an element of a matrix to a relation (i.e., a node to an edge). This issue is resolved in section 4.4.2.

So far there has not been any systematic research to generate a catalogue of structural characteristics. Equally, the differentiation between structural characteristics and structural metrics is still not clearly defined.

Some fields of research have developed indicators to measure the degree of complexity of a system as a numerical value. Yet little work has been done so far. While the metrics developed as part of this research are not exhaustive, they are meant to fill this gap (see section 5.2).

Network structures in practice often contain decision points that can only be represented with limits in a matrix. There has been work undertaken to overcome this problem [BELHE & KUSIAK 1995], which is not complete, however, as it does not allow logic operators to be included as part of the matrix description. Section 4.4.3 completes the existing approaches.

Common variant design is addressed through commonality issues, for example, [BRAUN & DEUBZER 2007]. The modeling of decision points and differentiation of co-existing solutions with a common base (e.g., one body with different component assemblies) still remains largely unsolved. In process management, process alternatives are often designed and compared to generate an improved process. However, there is no methodology yet to generate such structures based on a common matrix-based description.

MAURER introduces a way of excluding several cells of a matrix upon certain conditions to describe boundary conditions [MAURER 2007]. Yet, matrix notation is still unable to work with more complex conditional settings; even for static, non-conditional relations within a system, no notation supports the use of existing algorithms.

The evolution over time (or another axis) remains unsolved, although many problems represented in matrices undergo changes, e.g., team structures; however, no real mechanism of evolution of a matrix has yet been found. A first approach has been shown [EBEN et al. 2008] based on Delta DSMs that map how one DSM differs from another [DE WECK 2007].

The management of hierarchical decomposition within a cell is still difficult to consistently describe [DANILOVIC & BÖRJESSON 2001]. Often, it is necessary to go into detail for a few cells only; while it is possible to zoom into such a matrix within one cell (it is very similar to an MDM), no description of how to handle the multitude of relations from the zoomed matrix going into one single row/column in the higher-ranking matrix is available.

The intuitive and graphical representation of MDMs is still unsolved. DIEHL proposes a 3D-vizualization using several planes, where each plane represents a DSM, and the space in between is used to show how two planes interrelate [DIEHL 2009]. However, this is only applicable for small systems. Yet, there has been ample research for visualization[35].

Ultimately, no framework for a goal-oriented analysis has been proposed to operationalize the opportunistic and the requirements-driven approaches systematically [MAURER 2007, p. 93]. More generally, the potentials and limits of analyzing structural complexity in different contexts have not been given much attention. Section 6 proposes a framework to operationalize the selection of metrics for the context of engineering design processes.

In summary, different matrix-based dependency models were shown that enable the modeling and analysis of complex systems for which all entities and relations are basically known. Therefore, these models can be seen as deterministic models. In the context of analyzing a process chart, these models are, therefore, suitable to serve as a means to collect the entities and relations from any process chart in a common form, providing a basis to design structural metrics. In practice, however, these models are often not as definite as they seem, as generating a process model is essentially a process of consolidating different opinions as to how an as-is process is actually run. This, however, is not the focus here.

All modeling methods are paired with different analysis approaches that produce different structural characteristics which aid the analysis of the systems being modeled. Yet, a number of shortcomings still exist, some of which are highly relevant for process modeling, as section 2.2 will show. In particular, the modeling and analysis of logic operators in a structure, the systematic aggregation of

[35] Examples could be, for example, graph spectra [MAURER et al. 2004], linking it to common models [KARNIEL & REICH 2007] or various kinds of graphic representations [LIMA 2007].

different domains and their relationships, and the goal-oriented application of analysis in the context of process analysis have not yet been resolved.

### 2.1.4 Network Theory

Additional means of analyzing large network structures are provided by Network Theory. The combined use of approaches from Structural Complexity Management and Network Theory, therefore, provides a comprehensive toolset for the analysis of highly coupled structures; furthermore, both consider processes, among other fields of application, and thus provide empirical evidence for their use in process analysis. For a better understanding, the analysis tools provided by Network Theory and the relevant network models are presented here.

Network and Graph Theory are closely related. Whereas Graph Theory is focused on the formal modeling and analysis of the interaction of single nodes and edges of networks of limited size[36], Network Theory regards global properties of large networks. Thus, Network Theory makes extensive use of graphs, but with a different analytic approach[37] mostly based on statistics. Network Theory aims at creating viable models for large network structures, finding statistical properties to describe the networks, and making predictions about their behavior. The networks can be of any type. Commonly, four groups are regarded: Social networks (e.g., friendships, organizational structures in business, or email exchange), information networks (e.g., academic collaboration, websites and hyperlinks, or patent citations), technological networks (e.g., distribution networks, phone lines, or computer networks), and biological networks (e.g., metabolic pathways, genetic regulations, or the food chain) [NEWMAN 2003a, p.7]. To all these networks, the basic properties applicable to graphs (directed or not, weighted or not, etc.) shown in the previous section apply.

Network Science differentiates different models, as Table 2-3 shows. The work on random graphs is mostly focused on models, as in [ERDOES & RÉNYI 1959]; generalized random graphs can be found, for example, in the works of [BOLLOBÁS 1981] and [NEWMAN et al. 2001]. Small world networks are commonly referenced in [WATTS & STROGATZ 1998].

Having these models available, the basic properties of large networks as currently available can be accounted for. In this context, emergence is an effect particularly important to Network Theory, acknowledging essentially the fact that certain patterns originate from even random networks, and it is these structural patterns that govern the behavior of the overall network.

---

[36] Graph Theory is not limited to a certain network size; however, a direct analysis of nodes is pointless for networks with millions of vertices, as commonly encountered in Network Theory [NEWMAN 2003a, p. 2].

[37] There are several review publications on the state of the art available: [NEWMAN 2003a] is the most complete, while [STROGATZ 2001] provides more examples. [ALBERT & BARABASI 2002] is also considered very comprehensive. [DOROGOVTSEV & MENDES 2002], [HAYES 2000b], [HAYES 2000a], and [CAMI & DEO 2008] have reviewed the state of the art on Network Theory.

*Table 2-3: Overview of statistical models for large, static networks (based on [NEWMAN 2003a])*

| Model | Basic idea of model | Applicability and critique |
|---|---|---|
| Random graph | Poisson distribution model: Uniform probability for a node to connect to another node | Mostly suitable for the theoretical observation of network properties. |
| Generalized random graph | Configuration model: Distribution of degrees is given, nodes connect at random | Ability to overcome the unrealistic degree distribution of a random graph, but inability to describe transitivity (also called "the strength of the weak ties", see page 54) |
| | Power law degree distribution model: Degree of nodes is distributed according to power law | |
| | Directed graph model: Degree distributions of degrees of incoming and outgoing edges | |
| | N-partite graph model: Similar to directed graphs, distributions are given for each domain of nodes | |
| | Degree correlation model: The correlation of the prevailing degrees in the network is given | |
| | Exponential / Markov Graphs: Few models available, only for special cases | |
| Small world model[39] | Random "shortcuts" across the network are created to improve the possibility of a node to reach another node by a short path | Suitable for most real networks; shortcuts often overrated in comparison to real world networks |

The most common property is the **size** of a network, described by the number of nodes, the number of edges, the mean degree of a node, the mean distance of two nodes, and the diameter (i.e., the longest geodesic[38] in a network).

The **small world effect** formalizes the phenomenon commonly known as “six degrees of freedom”, an experiment undertaken in the 1960s by Stanley Milgram to assess how many people a letter would pass through to arrive at a recipient unknown to the sender [KLEINFELD 2002]. Using the small world effect, the shortest path between two nodes of the network can be calculated as well as the mean path length of all paths connecting any two nodes in the network. Obviously, the more “small world” a network is, the quicker information is spread or a common opinion is generated (see Figure 2-13). Unlike matrix-based methods in engineering, however, the network models used have statistical properties that do not occur in technical systems. Thus, this effect is not directly transferable.

[38] A geodesic is the shortest path between a given pair of nodes.

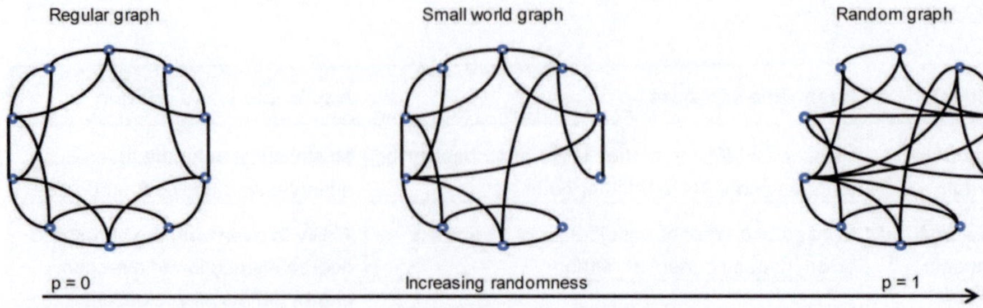


*Figure 2-13: Different network types[ WATTS & STROGATZ 1998]*

The **transitivity**[39] of a network addresses the fact that if node A is connected to nodes B and C, it is quite probable that nodes B and C are connected, too. Thus, nodes with a high degree tend to generate clusters around them, even if these are not explicitly present [WATTS & STROGATZ 1998].

**Resilience** is also called connectivity. The research interest in a network is how the average path length across the network changes if individual nodes are removed, and when the network falls apart into distinct groups; additionally, the size of these groups is used to characterize a network [ALBERT et al. 2000]. Typically, large networks are resilient against the removal of random nodes. This relates to the fact that common network structures consist of nodes with a low degree, thus being rarely involved in communication [NEWMAN 2003a]. The targeted removal of nodes with the highest degree, on the other hand, quickly breaks down the connectivity of the networks to such an extent, that with a few nodes removed, the network falls apart. This is especially true for scale-free networks (see next paragraph).

Networks in the real world possess different **degree distributions** [ERDOES & RÉNYI 1959] [BOLLOBÁS 1981]. While a completely random graph with equal probabilities *p* of a node being connected will generate a homogeneous network, a scale-free graph will turn to a hub-and-spoke-like structure (commonly measured using a histogram of the degrees in the network). See Figure 2-14 for the two cases. The application is highly relevant to judge the robustness of a network in terms of the random failure of a node and its target (i.e., an attack). While networks tend to remain generally intact if a node that is minimally connected drops out, removing a hub may cause the network to fail completely, as, for example, the failure of the power grid in the USA in 1996 showed [WATTS & STROGATZ 1998]. Scale-free networks have received particular attention, following a power-law degree distribution. In fact, most large networks, e.g., the internet, are scale-free networks, having dedicated hubs and few connected spokes [LI et al. 2005] [KIM et al. 2002] [BARABÁSI & ALBERT 1999].

---

[39] Also referred to as "clustering", which should not be confused with clusters in graphs or DSMs.

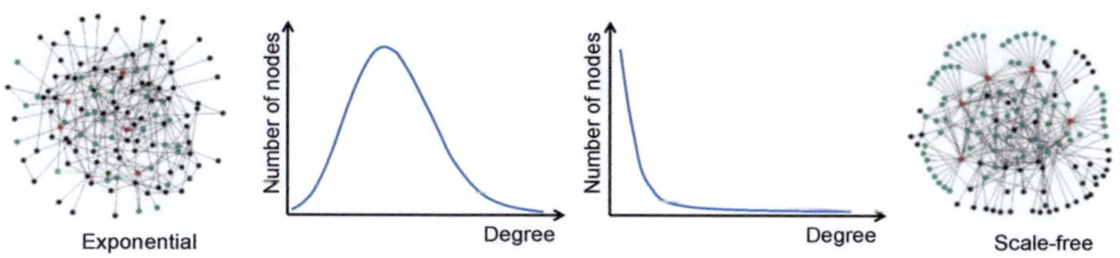


*Figure 2-14: Homogeneous exponential and inhomogeneous scale-free network after [ALBERT et al. 2000]*

In social networks especially, the characteristics of homophily or assortative mixing have led to research on **mixing patterns**. The interest is in how different nodes in a network establish groups of similarity of a kind that are not directly represented in the network [NEWMAN 2002] [NEWMAN 2003b]. A common example is a network of friendships in school. If the skin color of the children is introduced into the study, the network will sort the nodes in a way that distinct black and white subgraphs will appear [MOODY 2003]. A more special case is degree correlations; here, the pattern is about what constellations of degrees correlate [MASLOV & SNEPPEN 2002].

Social networks, in particular, show **community structures**, which are similar to the effect of clustering in DSMs (see Figure 2-11, right-hand side). A community is a group of nodes that are densely connected to each other and little connected to the outside network. To detect groups, cluster analysis is performed, which assigns a connection strength (much like the weight of an edge) to each pair of nodes [MOODY 2003] [NEWMAN 2003a]. However, more recent methods do not simply assign a community to a group of edges that are pairwise neighbors, but that are related via different geodesics across the group [GIRVAN & NEWMAN 2002].

Another feature of networks is the **navigation of the network**. While it is not only possible to find one's way across a large network even without knowing people explicitly ("a small world", [KLEINFELD 2002]), it is also common that individual paths across the network take shape.

**Centrality** is also a common property of individual nodes as well as networks [FREEMAN 1978]. Commonly, centrality is measured using the betweenness of a node, i.e., its position on a number of geodesics between all pairs of nodes in a network; the more geodesics a node is on, the more central it is because it is part of more communication paths across the network.

Finally, **motifs** have been recently researched. Motifs are parts of the network that, as a subgraph, are recurrent, i.e., they appear in a similar pattern in different places across the network [MILO et al. 2002].

In summary, Network Theory provides a set of structural characteristics that complement those provided by Structural Complexity Management. However, these structural characteristics are based on statistical network models that incorporate random network phenomena that do not necessarily occur in technical systems, such as shortcuts across the network, as found in Small World networks.

In fact, most networks that are regarded in Network Theory are networks from a social context. As processes are socio-technical systems, they do, in fact, exhibit these phenomena, too, to a certain extent. For example, an organizational structure in a process provides the basic communication structure; however, it does not elicit all communication channels available. The clustering coefficient that reviews the possible relations of a node is a good example, as it points to entities of the process that may be more importantly connected than shown in the actual network. Therefore, these structural characteristics provide a good complement to existing structural characteristics.

### 2.1.5 Other approaches to managing complex systems

**General Systems Theory** is interested in fundamentals, principles, and models valid for any kind of system, thus providing a general framework to describe how entities interact [VON BERTALANFFY 1968]. To this end, it applies differential equations. Systems Theory is based on the hypothesis of open systems, as opposed to many theories in physics, for example, i.e., systems that interact with their environment and are able to change their state through constant adaptation. General Systems Theory was later extended to New System Theory, differing mostly in the fact that the observer now was part of the system [PULM 2004, p.23]. It created the approaches of Self-Organization, Autopoiesis, and Dynamic Systems Theory. Self-Organization is, to some extent, related to emergence[40], i.e., how structure arises out of the relations and interactions of a system's entities. In particular, it regards how a system changes over time based on influences that originate from itself [DIETRICH 2001, p. 87]. Autopoiesis extends this concept to a level at which a system is able to reproduce itself based on the entities it is made of [MINGERS 1994]. Both approaches make use of four core principles inherent to open systems as proposed in General Systems Theory: Complexity (i.e., being a network of entities and their relationships), self-reference (i.e., the behavior of the system has an impact on the system itself), redundancy (i.e., entities that are in control cannot be separated from those that are being controlled), and autonomy (i.e., the behavior of the system can be detached – to some extent – from the environment) [PROBST 1987, p. 76].

**Cybernetics** [WIENER 1948] is closely related to General Systems Theory. Its main interest is in how complex, dynamic systems can be controlled to achieve a goal-oriented behavior [PROBST 1981, p. 7] regarding the control mechanisms among the entities of a system and the transfer functions that represent relationships. Cybernetics is an important foundation of management sciences [MALIK 2003, p. 80]. BEER bridged cybernetics and management science to create the field of Management Cybernetics [BEER 1972]. It is based on the concept of systems, and it incorporates the control mechanisms from cybernetics to model and determine how the actors in an enterprise are mutually interdependent and how they reach decisions by influencing each other [JACKSON 1991]. Cybernetics provides a paradigm where things are interdependent and certain organizational patterns directly impact the behavior of a system. This is also the basic understanding in this research.

---

[40] See section 2.1.4.

**Systems Engineering** is another discipline that uses the ideas from System Theory to better manage problems in engineering design. To do so, it extends the understanding of a system from the technical system under consideration to the project that creates the technical system [OLIVER et al. 1997, p. 85]. Systems Engineering makes extensive use of the network structure of the system and the control mechanisms that govern the system, and the context of Systems Engineering has given rise to a number of methodologies to model processes (e.g., IDEF, see appendix), Quality Function Deployment (see section 6.1.1), and various other models [INCOSE 2007]. A core concept is to link the behavior of a system (i.e., what a system does) to its structure (i.e., how a system is built), as proposed in this research [OLIVER et al. 1997, p. 21].

**System dynamics** equally regards systems, focusing on their internal network structure. However, it concentrates on the dynamics of this network to simulate its behavior in order to deduce improvement measures [FORRESTER 1977]. It is based both on quantitative models, mostly network-like flowcharts, and qualitative models to detect feedback loops that may either reinforce or balance the system.

**Operations Research** is interested in attributing resources to a problem under certain boundary conditions, using optimization algorithms to achieve an optimum solution. It applies methods such as linear programming, graph theory, scheduling algorithms, network theory, and different aspects of combinatorial analysis [FINKE 2008, p. ix]. Operations Research treats many problems that are similar in this research, and many algorithms for structural analysis have originated from Operations Research, especially regarding connectivity, shortest paths, matchings, and n-partite graphs [BIENA 2008].

**Information Theory** tries to quantify information, thus making possible many modern communication technologies. It applies a basic measure for the complexity of information, namely entropy, eliciting the average storage space necessary to recuperate a piece of information [SHANNON & WEAVER 1998, pp. 48-50]). It was later extended to Algorithmic Information Theory, combining it with the idea of determining the computability of an algorithm using a Turing Machine (i.e., a model of the computation of an algorithm). Tuning machines use the more formal Kolmogorov complexity as a measure for the computational resources necessary to describe a piece of information [BURGIN 1982].

To explain networks of economic transactions, **New Institutional Economics** is also related to the research presented here; however, it is mostly focused on explaining the rationale behind why two or more entities form a network. The Principal-agent problem regards, in particular, the "asymmetric" exchange of information within a network of potential partners, whereas Transaction Cost Theory focuses on the cost of exchanging information, and how this exchange can be organized most efficiently [KEIJZER 2007, p. 28].

### 2.1.6 Summary

With structure defined as the purposeful patterns that occur in the set of entities and relationships of a system, structural complexity prevails in different disciplines, creating different dependency models. Matrix-based models such as Multiple-Domain Matrices have only lately matured to a level able to describe large systems involving several domains and relationship types, and many questions remain unsolved. Yet, the method is able to embed many analyses that can be traced back to Graph Theory or Systems Theory.

*Table 2-4: Summary of features of structural analysis in different disciplines*

| | Structural feature | Meaning |
|---|---|---|
| Graph Theory | Adjacency | Immediate neighboring of two nodes |
| | Connectivity | Integrity of the overall network |
| | n-partite-ness | Existence of distinct, disconnected groups within the networks |
| | Paths | Channels of navigation through the network |
| | Cycles | Paths that end at their start node |
| | Reachability | Existence of at least one path to another node |
| | Planarity | Representation of network with no edges crossing each other |
| DSM | Sequencing | Ideal sequence of nodes in flow-oriented network |
| | Tearing | Iterations that inhibit an ideal sequencing |
| | Banding | Groups of independent nodes |
| | Clustering | Mutually related nodes |
| Structural Characteristics (MDM) | Isolated node | Node that is disconnected from the network |
| | Leaf | Node that is connected via only one edge |
| | Transit edge | Edge that lengthens a path without adding structure |
| | Transit node | Node that is transited by a path without adding structure |
| | Bridge node | Node that connects two structural characteristics |
| | Splits/joins | Node with few-to-many correlation of adjacent edges |
| | Bus | Combined split and join |
| | Hierarchy | Set of reachable nodes from a given node |
| | Similarity | Set of nodes similarly connected to rest of the network |
| | Biconnected component | Set of nodes that can be reached via at least two paths |
| | Spanning tree | Representation of minimum necessary reachability of a network |
| Network Theory | Size | Extent of network |
| | Small World Effect | Existence of shortcuts across network |
| | Transitivity | Probability of connectedness of neighboring nodes |
| | Degree distribution | Existence of hubs and spokes in network |
| | Mixing patterns | Relation of clustering to further attributes of the network |
| | Navigation | Relevance of shortcuts in Small World Network |
| | Centrality | Integration of a node into functioning of the overall network |
| | Motifs | Fractal patterns across different levels of abstraction |

Graph Theory and Network Theory have evolved in parallel, creating different techniques to systematically analyze structural characteristics, such as the centrality of an actor in a social network or planarity as a measure of understandability of a system. The recombination of the different appropriate structural characteristics from the different disciplines provides a comprehensive toolset for the analysis of any system, as shown in Table 2-4 (features appearing in several disciplines are listed only once). In particular, research in engineering and social sciences provides empirical evidence of the applicability of different structures in any kind of system and in processes specifically.

Furthermore, Multiple-Domain Matrices make it possible to create aggregate views that recombine different domains and relationship types into compact Design Structure Matrices used to analyze the long-range effects across several domains. The combination of Multiple-Domain Matrices with structural characteristics available in the different sciences, therefore, provides a comprehensive analysis tool for large complex systems. Yet, some gaps in modeling methodology still exist, especially the modeling of logic operators, the goal-oriented analysis and the aggregation of different domains. These issues will, therefore, be addressed later to complete the existing modeling methods.

## 2.2 Structural aspects of process management

Processes have been of interest for a long time. Process orientation was already proposed in 1934 [NORDSIECK 1934], yet process management did not really catch on until the 1980's [BECKER et al. 2005, p. 3]. A process-oriented organization is characterized by customer orientation, fewer interfaces, lower effort of coordination, clear responsibilities in terms of the results of the process, systematic improvement of process performance, process controlling, decentralized management, and organizational learning [SCHMELZER & SESSELMANN 2006, pp. 68-71].

In the following, the basics of process management are explained. In particular, those aspects that relate to the structure of a process are focused, i.e., dependencies and their patterns in process management. Quantitative analyses (e.g., towards lead time) are omitted.

### 2.2.1 Processes in Engineering Design

Process orientation has led to many different areas of research. In general, the more general notion of a process can be broken down into business processes and engineering design processes. Such a basic classification is, of course, not complete; there are many other taxonomies available, e.g., in primary (i.e., value-generating), secondary (i.e., doing the preliminary work for the value generation), and supporting processes (e.g., administration) [BECKER et al. 2005]. However, these are not addressed in this book, as the primary focus is on structure.

Table 2-5 lists common definitions of the term "process". Basically, a process is a set of interdependent tasks. Yet, a process is commonly characterized by its objective, its activities, its inputs and outputs, the events before and after it, its reference to time, and the resources used. Thus, many different aspects collectively enable a process, for example, resources or the inputs and outputs.

*Table 2-5: Definitions of the term "process" in literature*

| Reference | Definition |
|---|---|
| [BECKER et al. 2003, p. 4] | "A process is a completely closed, timely and logical sequence of activities which are required to work on a process-oriented business object. Such a process-oriented object can be, for example, an invoice, a purchase or a specimen." |
| [DAVENPORT 1993, p. 5] | "A process is simply a structured, measured set of activities designed to produce a specified output for a particular customer or market. It implies a strong emphasis on how work is done within an organization .... A process is thus a specific ordering of work activities across time and place, with a beginning, an end, and clearly identified inputs and outputs." |
| [HAMMER & CHAMPY 2003, p. 35] | "We define a business process as a collection of activities that takes one or more kinds of input and creates an output that is of value to the customer." |
| [HARRINGTON 1991, p. 9] | "A process is any activity or group of activities that takes an input, adds value to it, and provides an output to an internal or external customer." |
| [VAN DER AALST & VAN HEE 2002, p. 4] | "A process consists of a number of tasks that need to be carried out and a set of conditions that determine the order of the tasks. [...] A task is a logical unit of work that is carried out as a single whole by one resource. A resource is the generic name for a person, machine or group of persons or machines that can perform a specific task." |

The following definition for a **process** used for this research is based on [VAN DER AALST & VAN HEE 2002, p. 4], while additionally introducing the aspect of a *network* of tasks that are highly interdependent, which is of high relevance especially to processes in engineering [O'DONOVAN et al. 2005]. This definition is used, as it embodies the concept of a process as a multi-layered network (i.e., the first hypothesis of this research) and thus lays the foundation of assessing the structure of a process to deduce indications about its behavior.

> A process consists of interdependent tasks that exchange information via artifacts. The process is enabled and supported by the purposeful allocation of resources and time-oriented constraints. All of these entities are interrelated, on the one hand, via the input-output relationships among tasks along the principal process flow, and, on the other hand, via other relationship types that generate the overall process network.

In this context, a task is a logical unit of work that is carried out as a single whole by one resource over a period of time. A resource is the generic name for a person, machine, or group of persons or machines that can perform a specific task. All entities may have relationships among themselves.

A **business process** is "a special process that is directed by the business objectives of a company and by the business environment. Essential features of a business process are interfaces to the business partners of the company (e.g., customers, suppliers). Examples of business processes are the order processing in a factory, the routing process of a retailer, or the credit assignment of a bank" [BECKER et al. 2003, p. 4]. Such a process is, therefore, repeatable without the necessity to generate knowledge about the process execution.

An **engineering design process**, in contrast, is a process during which knowledge about an object is generated. As this object still necessitates designing, its nature is – at least in part – unknown. This generates uncertainty throughout the process that needs to be managed, and that causes an engineering design process to be much less deterministic than a business process. Table 2-6 distinguishes business processes and engineering design processes.

*Table 2-6: Difference between business processes and engineering design processes [VAJNA 2005, p. 371]*

| Business process | Engineering design process |
|---|---|
| • Processes are fixed, rigid, have to be reproducible and checkable to 100%<br>• Results have to be predictable<br>• Material, technologies, and tools are physical (e.g., in manufacturing) and / or completely described (e.g., in controlling)<br>• Possibility of disruptions is low, because objects and their respective environments are described precisely<br>• No need for dynamic reaction capability | • Processes are dynamic, creative, chaotic; many loops and go-tos<br>• Results are not always predictable<br>• Objects, concepts, ideas, designs, approaches, trials (and errors) are virtual and not always precise<br>• Possibility of disruptions is high because of imperfect definitions and change requests<br>• There is definitive need for dynamic reaction capabilities |

As can be seen, in engineering, design processes have the character of problem solving [LINDEMANN 2007, pp. 45-47], i.e., they cannot simply be processed but necessitate the generation of knowledge [HATCHUEL & WEIL 2003]. They thus represent a "wicked problem"[41] [RITTEL & WEBBER 1973]. Mostly, this is due to the high degree of novelty that is common for any product being designed. As a result, during the process, there is always a high degree of uncertainty present about the outcome – the earlier in the process, the more uncertainty there is in the process [LORENZ 2009, pp. 27-30]. In process management, this uncertainty takes shape especially in iterations, during which the design is reworked, improved and refined [WYNN et al. 2007] [ROELOFSEN et al. 2008]. Often, too, these iterations are not regularly cyclic, but they occur as leaps forward or backward in time [BADKE-SCHAUB & GEHRLICHER 2003]. While the process often creates quite

[41] A "wicked problem" refers to a problem that cannot be definitively described and that has no definite solution. Therefore, there are no optimal solutions, and solving the problem is hardly possible, only indications can be given.

erratic patterns due to this, artifacts, or rather the knowledge about their desirable properties, are characterized by their growing concretization during process runtime [KREIMEYER et al. 2008c], and the intermediate results that represent certain stages of this concretization determine the path of the process [VAJNA 2005] [PONN & LINDEMANN 2005], at times necessitating the re-planning of the process. At the same time, engineering design processes are often impacted by moving targets and late changes to the initial concept due to late learning during the design process. Overall, engineering processes, therefore, have a low degree of repeatability [VAJNA 2005], and they are difficult to model and plan [O'DONOVAN 2004]. Yet, their behavior follows basic patterns [EPPINGER 2001], namely the specific mutual dependencies between the organization, the process and the product architecture. These need to be well aligned and mutually adapted.

Engineering Design Processes are often carried out as **projects**, with the project organization bridging the organizational hierarchy and the common process [LINDEMANN 2007, p. 12 and p. 16]. A project, however, is understood as a "temporary endeavor undertaken to create a unique product, service or result" [PMI 2003, p. 5]. WYNN differentiates processes as mechanistic or projects as non-mechanistic [WYNN 2007, p. 84]; the interest in the process is more on the mechanism (or structural patterns, as in this research), while a project plan is more focused on the timeline.

In this context, the **control flow** is an important aspect. The control flow (also called control view) is the set of relations of the various entities of a project [SCHEER 1999, p. 102]; it is thus equivalent to the network of entities in the process definition applied here. The concept generally prevails in business process management. However, the term is used by other fields of science, as well.

**Process management** makes use of this understanding to analyze, design, implement, enact, monitor, and evaluate processes to improve value creation in the enterprise [ZUR MUEHLEN 2004, pp. 82-87], shown in Figure 2-15. The importance of the relations among the different entities of a process as a basis for the behavior of the process is commonly recognized; this has led to the understanding that improving the interfaces between different entities in a process provides the biggest leverage to obtain a more efficient process [RECHTIN 1991, p. 29] [WASSON 2006, p. 18] [FLURSHEIM 1977].

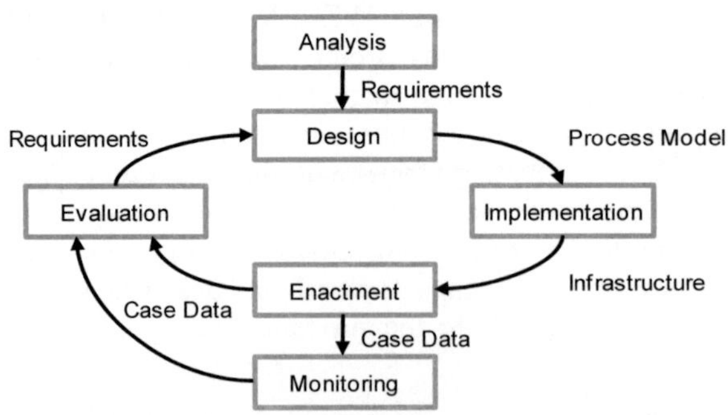


*Figure 2-15: Business process management lifecycle [MENDLING 2008, p. 5]*

Figure 2-16 visualizes the importance of how the paradigm, i.e., a certain perspective or understanding, drives the different activities during process management in engineering design [KREIMEYER 2008]; the model is similar to the Spiral Model in software development [BOEHM 1988].

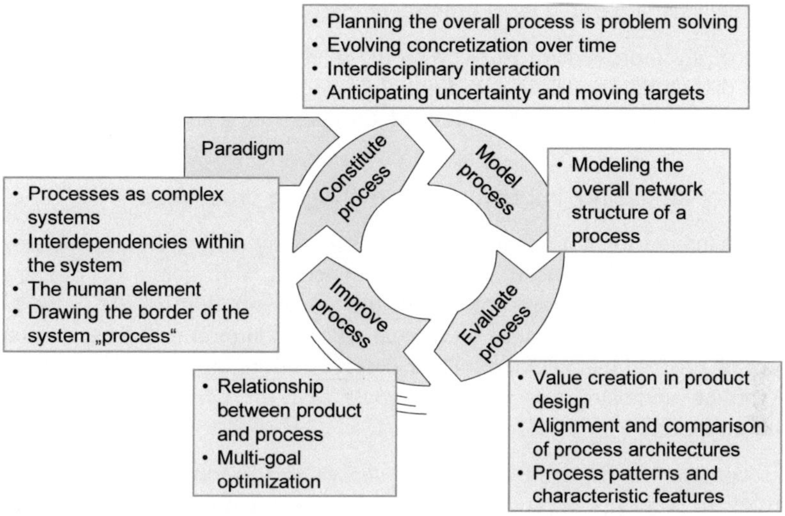


*Figure 2-16: Influence of the modeling paradigm and continuous process improvement [KREIMEYER 2008]*

The model works as follows: First, a model is prepared (the level of detail, the relevant views, the modeling language, and the access to information); then, problem areas of the process are defined and prioritized (i.e., the system border is drawn). Next, information is collected and models are created, and, last, the models are consolidated before they are analyzed for possible improvements [BECKER et al. 2005, 109-122].

In the context of this research, only structural aspects are of interest, and thus, the constitution only relates to collecting information about the relationships between the entities of the process. In the same manner, the model is a dependency model that then is evaluated, for example, for patterns that characterize the process's behavior, from which possible improvements are deduced.

As both figures show, the process model is the core component to process management [LINDEMANN 2007, p. 124], generating an overview about the current situation inside the company as the prerequisite for the improvement of a process. Of course, every process model is more or less simplified, made abstract, and reduced to the essentials. To model a process, usually the inputs and outputs, as well as the transforming tasks of the process, are captured, e.g., through workshops and interviews. The system boundary, in this context, describes departments, persons and facilities assigned to the system.

STETTER [2000] highlights his hypothesis that some weakness can be found in every design process model, for example, cost-intensive iterations in late phases, problems in finding and retrieving stored information, or problems caused by frequent product changes [STETTER 2000, p. 48]. Based on this hypothesis, it can be summarized that the identification of strengths and improvement potential in industrial practice not only consists of a search for strengths and weaknesses, but also includes the selection of the most prominent improvement potential.

The relevance of an improvement is given by the goals of a process improvement project, which dictate the modeling paradigm and which guide the later analysis of the model(s) created. The following section reviews possible goals more closely.

### 2.2.2 Goals of analyzing, improving and managing processes

Processes are managed for a number of reasons, satisfying different stakeholders, and there are various classifications of the concepts and goals of process management. Table 2-7 lists those aspects related to the structure of the process. These are adapted from the literature on typical errors, common problems, or the general intent of process management. Their categorization, as shown in the left-hand row, can be understood as common goals that for which processes are analyzed and improved.

The table is constructed from the references shown in the top row. From each reference, relevant concepts in process management were collected. In fact, some references directly address the goals of process management [KREIMEYER et al. 2008b] [BECKER et al. 2005, p.5, p. 30, p. 124], while others speak about the foci of process improvement in a more general manner [ZIMMERMANN 2008, p. 72] [SCHMELZER & SESSELMANN 2006, pp. 68-70] [IDS SCHEER 2007, p. 10] [GAITANIDES et al. 1994], and again others address typical problems in processes [BEST & WETH 2009, p. 77] [EUROPEAN FOUNDATION FOR QUALITY MANAGEMENT 1995]. All of these concepts were collectively classified with regard to their structural content, i.e., only those concepts that relate to the structure of a process to at least some extent were kept. In the context of this research, the concepts shown will be used as a framework to systematize a process in a goal-oriented manner. Section 6.1 will show common methods into which these concepts can be embedded.

*Table2-7: Different concepts in process improvement*

| Concepts of process management | [ZIMMERMANN 2008, p. 72] | [BEST & WETH 2009, p. 77] | [KREIMEYER et al. 2008b] | [BECKER et al. 2005, p.5, p. 30, p. 124] |
|---|---|---|---|---|
| **Planning** | • Long runtimes | • Long lead times | • Speed of design process | • Critical paths<br>• Time buffers |
| **Resource consumption** | • Effort for coordination<br>• Redundant tasks | • Redundant work<br>• Resource limits<br>• Resource availability | • Optimum between time, quality, and resources | • Cost reduction<br>• Obsolete processes<br>• Administration |
| **Quality** | • Redundancy to intercept errors<br>• Propagation of errors | • Fragmented tasks<br>• Errors in paperwork | • Quality | |
| **Flexibility** | | | | • Adaptation<br>• Flexibility and robustness<br>• Moveable activities |
| **Organizational decomposition** | | • Hierarchical reporting<br>• Coordination | • Design team coordination<br>• Setup of teams | • Adaptation of capacities |
| **Interfaces** | • Inefficient interfaces<br>• Insufficient supply of information | • Errors in transactions<br>• Disruptions (media, resources)<br>• Information: oversupply, outdated, incomplete | • Information exchange<br>• Operations / workflow management | • Integration of participants<br>• Optimization of interfaces |
| **Transparency of process** | | • Complexity of content<br>• High effort for maintenance | • Coordination of distributed design | • Process knowledge<br>• Standardization of workflows |
| **Planning** | • Repetitive tasks<br>• Iterations | • Expenses for coordination | • Optimization of internal processes<br>• Automatization of processes | • Speed |

*Table 2-7: Different concepts in process improvement (continued)*

| Concepts of process management | [EUROPEAN FOUNDATION FOR QUALITY MANAGEMENT 1995] | [SCHMELZER & SESSELMANN 2006, pp. 68-70] | [IDS SCHEER 2007, p. 10] | [GAITANIDES et al. 1994] |
|---|---|---|---|---|
| **Resource consumption** | | • High performance of process | • Efficiency<br>• Reduction of costs | • Use of methods<br>• Efficiency |
| **Quality** | • Consistency of information | | • Effectiveness | • Coherence<br>• Robustness (i.e., tolerance towards errors)<br>• Precision |
| **Flexibility** | | | | • Flexibility |
| **Organizational decomposition** | • Responsibility | • Distinct responsibility | | |
| **Interfaces** | • Interfaces within process | • Few interfaces | • Introduction of standards<br>• Harmonization<br>• Roles for collaboration<br>• Collaboration with external partners | |
| **Transparency of process** | | | | |

### 2.2.3 Process models and their structural content

Process models are an essential part of process management, as they help in understanding a process by representing the entities involved, their relationships, and the quality of their interactions. They are thus used for a variety of purposes which coincide with the different goals of process management, as shown before.

To assess process models for their structural content, it is necessary to understand to what extent each individual process model depicts a part of the structure of a process. In the interest of comparing what model is made for what purpose, BROWNING assesses different process models as to their focus and the different stakeholders interested in a process model [BROWNING 2009] [BROWNING 2008]. He concludes that while every model in his review is made for a different purpose, many models convey similar information.

To suit different needs, numerous methodologies for process modeling are available, and non-exhausted lists and comparisons are provided, for example, by

[BICHLMAIER 2000, pp. 43-61], [BAUMBERGER 2007, pp. 299-316], [SPATH & WEISBECKER 2008], [BROWNING 2009], [HEISIG et al. 2008], [LANGER et al. 2009], or [KARNIEL & REICH 2009]. [BROWNING & RAMASESH 2007] look more deeply into network-like process models, reviewing the literature published in the last decade. They conclude that, among other foci, the interaction of tasks and their impact on overall process improvement can be found in all existing models, yet needs more focus.

To be able to design a comprehensive approach that allows analysis of different process models in terms of their structural content, it is necessary to outline how process models can be compared and returned to one common denominator with respect to their structure. Both necessities are explained in the following section, starting with the latter aspect of reviewing how the interoperability of such models can be assessed to then analyzing models for their common structural content.

## Comparing and recombining process models

To develop an analysis method that suits not one but all common process models, a structured basis for its design is needed. To do so, a modeling framework needs to be created that incorporates the particularities of common process models, i.e., their specific modeling constructs, such as their semantics and semiotics. This section reviews the necessary methodology by looking more closely at the results of research on the interoperability of process models.

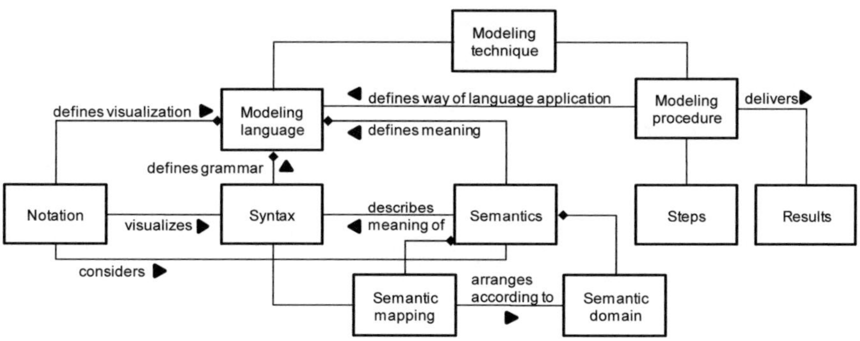


*Figure 2-17: Aspects of process modeling [KÜHN 2004]*

A process model involves many different aspects, summarized in Figure 2-17, for example, the semantics, their syntax, and the notation in which they are embedded. Existing process models vary in one or more aspects shown, and they do so to create an effective methodical support of one or several goals of process management. To do so, the semantics of the different modeling languages generally contain different aspects of, for example, scheduling, resource attribution or other domains. Other models, e.g., extended Event-driven Process Chains (EPC, see appendix 10.1.1), focus on the preparation and possible implementation of a process support through information technology and, therefore, integrate IT-systems and information objects into their semantics.

Again, YAWL (Yet Another Workflow Language, see appendix), for example, was designed to support the setup of the best possible workflow systems and, therefore, strongly focuses on formally correct decision logics, so-called workflow patterns, in the modeling scheme, while leaving out many other aspects that EPC integrates. However, if metrics are to be applied to estimate the complexity of a certain property of a process, it is necessary to have a basis to make these measures interoperable [CARDOSO 2005b].

As stated above, many process models are very similar to each other, following only marginally different foci, and a lot of work exists on the so-called **interoperability of process** or enterprise models.

To allow process models to be compared, BECKER & PFEIFFER [2008] suggest analyzing process models in terms of their semantics and syntax, and thus find two classes of conflicts that can arise between any two process models (language and ontology-based). To systematize the comparison of process models and to overcome these conflicts, HÖFFERER [2007] describes a meta-model-based approach to achieve a better interoperability between different process models by using ontologies[42] to compare how "close" one process model's meta-model is to the next [HÖFFERER 2007]. He thus proposes a terminological level at which models become comparable and which can be transferred from one model to another. The model is extended to a meta-level across the meta-models of different process modeling methodologies, referred to as a meta$^2$-model (see Figure 2-18).

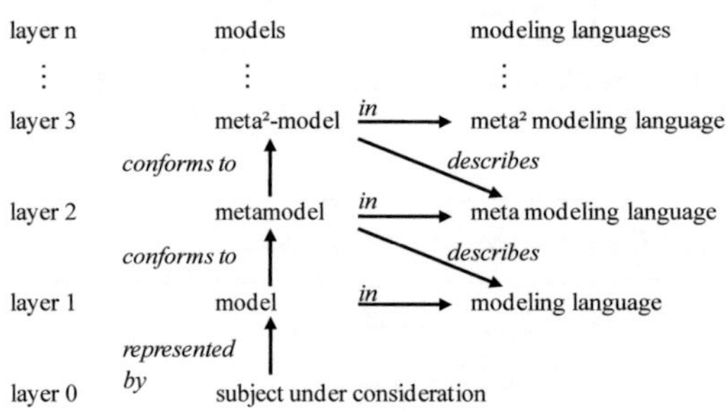


*Figure 2-18: Different meta-levels to compare process models [HÖFFERER 2007]*

Whereas modeling language-based conflicts limits the exchangeability of process models [BECKER & PFEIFFER 2008], the aspect of reviewing the semantics of processes at the level of the meta$^2$-model is of interest to see if different process models offer similar structural content, namely similar domains and relationship-types. This approach is quite similar to the SEMAT methodology which compares

[42] For the ontology-based comparison of process models, see also [GUIZZARDI et al. 2002], [PFEIFFER & GEHLERT 2005], and [SIMON & MENDLING 2007].

process modeling methodologies using a meta$^3$-model as a framework for the comparison [KLUTH et al. 2008]. In the following, the meta$^2$-level is chosen to compare the meta-models of different process modeling methodologies.

## Comparison of common process models and their structures

With the goal of analyzing processes and their structure, this section reviews common process models as to their structural content. Common models are, in this context, those models that are either considered standard in industry according to [IDS SCHEER 2007] or that serve as a common reference in research with [BROWNING & RAMASESH 2007] as the main reference. The interest of this analysis is to generate a meta-model later for structural process modeling able to accommodate common process models (see chapter 4 of this book), thus becoming an analysis adapter for existing process models.

Table 2-8 represents the most common process models used for engineering design process analysis. All models are described in detail in appendix 10.1 of this book. In the appendix, each model is described as a flow-oriented model according to the individual modeling conventions, and it is represented as a specific meta-MDM that shows all domains contained in the respective modeling language as well as the relationship types that can be found. Figure 2-19 shows the SADT meta-model on the left-hand side with four basic types of activities that generate four domains in the meta-MDM on the right-hand side. Those domains that are related can be seen in the flow-chart model; their relationship types are given in the meta-MDM to the right-hand side of Figure 2-19.

*Table 2-8: Common process modeling methodologies*

| Process modeling methodology | Acronym |
|---|---|
| Extended Event-driven Process Chains | eEPC |
| Object-oriented Event-driven Process Chains | oEPC |
| Business-Process Modeling (Integrierte Unternehmensmodellierung) | IUM |
| Unified Modeling Language | UML |
| Structured Analysis and Design Technique | SADT |
| Integrated Definition Method | IDEF0 / IDEF3 |
| Business Process Modeling Notation | BPMN |
| Yet Another Workflow Language | YAWL |
| P3 Signposting | |
| Petri-Nets | |
| Process Module Methodology | PMM |
| Program Evaluation and Review Technique | PERT |
| Objektorientierte Methode für die Geschäftsprozessmodellierung und -analyse | OMEGA |

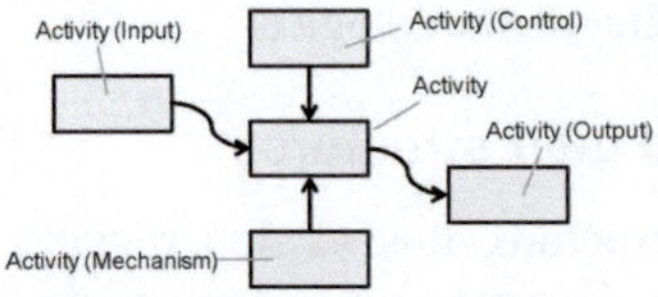


| | Control | Activity | Input / Output | Mechanism |
|---|---|---|---|---|
| **Control** | | • controls | | |
| **Activity** | • serves as control for | | • has output of | |
| **Input / Output** | | • is input for | • output of... is input for... | |
| **Mechanism** | | • supports | | |

*Figure 2-19: Example of a SADT as a flow-chart meta-model and as a meta-MDM to show the structural content*

Using the meta²-level, the process models from Table 2-10 were **compared concerning their structural content**. Figure 2-20 shows the general approach, using an example of two partial process models in EPC and IUM on the lower level, their meta-models on the middle level, and the regrouped domains on the top of the figure. The relationship types are not shown in the figure but are, of course, part of the analysis, as well. The analysis consists of two steps. First, common domains are collected to constitute the domains of a meta$^2$-model. Then, relationship types among these domains are established by collecting common relationship types of the individual meta-models for all process modeling methodologies.

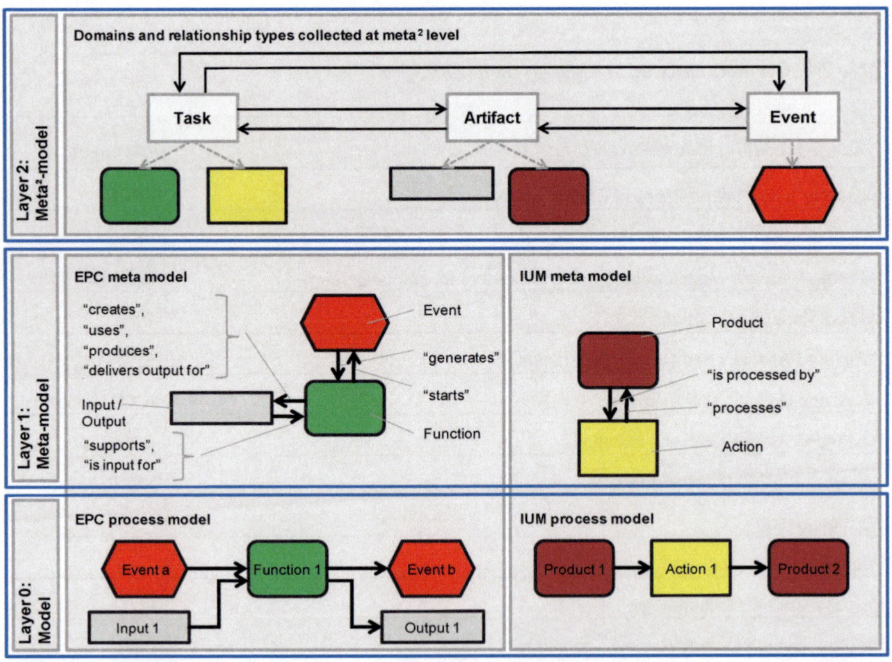


*Figure 2-20: Example of the meta²-model approach taken here to compare process models and their structural content, aligning an EPC and an IUM Model at level 3 of the model by [HÖFFERER 2007]*

Table 2-9 shows the domains of each process meta-model. Based on the descriptions of each model, these domains were regrouped into eight categories that represent common domains of process modeling. All descriptions and the necessary references are given in the appendix.

The table was designed like that of Table 2-7. Again, all individual classes of objects of the chosen process modeling methods were collected based on their descriptions in the references. For example, IUM lists "actions", "functions", and "activities" as descriptions for a task according to [MERTINS & JOCHEM 1998]. All classes of modeling objects were collectively classified. The resulting domains are listed in the top row of Table 2-9.

*Table 2-9: Regrouped domains of structural interest for all 13 reviewed process modeling methodologies*

| | Task | Artifact | Control | Event | Org. unit | Res. | Time | Product |
|---|---|---|---|---|---|---|---|---|
| eEPC | Function | Input / output | | Event | Org. unit | Resource | Milestone | (Output view) |
| oEPC | Method | Object | | Message | | Attribute (resource) | Attribute (object variable) | |
| IUM | Action | Product | | Order | | Resource | | |
| UML | Activity | | | Initial / final node | Respons-ibility | | | |
| SADT | Activity | Input / Output | Control | | | Mechan-ism | | |
| IDEF3 | Unit of behavior | Objects / object states | | Label | | | | |
| BPMN | Activity / gateway | Data object | | Event | Pool / Lane | Pool / Lane | | |
| YAWL | Task | Data | | Start / end node | | | | |
| P3 | Task | Parameter | | | | Resource | Time / duration | |
| Petri | Transition | | | Place | | | | |
| PMM | Work package | Input / output information | Informat. object / document | | Tool / method | | | |
| PERT | Task | | | Event, time | | | | |
| OMEGA | Activity | Business object | | | Org. unit | Technical resource | Milestone | External object |

There are specific modeling constructs that can be understood as additional domains, as Table 2-10 shows: P3 and PMM allow the specific decomposition of the tasks in the process, forming a domain of their own. Similarly, BPMN and YAWL introduce an individual domain to represent logic operations. BPMN furthermore models text annotations explicitly as a separate domain, although such annotations are common in all process models and, in fact, do normally not represent a specific domain of their own.

*Table 2-10: Further domains that occur among reviewed process modeling methodologies*

| | Decomposition | Logics | Annotation |
|---|---|---|---|
| BPMN | | (Gateway) | Text annotation |
| YAWL | | Condition | |
| P3 | Process | | |
| PMM | Process chain | | |

The same collection and classification can be done for the classes of relations that each modeling method provides. However, as the descriptions in appendix 10.1 show, many process modeling methods are not very specific about the different relationship types. For example, for SADT (Figure 2-20) no description is given at all; therefore the actual meta-model and the references only indicate an input/output relationship type [MARCA & MCGOWEN 1988]. The result of this collection of relationship types is not shown here but is taken up in section 4.3 to construct a meta-model for structural process modeling that entails all relevant domains and relationship types.

In summary, all process models contain aspects of the structure of a system to some extent, as they all consist of "boxes and arrows". While some models specify the structure very strictly, others leave more room for process modeling experts to adapt the model. Yet, it is possible to review existing models at a more abstract level to find a common denominator, both with regard to the domains involved and their relationships. Section 4.2 takes up that concept to generate a meta-model that later serves as a basis for the application and interpretation of structural metrics for engineering design process analysis.

### 2.2.4 Strategies to analyze design processes and models

Process analysis is a common buzzword and a wide field of research[43], and it has been so for many decades now. Generally, there are specific methods to analyze

[43] For example, [CHAMPY & HAMMER 2007], [DE BRUIN et al. 2000], [BECKER et al. 2005], [SCHMELZER & SESSELMANN 2006], and [GAITANIDES et al. 1994] provide an overview of general process management. [CLARKSON & ECKERT 2005] and [FAHRWINKEL 1995] review approaches specifically for engineering design.

and deduce improvements for each individual aspect of process management, as shown in Table 2-7. There are also two kinds of strategies: continuous improvement and radical reengineering [HAMMER & CHAMPY 2003]. Both, however, are based on modeling a process and the analysis of the process models.

Engineering design is closely connected to problem solving [LINDEMANN 2003] [LINDEMANN 2007]. In turn, a process must support the best possible problem solving, and, therefore, the structure of the process organization needs to be closely connected to the product architecture.

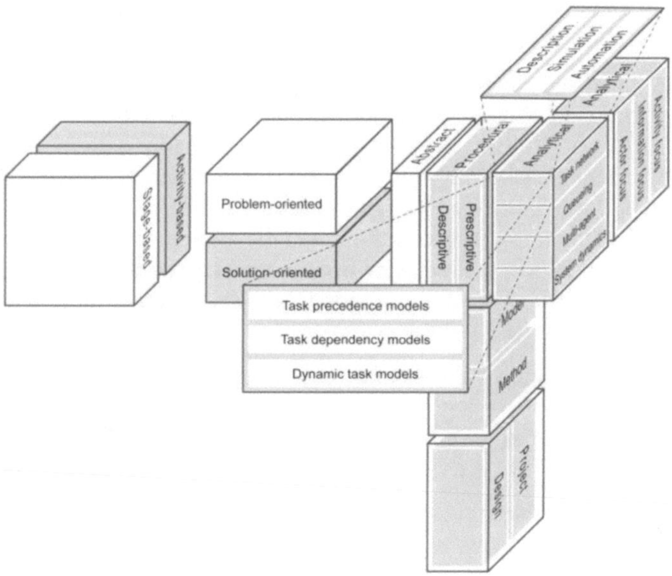


*Figure 2-21: Framework of organizing process models and their analysis [WYNN 2007, p. 60]*

This ambiguous borderline between process improvement and product design makes it difficult to systematize approaches to process improvement in engineering design. WYNN organizes process models, including the modeling and improvement intents in a framework consisting of several layers [WYNN 2007, pp. 16-60], as shown in Figure 2-21. The classes represent specific views of the design process, and each has created different models and improvement approaches.

As Figure 2-21 shows, the structure of a process is only addressed implicitly, which reflects the fact that there has been minimal attention in research on process improvement so far. In Wynn's system, structure is a part of the analytical approaches, and is addressed to activities, information transfer, or the actors. To this end, different strategies exist (task networks, queuing models, multi-agent simulations, system dynamics), but there is no overall approach that supports the selection of any of these quantitative, detailed and labor-intensive methodologies from a more qualitative point of view.

Generally, different methods of analyzing and improving processes are available [FREISLEBEN 2001, pp.71-74]: Experiential, analysis of historic structures, identification of duplication of work, analysis of interfaces, capability of communication, calibration of milestones, and comparison to reference models. However, these do not relate to structures directly except for the analysis of historic process models, which are the focus of this research.

The structure of processes has generally been addressed from a semantic point of view, having created many procedural models[44] to guide the design process; however, these remain at a very rudimentary level. At a more sophisticated level of detail, a DSM is generally used to interrelate the tasks of a design process to improve sequence, communication, or synchronization [BROWNING 2002] [KUSIAK et al. 1995] [STEWARD 1981].

Other research has enriched the simple dependencies in these models to include, for example, the possibility to overlap tasks [KRISHNAN & EPPINGER 1997], the uncertainty inherent in a task [CHALUPNIK et al. 2008], decision options and scenarios [CLARKSON & HAMILTON 2000] [WYNN 2007], the different aspects of behavior of the actors of the tasks on a strategic level, such as coordination techniques, cooperation techniques, organization models, project management, and others [WHITFIELD et al. 2000], the behavior of actors on an operational level, e.g., resource attribution, scheduling, and others [COATES et al. 2000], and complex simulation [KARNIEL & REICH 2009].

Besides these approaches, which are focused mostly on the principal process flow as a set of tasks and their supportive entities (e.g., resources), another stream of research has focused on how the design process can be best aligned for the product architecture [SOSA et al. 2004b] [KREIMEYER et al. 2007c] [DANILOVIC & BROWNING 2007].

Little work has been done outside these basic metric designs, which relate mostly to product complexity. However, individual work on, for example, customer integration [KAIN et al. 2008], the role of social networks [LIBERATI et al. 2007], or enterprise architectures [WALDMAN & SANGAL 2007] have shown the need to manage the structure not only among the tasks but across all entities and domains of a design process [EPPINGER 2001].

So far, no comprehensive approach is available for the structural analysis of a complete process, and there seems to be no framework for the systematic improvement of a process's structure. There are, however, many fragmented methods and algorithms available that support the analysis and improvement.

### 2.2.5 Summary

A process can be understood as a system made up from different domains and relationship types, and it forms a network structure which only serves its purpose as a whole. While common business processes have already become highly

---

[44] Typical procedural models are, for example, the VDI 2221, the Munich Procedural Model [LINDEMANN 2003], the SPALTEN model [ALBERS et al. 2005], and the V-model of the VDI 2206. An overview can be found in [BAUMBERGER 2007, pp. 72-77].

complex, in engineering design this complexity is even greater through growing specialization and the distributed generation of knowledge.

Process management offers different methodologies to improve such processes, and improvements follow certain goals. Those goals relevant to structural process improvement were reviewed and consolidated to serve as a basis to set up a framework to organize structural metrics for process analysis.

An important foundation of process management is the modeling of processes. Each modeling method brings with it different domains and relationship types for the set of common aspects regarded by process management. Using a meta$^2$-model, it was shown that process models in research and industry have common structural content that can be used to access the process structure embedded in these models; furthermore, different process models can be combined and/or compared that way with regard to the structure of the process. Nevertheless, there is no comprehensive approach to analyze the structure of a process in a goal-oriented manner. Although different methods for process analysis exist, these remain largely unconnected.

## 2.3 Metrics to analyze the structure of a process

This section reviews metrics as a means of the systematic analysis of large systems[45]. First, the foundations of measuring are introduced; then the different aspects of good measurement of the complexity of a structure in network, software, processes, and engineering design, are reviewed. In particular, existing structural metrics are reviewed in detail to increase the understanding of the existing basis for developing metrics able to characterize different structures in an engineering design process.

### 2.3.1 Basics and measurement foundation

**Metrics**[46] are a means of representing a quantitative or qualitative *measurable* aspect of an issue in a condensed form [HORVÁTH 2003]. This "measurement is a mapping of properties of empirical objects to formal objects by a homomorphism" [ZUSE 1998, p. 92]. As such, a metric is therefore intended to depict an actual situation in a reduced and efficient manner.

**Measurement theory**[47] [ZUSE 1998] provides the basis for the design of such metrics. It describes how a phenomenon can be measured by establishing mapping

[45] "Large" will not be specifically defined, as it addresses essentially the fact that a system has many entities and relationships that are complex and thus difficult to handle.

[46] Measurement theory also refers to scales and measures, which, in this context, will be used synonymously. In economics, the term "performance indicator" is used, as well; however, it is not applied here, as it implies rating a good or bad performance rather than a basic metric [KAPLAN & NORTON 1992].

[47] Measurement Theory is succinctly reviewed in [SUPPES & ZINNES 1963] (available at http://suppes-corpus.stanford.edu/article.html?id=43, accessed 9.8.2009) and in [LUCE et al. 1988]

of an empirical concept to a mathematical concept. Measurement foundation addresses three common problems [STEVENS 1946]:

- The *representation* problem addresses the fact that a numerical scale should represent the relations that prevail in the real world.
- The *uniqueness* problem assesses the invariance of a metric to basic mathematical transformations.
- The *meaningfulness* problem allows the possibility of drawing conclusions from measured value.

Besides these foundations, the *validity* of a metric is, of course, of particular interest. The goal is to see if a metric actually represents what is supposed to be measured. The validity of a metric, therefore, largely relates to the meaningfulness problem. It can be broken down into three aspects [MENDLING 2008, p. 106], visualized in Figure 2-22:

- *Content*: Is the full range of possible meanings of the object represented?
- *Criterion*: Is the measured aspect the correct one to represent the topic of interest?
- *Construct*: Is the criterion in line with theoretical reasoning?

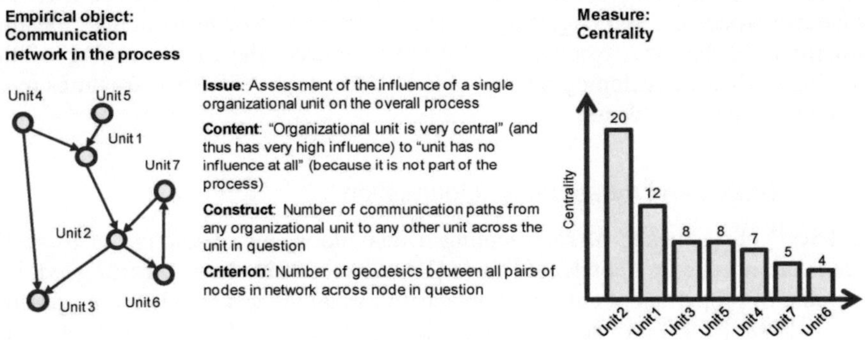


*Figure 2-22: Example of a structural metric and the different aspects of its description*

These problems are approached by different kinds of metrics. The scale hierarchy [STEVENS 1946] classifies metrics as nominal, ordinal, interval, and ratio scales. Identification numbers, for example, are *nominal* values, attributing a name to an empirical issue. *Ordinal* scales, such as the weight of an edge that relates to, for example, the intensity of the use of the communication channel represented by the edge, relate directly to the proportions in the empirical domain. *Interval* scales preserve only the relative distance between two empirical observations. *Ratio* scales, ultimately, extend the interval by a zero that expresses the absence of an empirical observation.

Metrics can also be classified as fundamental and derived [ZUSE 1998, p. 95]. Fundamental metrics commonly emerge at an initial point of research; derived measures aggregate one or several fundamental measures.

There is ongoing discussion about what kind of metrics can be seen as "real metrics" [AICHELE 1997, p. 74]. Some researchers argue that only ratios are real measures, as they not only yield a result but also a scale provided by the baseline of a fraction. Other researchers see the problem pragmatically, arguing that any metric that is able to express an empirical problem in a meaningful way is a useful measure. In this research, the latter view is followed.

Overall, two basic strategies to generate specific measures are possible: Either, well-understood fundamental metrics can be applied to a comparable context to generate new derived measures, or new fundamental metrics can be used. In this research, as presented in the following chapters, mostly new derived metrics are developed that use pre-existing empirical foundations.

A set of related metrics is commonly organized as a **measurement system**[48]. The goal of establishing such a system is to organize metrics concerning their number, precision, and appropriate allocation based on their commonalities and relations among each other [AICHELE 1997, p. 79]. A measurement system can thus be defined as an "ordered set of metrics that are semantically related, that complement each other and that – as a set – are intended to represent an empirical issue in a well-balanced and complete way" [LACHNIT 1976].

A measurement system, therefore, is intended to structure a complex issue[49] in the real world in a condensed way, while making it possible to detail individual aspects as needed. Thus, an issue can be accessed in a structural manner, while it also serves as a framework to compare different objects under observation [SCHÜRRLE 1995, p. 14]. Like metrics per se, such a system intends to be homomorphous containing aspects that are of relevance in the real world. A measurement system can be structured in four different basic ways [AICHELE 1997, p. 81]:

- A *mathematical system* relates metrics by calculating combined or derived metrics from fundamental ones. This way, hierarchies of metrics, like the DuPont-System of Financial Control, are set up.
- *Practical systems* apply factual logic to relate metrics; relations are usually empirically established, such as the RL-System [AICHELE 1997, p. 81].
- A *heuristic system* is more focused than a practical system, developed explicitly to solve a specific issue and relate metrics to this issue. The GQM-approach shown in section 6.1.2 is an example that will be used later.

[48]Also referred to as "scorecard", "metrics suite", or "ratio system" [REICHMANN & LACHNIT 1977]

[49] Measurement systems are most commonly found in economics, e.g., the balance sheet analysis, the DuPont-System of Financial Control, or Tucker's Managerial Control Concept. Compare, for example, [AICHELE 1997, pp. 84-109] for an overview of measurement systems in economics or [GEIGER 2000, pp. 129-133] for such systems in engineering management.

- An *empirical system* is focused on statistically significant metrics that have originated from empirical observation. In contrast to a heuristic system, the attribution of a metric to an issue is considered more objective.

In practice, measurement systems are used as early-warning instruments, as means of analysis (e.g., for benchmarking processes or spotting improvement potential in a process), or as management tools to plan and control a complex system [GEIGER 2000, p. 99].

With the above definitions, TSAI and KERNLER summarize the **requirements of metrics** as follows [KERNLER 1996, pp. 35-38] [TSAI et al. 1986]:

- *Purposefulness*: Metrics should provide sufficient informational content to describe the issue in question.
- *Homomorphism*: Metrics should be designed to be as homomorphous with the behavior of original data as is possible and purposeful.
- *Simplicity*: A user who should be able to understand the metric easily.
- *Consistency*: A user should produce the same result when measuring the same process.
- *Automation*: The metric should be suitable for process automation.
- Metrics must be *additive*: If two independent processes are put into a queue, the value of the complexity metric should be at least the sum of the single values.

In addition, Weyuker's properties have become the established reference for metrics design, especially in software engineering [WEYUKER 1988]. They provide a set of properties that any good metric should fulfill. The set is, however, rather generic and is criticized for this [MENDLING 2008, p. 117]. In fact, a metric that is correct in terms of Weyuker's properties can still be meaningless, i.e., Weyuker's properties do not acknowledge the basics of measurement foundation [CHERNIAVSKY & SMITH 1991]. Secondly Weyuker's properties deny the fact that a single metric cannot capture complexity in all its facets [ZUSE 1998]. Still, the properties are commonly applied to define metrics [CARDOSO 2005a].

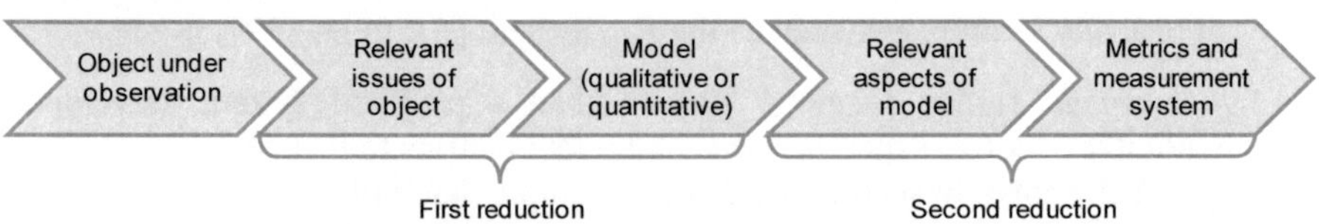


*Figure 2-23: Modeling process of representing a system as a measurement system [GEIGER 2000, p. 95]*

The **procedure to design metrics** commonly involves two stages of reducing a system [GEIGER 2000, p. 95]. In the first stage, the system from the real world is reduced to its relevant issues and then quantitatively or qualitatively modeled. In the second step, the relevant aspects of this model are reduced and quantified. A similar model is proposed in [MUTSCHELLER 1996, pp. 63-83]; the author, however, extends the procedure to include the implementation and review of the applicability of the measures. In this research, essentially, the second reduction is the focus, as the metrics are used to analyze existing process models.

### 2.3.2 Metrics to describe networks

Metrics to describe network structures are generally derived from applied graph theory and network theory, as described in sections 2.1.2 to 2.1.5. They are usually grouped into three categories [BRANDES & ERLEBACH 2005]:

- *Element-level metrics* assess the position of a single element within the overall network.
- *Group-level metrics* regard the constitution and quality of groups[50] of elements within the overall network.
- *Network-level metrics* characterize the properties of the overall network.

MENDLING lists the most common metrics that are commonly used to describe a network structure [MENDLING 2008, pp. 107-109]:

- The *degree*, i.e., the number of edges connected to a node represents the connectivity of a single node.
- The *density* of a network, i.e., the ratio of existing edges to the maximum number of possible edges measures the cohesion of the overall graph.
- The *centrality* of a node, based on different methods [FREEMAN 1978], represents the cohesion of a network around a central node.
- The *connectivity*, i.e., the number of nodes that need to be removed for the graph to be unconnected, measures the level of homogeneity of the network.

Network metrics thus incorporate different aspects from Graph Theory; in fact, many of these metrics only take shape in different applied sciences, which will be reviewed below.

### 2.3.3 Metrics in software engineering

Software metrics are highly relevant to process management, as a software program and the control-flow graph of a process are very similar; several authors have drawn attention to the fact that executing a software is much like running a workflow or a process [CARDOSO 2005b] [GRUHN & LAUE 2006a] [ROLÓN et al. 2006a].

In software engineering, metrics are a popular basis for quality assurance[51]; they are employed to measure the degree of complexity in software to estimate the

[50] A module is a group of highly interconnected entities, whereas these high interconnections have been planned by the structural organization and are therefore predefined.

[51] An overview of the foundations is provided in [NAVLAKHA 1987]; the recent state of the art can be found in [ZUSE 1998] and [DUMKE & LEHNER 2000]. An annotated bibliography is

level of error that the software is likely to encounter and to design test methods adapted to a specific new software development [ZUSE 1998]. Typically, thus, the metrics are selected top down [MENDLING 2008, pp. 110-111], i.e., the metrics are applied to measure certain aspects of a software program that are part of the quality assurance. This is why they are generally regrouped using the Goal-Question-Metric measurement system that will be shown in section 6.1.2.

*Table 2-11: Common structural metrics for procedural programming paradigms*

| Metric | Description |
|---|---|
| *Lines of code* [AZUMA & MOLE 1994] | Lines of code measure the size of a program. This size varies substantially among different programming languages; thus the measure is of limited informational value (compare a dictum by Bill Gates: "*Measuring programming progress by lines of code is like measuring aircraft building progress by weight.*"). |
| *Cyclomatic number* [MCCABE 1976] | The Cyclomatic number provides the number of paths through a program to visit all nodes, thus assessing the number of decision points. It is mostly used to define test cases. The original publication mentions a threshold of 10, beyond which the complexity should be justified. Microsoft's Visual Studio 2008 reports a violation for values exceeding 25. |
| Halstead's metrics: *length, vocabulary, volume, difficulty, effort* [HALSTEAD 1977] | Halstead disconnects the complexity of a program from the complexity of the language (number of distinct operators) the program is written in (total number of operators used); he also differentiates – in the same way – the variables (operands). Yet, only lexical complexity is measured. |
| *Information flow metric* [HENRY et al. 1981] | The information flow metric measures the fan-in and fan-out of a module, i.e., the degree of a connection of a module to the outside world. The metric is commonly used to estimate the likelihood errors. |
| Oviedo's *data flow complexity* [OVIEDO 1980] | The data flow complexity displays the degree to which a program can be broken down into distinct blocks that are executed as a unit and only have relations at a definition-level. It shows, thus, the potential for decomposition. |
| *COCOMO metric* [BOEHM 1981] | COCOMO is a cost estimation approach that uses the programming effort estimation metric as a basis for cost estimation. This metric is based on the size of a program (i.e., lines of code) and the programming language. |
| *Defect density* [ZUSE 1998, pp. 36-40] | The defect density represents the number of defects found in each module of a program. It is most helpful to locate error-prone parts of the software and to concentrate programming effort (as outliers among all modules). |

Here, only basic metrics are presented, which will be of interest later. Commonly, there are two kinds of metrics, as there are two fundamentally different programming paradigms: Procedural programming uses a series of function calls, plus connecting split and join operations (e.g., goto, for…then, etc.), to constitute the control-flow of a program, whereas object-oriented programming uses classes

available at http://ivs.cs.uni-magdeburg.de/sw-eng/us/bibliography (viewed on 15 February 2009).

to define and instantiate objects, which then use methods to transform data. Thus, their control-flow graphs vary substantially.

Table 2-11 lists common metrics for procedural paradigms from the literature. They are typically based on counting function calls along the control flow; this is in line with the programming paradigm, as the metrics are generally oriented to mimic the execution of the program.

For object-oriented programming, the cohesion (i.e., the degree of the functional relationship of one entity to another entity) and the coupling (i.e., the interdependence between two entities) are important [WAND et al. 1990] and drive the metrics as shown in Table 2-12.

There are many more metrics available in software engineering. However, most of them are more conceptual and have not made their way into design practice. Two other measures, however, more abstract metrics have influenced many other metrics substantially. The *entropy* of information is part of information theory [SHANNON & WEAVER 1998] and measures the degree of disorder in a system; it was developed as part of the development of modern telecommunication facilities and has impacted the measurement of runtime complexity decisively. *Kolmogorov complexity* is a measure for the shortest program that can output a given string [CARDOSO 2006]; it plays an important role in the formalization of software code and has, as such, laid the foundation for the formulation of effort calculation as shown above, for example, in Halstead's metrics.

*Table 2-12: Common structural metrics for object-oriented programming paradigms*

| **Metric** | **Description** (for details, see [WAND & WEBER 1990]) |
|---|---|
| *Methods per class* | The metric methods per class sum of all method-related metrics (all others) to estimate the overall complexity of a program. |
| *Depth of inheritance tree* | The depth of the hierarchy measures which attributes are inherited from the original class and estimates the impact and propagation of possible changes (i.e., how many other classes can be reached by a change). |
| *Number of children* | The number of successors of a class estimates the impact that changes within the program have on a particular class (i.e., the reachability). |
| *Coupling between object classes* | This metric enumerates the number of other classes to which a class is related to estimate errors and clear the structure of a program. |
| *Response for class* | This metric defines the size of the set of methods of a class that can respond to a specific message to estimate the run-time complexity of a program. |
| *Lack of cohesion* | This metric provides a comparison of variables and methods, and estimates thereby the necessity to split a class into two or more classes. |

In summary, metrics from software management can be classed as highly relevant for process analysis, as software is similarly characterized by many decisions that cannot be analyzed in a deterministic manner. Software engineering offers several metrics, including empirical foundations, because of their relevance in software design. The transfer of these foundations to process management is highly possible, as the following sections will show.

### 2.3.4 Metrics in process management

Structural metrics for process assessment are a recent development [GHANI et al. 2008]; while quantitative measures have been used for a long time, the pioneer work on qualitative metrics occurred only in the 1990's with the assessment of Petri nets of the first structural and dynamic metrics [LEE & YOON 1992].

Today, a number of metrics exist[52]. Yet, these metrics have neither been consolidated nor compiled into a coherent body of knowledge; so far, the knowledge on qualitative assessment of processes and their structure is still fragmented [MENDLING 2008, p. 114].

Generally, two kinds of metrics are in use: those that are mainly used for the prediction of the behavior of a process, and those that are used to estimate errors in a process model [GRONBACK 2006] [GRUHN & LAUE 2006a]. However, the border between the two kinds cannot be clearly defined, as often process models are designed in a way that is not completely free of errors, while deviations from a semantically and semiotically correct model are intended to transmit a certain purpose or meaning [MENDLING et al. 2007]. Table 2-13, therefore, does not differentiate the two kinds.

Overall, many metrics are similar and use comparable concepts that have been, in part, embodied independently from each other. At the same time, many approaches remain conceptual and without empirical evidence. Lastly, few authors provide detailed algorithms that can be used to compute[53] the metrics.

Metrics for business processes represent thus the main contribution that is used to answer the research question behind this book. Although many were developed for business processes and workflows, they can be transferred to engineering design processes without limit, as engineering design processes are basically a subclass of a business process, being more complex and less deterministic in their overall behavior (compare Table 2-6).

---

[52] A detailed overview of metrics with a structural focus on workflows and business processes is provided in [MENDLING 2008, pp. 110-117]; the earlier work is also comprehensively summed up in [CARDOSO et al. 2006], and [GRUHN & LAUE 2006a].

[53] As complexity metrics are difficult to compute, tool support is necessary. Currently, there are three software tools available that embody a wide set of metrics: KOPeR [NISSEN 1998], EPCMetrics [GRUHN ET AL. 2006], and STAN [BECK & STUHR 2008].

*Table 2-13: Overview of common metrics in process management (based in part on [MENDLING 2008]*

| Metrics | Description |
|---|---|
| *Distinct paths, hierarchy levels, cycles, diameter, parallelism* [NISSEN 1998] | As one of the first concepts for metrics to describe a process, Nissen's metrics are meant to facilitate the (re)design of a process. Nissen counts the number of independent paths, the levels of hierarchy found in the model, the number of independent cycles, the overall diameter (i.e., the longest path), and the degree of parallelism among activities. |
| *Simplicity, flexibility, integration, efficiency* [TJADEN et al. 1996] | Tjaden et al. developed their approach focusing on establishing a balance between the four metrics they designed. Simplicity represents basically the size of the process; the flexibility and integration level are not actually calculated but scored using a checklist-like function point approach. Tjaden et al., however, do not provide a detailed concept of efficiency. Therefore, their approach is criticized as too abstract [BALASUBRAMANIAN & GUPTA 2005]. |
| *Size, length, structural complexity, coupling* (*concurrent* and *sequential contribution, arcs of a subnet*) [MORASCA 1999] | Morasca bases his metrics on a detailed, formalized theoretical foundation to assess Petri-net process models. As size, he uses the number of places and transition, as length the diameter of the network, as structural complexity the number of distinct paths, as concurrent contribution a count of edges, as sequential contribution a count of places, and finally an assessment of modules and their incident and outgoing edges as a measure for modularity. However, his approach is purely theoretical and not empirically supported. |
| *Network complexity, Cyclomatic number, reduction coefficient, restrictiveness, number of trees* [LATVA-KOIVISTO 2001] | Latva-Koivisto uses established foundations to build a set of metrics that describe a process in a complete manner; however, his approach is criticized as incomplete, as it does not differentiate different entities in the process model, and it lacks more tree-related metrics [MENDLING 2008, pp. 115]. The network complexity is similar to the density of a network, the cyclomatic number is similar to that by [MCCABE 1976], the reduction coefficient calculates the number of steps to reduce the network to a maximum path length of 1, the restrictiveness estimates the number of possible sequences in a process, and the number of trees assesses the existence of hierarchies that permeate the process. |
| *Cohesion of functions, events, and data objects* [DANEVA et al. 1996] | As a toolkit for EPC, the cohesion of functions among each other represents the intensity of involvement of a function in the control-flow; events and data objects in the control-flow are assessed similarly. The authors conclude that their measurement suite is mainly suitable to identify error-prone fragments of the process. |
| *Cycle Sets* (*number of cycles, entry nodes into cycle, exit nodes out of cycle*), *structuredness* (*existence of deadlocks*), *nesting depth* [GRUHN et al. 2006] | The authors extend, essentially, the count of cycles as in [NISSEN 1998]: They integrate the count to those nodes that serve as an entry point or an exit to cycles. Equally, they assess process models for the existence of deadlocks (which later is completed by [MENDLING 2008]), i.e., joins and splits that are not harmonized. Ultimately, they introduce a measure for the nesting depth, i.e., the maximum number of splits in a given path across the process. |

*Table 2-13: Overview of common metrics in process management (continued)*

| Metrics | Description |
|---|---|
| *Control-flow complexity* [CARDOSO 2005a] | The control-flow complexity is an adaptation of McCabe's Cyclomatic number, integrating the characteristics of how the process can continue after each decision point that splits the control-flow. Well validated, it is not designed to predict errors. |
| *Log-based complexity* [CARDOSO 2007] | Log-based complexity extends Cardoso's control-flow complexity, having a structural focus, to a run-time measure that assesses the process complexity based on the log data of how the process is executed; it is, thus, only suitable for formalized workflows. |
| *Maintainability (analyzability, understandability, modifiability,* all using the *size, complexity, coupling* of a process) [ROLÓN et al. 2006a] | Rolón et al. estimate the maintainability of a process, breaking it down into three concepts: The analyzability addresses the probability of discovering errors in the process, the understandability estimates the ease of comprehension of the model, and the modifiability is a measure for the probability of making the correct changes to erroneous models. The metrics are based on size (counting all entities in the process), complexity (based on the number of dependencies), and coupling (between activities). |
| *Degree of automatization, activity automation, role integration, role dependency, activity parallelism, delay risk* [BALASUBRAMANIAN & GUPTA 2005] | This set of metrics supports process planning by supporting the evaluation of different alternatives for a given process. To this end, they extend the existing pool of metrics by a few more metrics that address the possibility of automated process execution and decision making where possible, the purposeful integration of resources, the measurement of the degree of parallelism among tasks, and the identification of structural bottlenecks that possibly cause delays. |
| *Cognitive weights* [GRUHN & LAUE 2007a] | The key idea of their approach, being based on [SHAO & WANG 2003], is to assign a cognitive weight to basic software control structures: the more difficult a control structure is to understand, the greater its cognitive weight. The approach is based on empirically funded coefficients for basic patterns in the control-flow. |
| *Cognitive cross-connectivity* [VANDERFEESTEN et al. 2008] | The proposed metric measures the strength of the links between pairs of entities in the process based on the hypothesis that encompassing "the structural relationship between any pair of model elements" is a complex mental operation. |
| *Understandability* [GHANI et al. 2008] | Besides other common complexity measures, Ghani et al. introduce the understandability of a process model as a contribution to its maintainability. They propose calculating it based on anti-patterns, for which, however, they do not provide empirical evidence. |
| *Size, density, partitionability, connector interplay, cyclicity, concurrency* [MENDLING 2008, pp. 117-130] | Mendling collects and extends metrics and their empirical foundations to verify process models and predict possible errors. These metrics make use of the strong formalism that guides the EPC process modeling language, and the different metrics embody these formalisms to assess EPC models for their formal correctness. |

### 2.3.5 Metrics for engineering design processes

Generally, metrics in engineering design processes are meant to serve three purposes: estimation, monitoring, and performance measurement. Yet, there is generally little specific work on metrics for engineering design processes available[54]. This is mainly because product development has the nature of "a mental exercise" and because of "a lack of easily identifiable items to measure" [BASHIR 1999]. It is true that the existing metrics, therefore, remain either highly specialized, or they are conceptual and hard to apply.

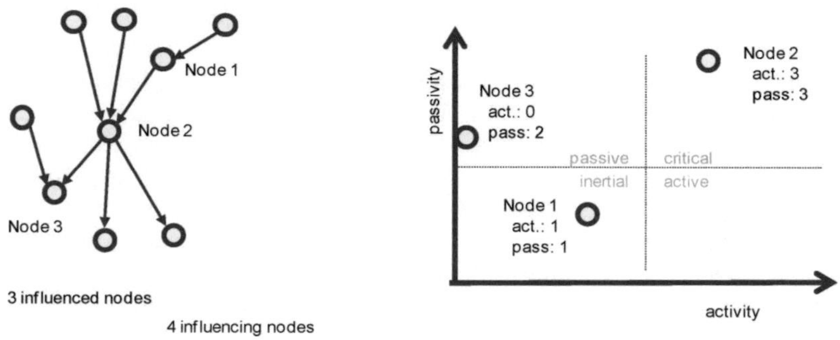


*Figure 2-24: Activity and passivity of the elements of a system*

The most common metrics[55] are very simple. Having originated from systems engineering, the measures of *activity* (also called active sum) and *passivity* have a strong structural focus. They compare the immediate impact of neighboring entities on each other and thus are similar to the degree-measure that is a common basis in graph theory [DAENZER & HUBER 2002, pp. 558-560 ] [LINDEMANN 2007, pp. 73-76].

Other metrics in engineering design incorporate the numerical evaluation of the product or process architecture to some extent. Mostly, these metrics are designed as effectivity and efficiency measures, for example, for attributing the necessary manpower [NORDEN 1964] or for estimating the *degree of efficiency* of a design process [O'DONNELL & DUFFY 2005, pp. 70-79]. Similarly, there are metrics to assess the impact of project characteristics on process planning [CLARK 1989]. These involve, for example, the level of risk in new product development, which

---

[54] [BASHIR 1999] provides a sound overview of the state of the art of metrics in engineering design at the time of writing, while [HORNBY 2007] provides a recent overview of common and more specific measures (which are not further regarded here).

[55] Score evaluations and similar methods are common in engineering design, e.g., in risk management [LINDEMANN 2007, p. 276]. As they have no explicit focus on structure, they are not considered here.

is broken down into metrics for *product innovation*, *product complexity*, *design maturity*, and *schedule pressure* [ZURN 1991]. Also, lead time can be broken down into the driver's *product complexity*, *management complexity*, and *amount of charge* [GRIFFIN 1993]. LIU ET AL. establish a metrics-based process review from a structural point of view [LIU et al. 2003]. They apply theoretical measures to assess the *critical degree* (a weighted measure of the task precedence in the process), the *likelihood of error occurrence* (based on an estimation of the novelty of a product and the availability of knowledge in the company), and the *spread degree of a task* (a weighted measure of the reachability of subsequent tasks) to support the planning of a process based on other process reviews.

As can be seen from these few process-based approaches, the estimation of the design complexity[56] (i.e., a measure of how complex a product is) is the focus in all approaches, as the product complexity necessarily drives the process complexity. Early works measure design complexity by the *coupling between the design targets and their variables* [DIXON et al. 1988]. SUH develops different *metrics for function coupling* [SUH 1999], and other authors introduce *estimators of design complexity* [BASHIR & THOMSON 1999] [SUMMERS & SHAH 2003]. HORNBY ultimately introduces *modularity* (based on the number of modules and the degree of their interaction with the overall system), *reuse* (measuring the repeated occurrence of entities within the design) and *hierarchy* (similar to the nesting depth in process management) as classifying measures for product complexity [HORNBY 2007].

A measure that focuses purely on process complexity is illustrated by [SCHLICK et al. 2008]. Here, a numerical DSM is used to model project dynamics; the approach is able to cope with large teams "who make at least partially autonomous decisions on product components but also strongly interact in their impact on project performance".

There are, of course, many other measures available that, however, do not relate to structural complexity but are measures used in, for example, benchmarking projects, process audits, or performance measurement for management and organizations (mostly financial and operational measures in project management [PMI 2003]).

In summary, although engineering design science strongly focuses on model building, there is virtually no work on complexity metrics available [BENSON 2007], even less so for processes. This can be attributed in part to the fact that engineering design processes can be treated like business processes concerning their analysis; however, the specific interpretation basis for such processes cannot be directly transferred, and there is a large gap in science about the meaning of the available structural metrics for processes.

---

[56] Compare [AMERI et al. 2008] for a detailed comparison of existing measures of product complexity.

### 2.3.6 The limits of using metrics in an organization

GRIFFIN points out that measurement is an important first step towards process improvement [GRIFFIN 1993], as it is important to have a sound overview of the initial state, the needs and development potential, and a final comparison after improvement measures have been implemented.

Yet, metrics are no remedy for any problem. "What you measure is what you get" was the driving dictum for the research of the Balanced Scorecard, at the time revolutionary, as it allowed a wider, balanced picture that substituted a former management that was only guided by financial goals (see section 6.1.3 for more details). To this end, metrics already show their biggest disadvantage: While reducing the complexity of the representation of an issue, they tend to oversimplify or omit dependencies of an issue, thus making the representation incomplete [KAPLAN & NORTON 1992]. It is, therefore, necessary to select a group of metrics to represent a problem in a balanced way.

Secondly, metrics directly impact the behavior of personnel in a company. In particular, in the management concept "Management by Objectives", metrics are a common basis for the evaluation of the performance of personnel [DRUCKER 2007, p. 261]. The concept is based on goals for each member of a company, whose fulfillment is measured to assess the individual performance [ODIORNE 1980, p. 82]. As part of this measurement, the motivation of an employee is directly related to the results of the measurement; in turn, a measurement influences the behavior of an employee in a positive way, but it can equally demotivate [MUTSCHELLER 1996, p. 61]. Thus, measures need to be chosen carefully to ensure they are not influenced by staff out of fear or personal ambition. This leads to two consequences: One the one hand, staff needs to be integrated early into the process of installing measures in the company to achieve a transparent measurement system that is favorably accepted and thus maintained; on the other hand, metrics need to be changed at regular intervals to ensure that the staff also considers other relevant specifics of a situation that are not part of the measures in place.

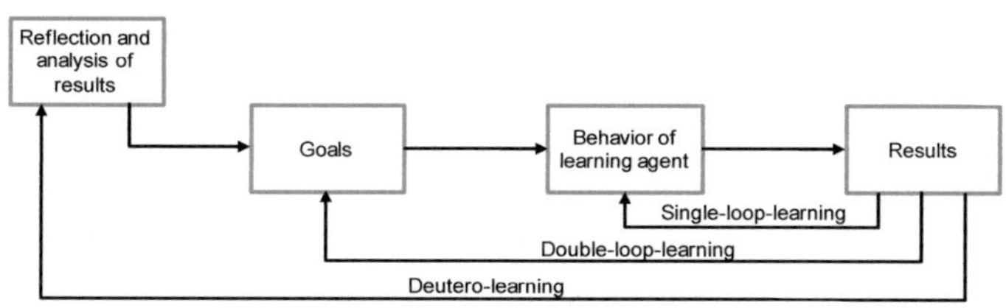


*Figure 2-25: Three types of organizational learning*

Organizational learning similarly stresses the fact that "maps and images" guide the behavior of individuals in an organizational setting [ARGYRIS & SCHÖN 1978, pp. 17-19]; metrics can serve as such images that are considered attainable and therefore guide individuals without considering other aspects. In this context, individuals take on the role of a so-called learning agent, whose learned behavior is guided by the impact of their actions [ARGYRIS & SCHÖN 1978, p. 29]. Three types of learning are common, which involve more reflection of the third type, the deuteron-learning (see Figure 2-25). Generally, single- and double-loop learning is guided by the results of the behavior, upon which measurement of the results can have a detrimental influence if metrics are designed in a way that neither stipulates reflection nor is set up in a holistic way to avoid one-sided behavior.

In engineering practice, metrics are mostly used for project controlling; they are rarely used for concept and test design [FINK & HAMPP 2005]. An empirical study in the software industry revealed five common strategies as to how engineers commonly cope with metrics in a company [SIMON & SIMON 2005]:

- Optimism-strategy: Personnel disapprove of the metrics, perceiving them as implicit criticism and a constraint on their professionalism.
- Delegation-strategy: The results of metrics are attributed to external reasons that are not related to an individual engineer's work, and thus responsibility is denied.
- Automatism-strategy: Problems that surface through the metrics are blamed on tool-support (automated workflows, or, in software engineering, code generators).
- Particularity-strategy: As design in problem solving, metrics are not recognized as relevant to the specific issue, and their validity for common situations is denied.
- Tortoise-and-Hare-strategy: Refers to the fact that an issue was already improved before metrics were introduced, and engineers tend to turn away from a problem.

Generally, it can be stated that metrics provoke a lack of emphasis on the environment in terms of a goal (and its respective measure) due to a lack of understanding of the actual system [DEMING 1994]. Three "traps" are inherent in such measurements:

- Common and special situations are little differentiated by such measurements, although criteria are not universally valid. Typically, measures depend on other external influences; deflections thus must exist.
- A single measure is not the best criterion to judge an issue, and often the focus is misplaced. A more comprehensive set of metrics helps avoid single-sided improvements. The overemphasis of individually measured aspects can, in some cases, lead to "bogus metrics", i.e., metrics that mislead the company [BOLLINGER 1995].
- The common assumption that improving a single measure improves the overall system is wrong, as in most cases a more holistic view is needed to correctly assess a system.

It is possible that a set of metrics is misleading in the long term. It is generally recommended that measures be alternated from time to time and there be overlapping measures that exercise a certain control over each other, which, at the same time, lowers the risk of manipulation of an individual measure [KAPLAN &

NORTON 1992]. The goal is to lower the risk of simply seeing the measure and not seeing the object behind the measure. Furthermore, a balanced set of several measures works best to achieve a comprehensive picture.

Metrics for analysis should, therefore, not be applied in isolation, as they do not describe "goodness" per se. Rather, they point to possible quality issues that need to be further evaluated [BOLLINGER 1995], which in this research is targeted by assessing outliers to show how a process improvement project can be prioritized and where possible improvements can be expected.

### 2.3.7 Summary

As the most reduced model possible, metrics are able to represent a system in a condensed form to show important characteristics and to point to important aspects. In process analysis especially, metrics are a tool for the identification of weak spots and their conditions [BENSON 2007]. As such, metrics can introduce the risk of reduction past a meaningful limit (reductionism).

At the same time, metrics need to be used carefully in a company, as they necessarily influence the behavior of staff, especially in the context of management by objectives, where objectives are related to measures. They should, therefore, be used mainly to focus on further investigations [BENSON 2007].

There is a substantial body of metrics available that is able to assess the structural complexity of a system with a view to different patterns. These metrics are used, on the one hand, to discover modeling errors, and, on the other hand, to better understand a system's behavior through the measurement. Many different metrics for the analysis of network structures in all kinds of disciplines exist and can be transferred to process management; yet no comprehensive compilation is available. At the same time, the transfer to the specifics of engineering design processes, i.e., what behavioral aspects relate to what structural characteristic evaluated in a metric, remains unsolved. There is, especially, no systematic listing of the significance of the available metrics against the domains and relationship types common to process management.

Commonly, metrics are not independent of each other but can be organized in a measurement system (according to, for example, focus, goal, granularity). This enables the systematic and goal-oriented employment of metrics. This is especially important in regards to the structural analysis of a process, as a metric can only be purposeful in the context of a goal and the related semantics; metrics, therefore, cannot be designed without a meta-model that provides a semantic context to later interpret the metric.

*Table 2-14: Available metrics (see section 2.3) and structural characteristics assessed by metrics*

| | | Graph theory | | | | | | | DSM | | | | Structural complexity mmgt. | | | | | | | | | | Network theory | | | | | | | | |
|---|---|---|---|---|---|---|---|---|---|---|---|---|---|---|---|---|---|---|---|---|---|---|---|---|---|---|---|---|---|---|---|
| | | Adjacency | Connectivity | n-partite-ness | Paths | Cycles | Reachability | Planarity | Sequencing | Tearing | Banding | Clustering | Isolated node | Leaf | Transit edge / node | Bridge node | Splits / joins | Bus | Hierarchy | Similarity | Biconnected comp. | Spanning tree | Size | Small World effect | Transitivity | Degree distribution | Mixing patterns | Navigation | Centrality | Motifs | none |
| Procedural prog. | Lines of code | | | | | | | | | | | | | | | | | | | | | | X | | | | | | | | |
| | Cyclomatic Number | | | | X | | X | | | | | | | | | | | | | | | | | | | | | | | | |
| | Halstead's Metrics | | | | | | | | | | | | | | X | | | | | | | | X | | | | | | | | |
| | Information flow | X | | | | | | | | | | X | | | | | | | | | | | | | | | | | | | |
| | Data flow complexity | | | | | | | | | | | X | | | | | | | | | | | | | | | | | | X | |
| | COCOMO metric | | | | | | | | | | | | | | | | | | | | | | X | | | | | | | | |
| | Defect density | | | | | | | | | | | | | | | | | | | | | | | | | | | | | | X |
| Object-orient. | Methods per class | X | | | | | | | | | | | | | | | | | | | | | | | | | | | | | |
| | Depth of inherit. tree | | | | | | | | | | | | | | | | | | X | | | | | | | | | | | | |
| | Number of children | | | | | | | | | | | | | | | | | | X | | | | | | | | | | | | |
| | Coupling btw. classes | | X | | | | X | | | | | | | | | | | | | | | | | | | | | | | | |
| | Response for class | | | | | | | | | | | | | | | | | | | | | | | | | | X | | | | |
| | Lack of cohesion | | X | | | | | | | | | | | | | | | | | | | | | | | | | | | X | |
| Process management | Distinct paths | | | | X | | | | | | | | | | | | | | | | | | | | | | | | | | |
| | Hierarchy levels | | | | | | | | | | | | | | | | | | X | | | | | | | | | | | | |
| | Cycles | | | | | X | | | | | | | | | | | | | | | | | | | | | | | | | |
| | Diameter | | | | | | | | | | | | | | | | | | | | | | X | | | | | | | | |
| | Parallelism / concurr. | | | | | | | | | | X | | | | | | | | | X | | | | | | | | | | | |
| | Simplicity | | | | | | | | | | | | | | | | | | | | | | X | | | | | | | | |
| | Flexibility | | | | | | | | | | | | | | | | | | | | | | | | | | | | | | X |
| | Integration | | | | | | | | | | | | | | | | | | | | | | | | | | | | | | X |
| | Efficiency | | | | | | | | | | | | | | | | | | | | | | | | | | | | | | X |
| | Size | | | | | | | | | | | | | | | | | | | | | | X | | | | | | | | |
| | Length | | | | | | | | | | | | | | | | | | | | | | X | | | | | | | | |
| | Structural complexity | | | | X | | | | | | | | | | | | | | | | | | | | | | | | | | |
| | Coupling | | | | | | | | | | | | | | | | | | | | | | X | | | | | | | | |
| | Network complexity | | | | | | | | | | | | | | | | | | | | | | X | | | | | | | | |
| | Reduction Coefficient | | | | X | | | | | | | | | | | | | | | | | | | | | | | | | | |
| | Restrictiveness | | | | | | | | X | X | | | | | | | | | | | | | | | | | | | | | |
| | Number of trees | | | | | | | | | | | | | | | | | | X | | | | | | | | | | | | |
| | Cohesion | | X | | | | | | | | | | | | | | | | | | | | | | | | | | | | |
| | Cycle sets | | | | | X | | | | | | | | | | | | | | | | | | | | | | | | | |
| | Structuredness | | | | | | | | | | | | | | | | X | | | | | | | | | | | | | | |
| | Nesting depth | | | | | | | | | | | | | | | | X | | X | | | | | | | | | | | | |
| | Control-flow complex. | | | | | | | | | | | | | | | | X | | | | | | | | | | | | | | |
| | Log-based complexity | | | | | | | | | | | | | | | | | | | | | | | | | | | | | | X |
| | Analyzability | | | | | | | | | | | | | | | | | | | | | | X | | | | | | | | |
| | Understandability | | | | | | | | | | | | | | | | | | | | | | X | | | | | | | | |
| | Modifiability | | | | | | | | | | | | | | | | | | | | | | X | | | | | | | | |
| | Degree of automat. | | | | | | | | | | | | | | | | | | | | | | | | | | | | | | X |
| | Activity automation | | | | | | | | | | | | | | | | | | | | | | | | | | | | | | X |
| | Role integration | X | | | | | | | | | | | | | | | | | | | | | | | | | | | | | |
| | Role dependency | | | | | | X | | | | | | | | | | | | | | | | | | | | | | | | |
| | Activity parallelism | | | | | | | | | | | | | | | | | | | | | | X | | | | | | | | |
| | Delay risk | | | | | | | | | | | | | | X | | | X | | | | | | | | | | | | | |
| | Cognitive weights | | | | | | | | | | | | | | | | | | | | | | | | | | | | | | X |
| | Cognitive cross-conn. | | | | | | | X | | | | | | | | | | | | | | | | | | | | | | | |
| | Density | | | | | | | | | | | | | | | | | | | | | | X | | | | | | | | |
| | Partitionability | | | | | | | | X | | | | | | | | | | | | | | | | | | | | | | |
| | Connector interplay | | | | | | | | | | | | | | | | | | | | | | X | | | | | | | | |
| | Cyclicity | | | | | X | | | | | | | | | | | | | | | | | | | | | | | | | |

## 2.4 Directions from the state of the art

The management of engineering design processes is a wide field of research; there are many different models and methods available, few of which, however, are designed to cope with the structure of a process in a comprehensive manner. Nevertheless, the management of structural complexity and related fields of science have provided different means of understanding, modeling, and managing relationships in a complex system.

Both process management and the different sciences related to structural complexity have created a number of different metrics that evaluate the complexity of a system based on various structural characteristics; many of them are empirically validated and recognized in research and application, and the transferability to the management of processes is generally confirmed. Table 2-14 summarizes existing metrics and the structural criteria they assess: A mark in a cell relates a metric and a structural characteristic that the metric focuses on; however, it is possible that a metric also includes other structural characteristics that are not essential to its function. The table was generated by reviewing every metric (as shown on page 80 and the following pages) for its basic structural focus, as described in the literature.

As the table shows, some structural characteristics are not evaluated yet; however, a substantial toolset exists that is suited for the structural analysis of a process. In particular, n-partite-ness, isolated nodes, leafs, bridge nodes, biconnected components, spanning trees, the Small World effect, transitivity, degree distributions, navigation, and centrality are not evaluated as recognized metrics, although some of these structural characteristics have, in fact, the character of metrics themselves. These need to be reviewed in detail for their use in extending the set of means of analysis for processes.

Furthermore, process management and the analysis of processes is generally a goal-oriented procedure, which works to improve a process for one or another concept. As a review of the literature shows, the common goals of process management (planning, resource consumption, quality, flexibility, organizational decomposition, interfaces, and transparency of process, see page 66 and the following pages) and the typical properties of design processes (dynamics, creative nature, loops, leaps, iterations, results that are not predictable, changes, imperfect definitions, uncertainty and risk, maturity of artifacts, process path not determinable, and the involvement of many stakeholders, see page 61 and the following) and the means of analyzing a structure have not been systematically used in conjunction, and, therefore, no mapping between them is available. However, as individual works show, the goals can be related to certain structural patterns via the individual process patterns; for example, the triangularization of a DSM of tasks relates directly to the reduction of leaps and loops by proposing an improved sequence of tasks, thus contributing to better process planning.

These structural characteristics may possibly occur for all domains and relationship types that exist in process models. In fact, it is necessary to know the specific semantics of a process model to give meaning to the structural characteristics and to the structural metrics, as only the semantics of the nodes and edges of the underlying network structure allow indications about the behavior to be deduced. However, the specific attribution of all structural characteristics to all relevant domains has yet not been researched. Thus, no dedicated means of systematical analysis exists, but only fragmented parts thereof.

As common domains, tasks, artifacts, events, organizational units, resources, and time, were identified as basic domains of process management (Table 2-9). These can serve as an approximation to interpret the structural metrics and give them significance, which, however, will be precise if the relationship type is also considered in the second step. However, in many cases, this procedure will be too complex to be handled. To facilitate this procedure, principal relationship types were identified.

# 3. Concept of an integrated set of complexity metrics

This chapter provides an overview of the following three chapters that delineate different aspects of the solution and how they were developed. The reason for this layout is because developing the solution first requires detailing the modeling method, then the metrics, and lastly the contextualization of the metrics in an overall scheme. However, when using this approach, the elements to the solution are reversed. Therefore, this section provides a concise preview of the elements of the solution and their dependencies.

In general, the measurement system is intended to provide a method to analyze a process chart by drawing inferences about the process's behavior from the structure of its entities, as modeled in the process chart. Therefore, it is necessary to identify possible constellations of nodes and edges as basic constituents of structural characteristics to develop structural metrics independently of domains and relationship types. In a second step, the metrics are combined with common domains and relationship types to evaluate the particularities of engineering design processes and, thereby, give the structural metrics a process-focused meaning.

## 3.1 Solution design process

The solution was developed successively by collecting requirements from industry and combining them with further requirements and existing solutions in the literature. Potential concepts for the solution were compared each time and recombined to provide the best possible solution to the requirements identified. At each step extending the solution, the results were verified using industrial case studies, some of which are shown here.

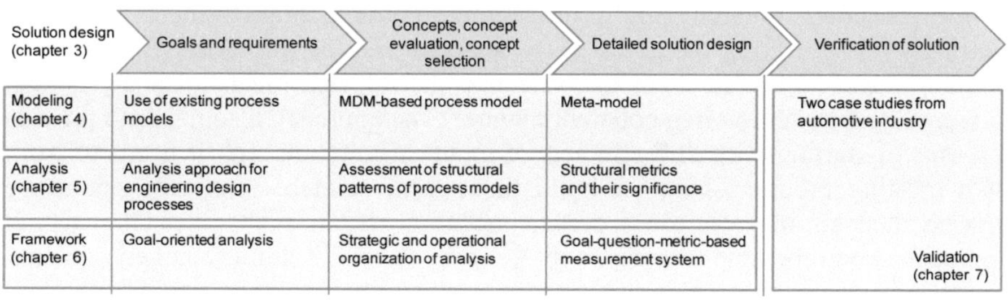


*Figure 3-1: Solution design process*

As Figure 3-1 shows, these steps were run parallel in order to develop the modeling method, the analysis method, and the framework. Of course, these three strands of the solution design process are not independent of each other, but the modeling method serves as a basis for the analysis, and the overall analysis approach is systematized in the framework. For this reason, the following chapters are ordered accordingly. To prepare these three chapters and the successive case study, the requirements and directions from the state of the art are collected into a general concept, which is detailed below.

## 3.2 Requirements for the solution design

Throughout the previous chapter, the different fields of science and their contributions to a structural measurement system were reviewed. Based on this, the definition of the problem can be refined with the following sub-problems:

- No high-level analysis procedure is available to systematically screen all entities and relationships in a process chart for possible weak spots.
- No overall modeling method exists that represents the structure of a process including domains and relationship types common to process management. Current dependency models are unable to cope with Boolean Operators that are commonly employed in process models.
- No comprehensive set of analysis tools exists to draw inferences about the behavior of a process from its entities and their mutual relations.
- No framework is available that orients the analysis of a process towards goals that are commonly followed in process management.

To provide a solution to this problem, many constituents of the solution are already available, as chapter 2 showed. An important part of the solution design is, therefore, to assemble the existing parts and fill the remaining gaps in a way that the solution is correct, complete, consistent, and clear. In the following section, the requirements for each of the sub-problems are presented.

The **overall approach** is required to systematically guide a user through all phases of the procedure of analyzing a process for possible improvement potentials, starting with specifying the goals of the analysis and the questions relevant to it, building the model to be analyzed, selecting the relevant means of analysis, and interpreting their results in a coherent manner. The approach also needs to provide a means of starting with different sets of input information, i.e., it needs to work with existing process charts or build them from scratch. Similarly, not every process analysis may follow a certain set of goals; in such cases, the overall approach should be able to analyze the process at hand in a more generic manner to obtain basic insights that help refine the analysis approach and possibly re-run it with regards to certain details.

The **process modeling method** provides the semantics of common process models to ensure that the analysis is able to work with different existing process models that serve as input. To do so, the model needs to encompass the domains and relationship types that are relevant to the goal-oriented analysis of a process as well as for generic screening for potential improvement. Thus, the process model needs to serve as an adaptor to filter relevant dependency data from existing process models and make it available for analysis. To do so, the model needs to be able to represent all modeling constructs that are relevant to process management, i.e., entities, relationships, attributes to entities and relationships, and logic operators. Furthermore, to be suitable for more extended analyses, it should provide an interface to other models, especially the product architecture. For complex process models, the process modeling method also needs to be able to create condensed, aggregate views of the specific domain being analyzed to ensure the efficient handling of the dependency data required for the analysis of the structure. The model is referred to as the **Structural Process Architecture**.

The **analysis method** needs to provide a comprehensive toolbox to assess the structure of a process in terms of the existence and impact of all relevant patterns that may occur among the entities and relationships in a process chart. The patterns need to be connected to their structural significance for all domains and relationship types as defined for relevant existing process models, based on empirical evidence. It is furthermore necessary to reduce these patterns into metrics to evaluate their occurrence for every single entity, group, or network (depending on the scope of the pattern) to ensure that all entities and relationships are assessed regarding their contribution to the behavior of the process. Using metrics for this reduction, the metrics need to adhere to measurement foundations (representation, uniqueness, meaningfulness) and to Weyuker's Properties. To further narrow the focus of the analysis and to allow a hierarchy of the results to be obtained through the metrics, the results of the metrics should be ranked by their degree of distinctiveness in the process. To do so, different means of identifying entities that "stick out" are necessary. This analysis should be intuitive and automatable for processing large process charts, as well. It takes shape as the **Structural Measurement System** (SMS)

Last, the overall **framework** needs to connect the solution elements consistently by linking the goals of the analysis to an operational layer that connects them to the patterns (and their evaluation through metrics) as well as to the relevant domains and relationship types in the process chart. The framework should, as a simplified access to the analysis, provide goals common to process management as primary points of entry to a process analysis. It is presented later as the **Structural Goal Question Metric** (S-GQM) approach.

As the development of a structural measurement system presents a method design, the method description follows the basic needs of the Munich Method Model [LINDEMANN 2007, p. 56]. To facilitate its application, it should be as concrete as possible to enable an "as is" deployment in industry [WALLACE et al. 2003]. A close relationship between the method developed and its industrial application is, therefore, critical. Ultimately, it needs be flexible in its application to accommodate varying needs and boundary conditions [LÓPEZ-MESA et al. 2004]. Thus the method development needs to be as modular as possible.

## 3.3 Constituents of the solution

Overall, the method to be designed needs to provide methodical circumnavigation of the complexity barrier at an abstract level (see page 22) to cope with the complexity of a process chart. To do so, the procedure shown in Figure 3-2 is followed. It starts from a set of goals of the analysis that can be concretized using different questions; these questions are those that a process analyst might be interested in finding answers to. To address these goals, the procedure needs to link each question to a set of metrics, domains, and relationship types that will provide answers. Similarly, it needs to help the user interpret the results obtained in order to collect indications about the process's behavior, and possibly potential improvement. To prioritize the results, structural outliers can be sought that indicate the specific particularities of a process from a structural point of view.

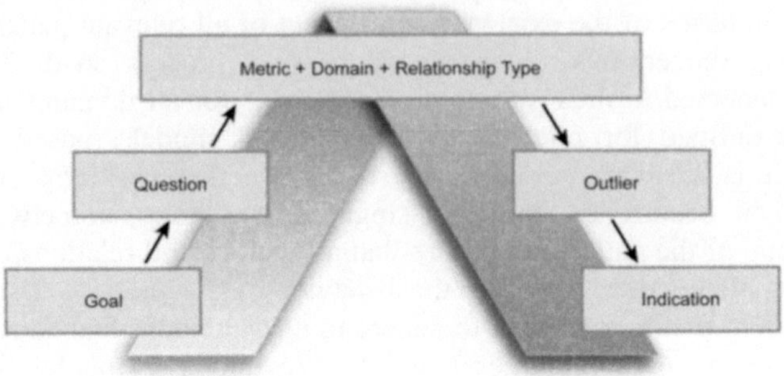


*Figure 3-2: Basic procedure of the goal-oriented analysis of a process structure*

The basic procedure is, therefore, straightforward. After an initial goal has been selected, one or more questions are suggested by the framework to allow a better focus of the analysis. Metrics are allocated to each question to help answer the selected questions. At the same time, a question commonly only relates to one domain or a few domains, but not all; thus, relevant domains are chosen at the same time. Each is accompanied by its principal relationship type[57]. Using these tools, a process can then be analyzed to identify possible structural outliers. The structural significance each metric provides can then be used to investigate the nature of the outliers to answer the initial questions and guide further improvement measures.

Figure 3-3 concretizes the elements of the overall solution and their interdependencies. It shows the necessity of the modeling method (inner box) to develop an analysis approach (middle box) that is framed through the attribution to goals and questions (outer box).

The individual selection of the models, methods, and tools used for each of these three constituents is argued in the respective chapters. The overall analysis procedure is detailed in the following section.

[57] In theory, the selection of the appropriate relationship type depends on the question, the metric, and the domain simultaneously; the selection of the principal relationship type, therefore, is a certain simplification, as it does not incorporate all influences that determine the relationship type. However, the principal relationship types, as part of the meta-model (section 4.2), are described in a way to be as generic as possible without losing the nature of common relationship types applicable to a domain. Therefore, only a small error is introduced. At the same time, the framework is significantly simplified, as one additional variable aspect depending on three other inputs is removed.

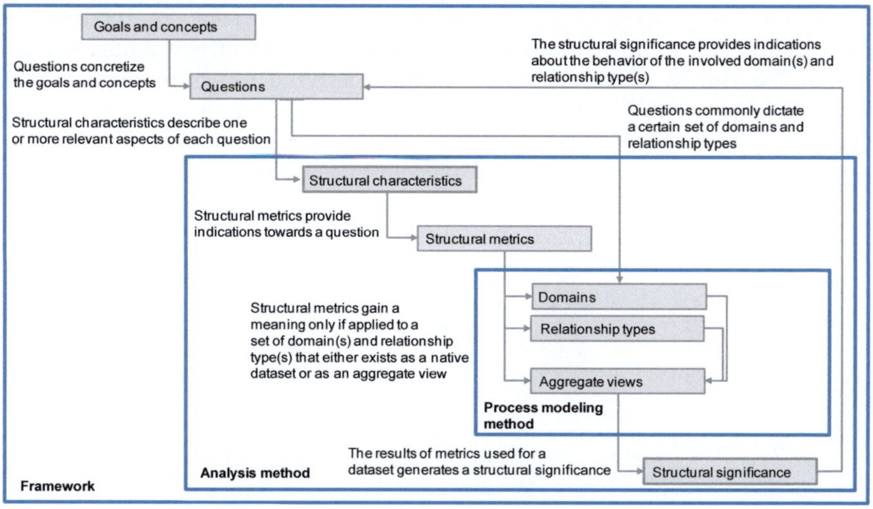


*Figure 3-3: Constituents of the overall solution and their principal dependencies*

## 3.4 Overall concept: Analysis procedure

To enable a goal-oriented analysis of the structure of the system "process", several steps have to be carried out. There are different procedural models available to support an analysis in a structured manner. The Structured Analysis approach views systems from a data flow perspective [DEMARCO 1978, pp. 1-7 and 37-44]. In Systems Engineering, the systematic analysis is more holistically focused, using a problem solving procedure [DAENZER & HUBER 2002, p. 96]. The information structure approach to analyze DSMs is more focused on DSMs, including building and analyzing a model for task interdependencies [YASSINE et al. 1999]. In structural management, these different models together are used to design a procedure for the analysis of a systems' structure of the five latter steps, as shown in Figure 3-4 [MAURER 2007, p. 69]. In the context of process analysis, this procedure is extended by an initial step, as setting goals is highly relevant for a purposeful analysis. This phase is optional if the process is not analyzed to answer a specific question but only to generate a general picture of a process.

In Figure 3-4, the different elements of the overall solution are depicted. The different aspects are explained in the following paragraphs lists how each of these phases relate to the structural analysis of a process and what deliverables are gained. Initially, the **goals** of the analysis project are set, and each can be detailed by various questions. During the **system definition**, the overall scope of the process analysis is defined, including what process is going to be regarded and which aspects of this process are relevant to the analysis; to this end, the basic goals need to be determined to purposefully set the system border, i.e., the relevant domains and relationship types are defined. Likewise, the metrics needed to answer the questions from the initial phase are selected. All of these elements can be obtained using the questions from the first phase. The **information acquisition** then generates the structural datasets of relevance, i.e., entities and relationships that are then **modeled** into an overall dependency model using a Multiple-Domain Matrix. This matrix is then **analyzed** using structural metrics that are selected according to the goals of interest, as defined in the first phase. Ultimately, the results from the metrics are interpreted in order to prioritize possible weak spots in the process and to deduce measures.

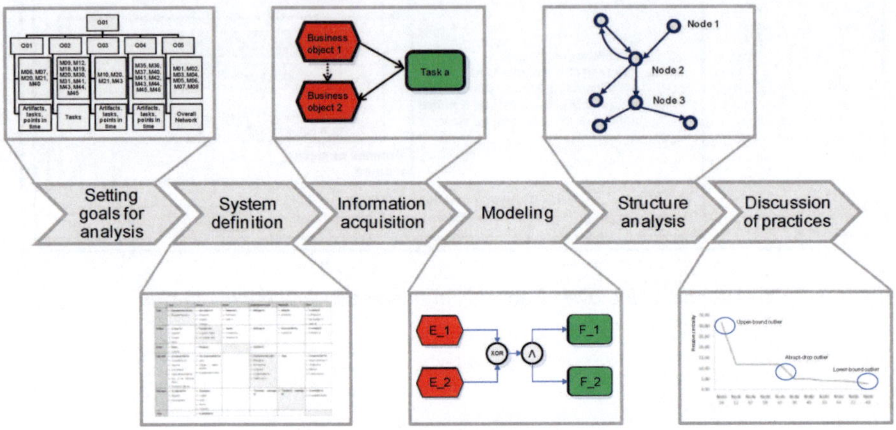


*Figure 3-4: Basic procedure of structural analysis based on [MAURER 2007, p. 69]*

## Setting goals for analysis

To enable a goal-oriented process analysis, in the initial step the Goal-Question-Metric (GQM) scheme is used to guide the analysis. From a list of eight different goals, an appropriate goal can be chosen and further detailed by selecting appropriate questions. The intent is to point the user to relevant aspects of the chosen goal of process improvement. To this end, the common goals of process analysis as described in the state of the art are used and extended to accommodate possible needs of structural process analysis. Chapter 6 details the framework that embodies the goals and questions.

The framework is, by no means, meant as a rigid or prescriptive system. It can be extended and adapted to suit individual needs of process analysis just as well. Also, it is possible to omit this step and run a more individual analysis that works to characterize a process in general, looking for possible improvement potentials without following a particular goal.

*Table 3-1: Focus and deliverables of each phase during the structural analysis of a process*

| **Phase** | Setting goals for analysis | System definition | Information acquisition | Modeling | Structure analysis | Discussion of practices |
|---|---|---|---|---|---|---|
| **Focus** | Definition of scope of analysis | Definition of involved aspects of process | Process recording | Compilation of structural process model using MDM | Goal-oriented analysis based on structural metrics | Interpretation of the results of structural metrics and deduction of measures |
| **Output** | Goals of analysis, relevant questions to be answered | Domains, relationship types, set of metrics | Entities and relationships in partial DSMs and DMMs | Overall MDM | Results of structural metrics | Indications of potential to improve process |

## System definition

If questions have been selected, metrics can be chosen in the next step. Those structural metrics that are able to provide answers are allocated to each question (section 6.2). If no particular questions are asked, metrics can be selected individually from their descriptions or from different classifications that are available. These classifications are listed in section 5.2.4.

The metrics then need to be assigned to the datasets that are to be acquired later or that are already available. If questions guide the analysis, the framework provides relevant domains and their relationship types. In other cases, either individual information needs or the available datasets guide the selection of domains and relationship types.

Datasets for the analysis need to be available as Design Structure Matrices (DSM) for almost all structural metrics. Two kinds of input datasets are, therefore, possible. Either, the native data is available as a DSM, or an aggregate view can be computed for relations that span at least one intermediate domain. Section 4.5 introduces these different ways of building the model.

More generally, the domains and relationship types involved can be selected from a structural process meta-model that is used to introduce common aspects of process modeling into the analysis. This meta-model is also used as a reference for the development of the structural metrics. It is detailed in section 4.2.

## Information acquisition

In all cases, the input data needs to be acquired; this can be done either through workshops or by parsing and converting existing process models. Section 4.5 briefly addresses these issues. The structural content of common process models is used to set up the meta-model so that it can integrate all kinds of process models into one overall structural model. The partial models acquired are modeled as dependency models of any kind that can be converted to partial Design Structure Matrices (DSM) or Domain Mapping Matrices (DMM).

### Modeling

The different partial models are then assembled into one overall Multiple-Domain Matrix (MDM) within the frame that is spanned by the meta-model. To integrate all particularities of common process models, section 4.4 introduces new aspects into the MDM, such as the integration of the product architecture, the modeling of different attributes, and the integration of logic operators to model decision points.

### Structure analysis

The analysis of the model which is then available is undertaken using structural metrics that are developed and described in detail in section 5.2. These metrics are based on the numeric evaluation of structural characteristics as found in different disciplines that focus on the management of structural complexity. Section 5.1 collects these different approaches and proposes a procedure to develop structural metrics from these structural characteristics.

### Discussion of practices

A major part of the development of the metrics is the description of their significance. Section 5.2.2 details their meaning in reference to all domains of the meta-model. Therefore, the results that are obtained from the application of the structural metrics to a process model can systematically be interpreted to provide detailed and comprehensive insight into the question that is being analyzed and to deduce possible measures for improvement.

# 4. Modeling the structure of design processes

In this section, a model representing the structure of a process is laid out to serve the following purposes:

- Serve as an interface to other process models.
- Provide the semantics of common process models as a basis for the development of metrics that are independent of a specific process model.
- Represent all modeling constructs that are relevant to process management from a structural point of view (entities, relationships, attributes to entities and relationships, logic operators)
- Provide an interface to the product architecture.
- Support the creation of aggregate views onto a specific domain.

As a modeling technique, Multiple-Domain Matrices (MDM) are chosen as an apt way of representing and manipulating a network structure consisting of different domains and relationship types. The argument for this choice is given in section 4.2. First, the use of an MDM is argued and explained; then, based on the different process models reviewed in the state of the art, an MDM-based Structural Process Architecture (SPA) is developed that serves two purposes. First, it is meant to enable assembling a structural process model from different partial models that may be available in a process improvement project; secondly and more importantly, the meta-model is needed to provide a reference for the interpretation of the metrics and the measurement framework developed to systematically access the metrics. Finally, MDM is extended to include attributes to edges and logic operators, which, up to now, was not possible in this notation. Also, the linkup of the process structure to the product architecture is explained.

## 4.1 Design processes as a multi-layered network

The focus in this research is on engineering design processes. There are, in fact, many different kinds of processes that are common to engineering design, for example, the planning process, the technology development process, the purchasing process, and others[58]. Here, the focus is on the primary process, i.e., the process of generating new or adapted technical designs [BECKER et al. 2005, p. 7] [SCHMELZER & SESSELMANN 2006, p. 55].

As the introductory case study in section 1.1 showed, engineering design processes are, in fact, a network of multiple domains that coexist to enable the development of a product. Each of these domains is networked in itself, commonly, in many different ways, and the different domains are networked among each other. Figure 4-1 shows an example of three domains that make up the network layers of a process; all are mutually linked and coupled.

[58] See [BAUMBERGER 2007, pp. 123] for an overview of different kinds of processes.

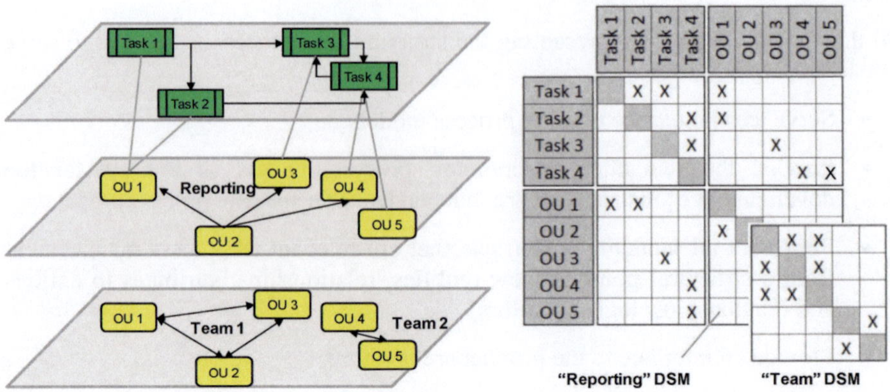


*Figure 4-1: Structural equivalence of process model in flow-oriented notation and MDM notation*

Inevitably, the behavior of the process depends, to a large extent, on this network and its structure, as, for example, communication among organizational units is only possible if they are related to each other. Thus, the specific setup of entities and their relationships constitutes the value of an actual process.

To model the structure of an engineering process comprehensively and to gain a deeper understanding of it, it should therefore be understood as the multi-layered network it actually is, i.e., it should comprehend every part of the company organization that is actually necessary to enable it. To analyze a process, in turn, it is important to select and relate all domains that are relevant to such a specific analysis in an integrated manner, which simultaneously enables and facilitates systematic and comprehensive analysis.

## 4.2 MDM-based modeling of the structure of a process

There are many different modeling methodologies that could be applied to represent such networks; in fact, there are plenty of dependency models available, as section 2.1 showed, e.g., graphs or incidence lists. [BELHE & KUSIAK 1996] compare different dependency modeling methods for process models. They conclude that all major models can be converted among them, using adjacency matrices to represent the dependencies. Such matrices are similar to DSMs and, therefore, to MDMs.

Multiple-Domain Matrices (MDM) allow multiple network structures to be represented, both within a single domain (e.g., tasks) and across domains (e.g., the attribution of organizational units to individual tasks). Equally, an MDM is able to capture different relationship types that coexist concurrently. This makes it an ideal tool for modeling the structure of design processes. Figure 4-1 visualizes the concept. In this example, two different networks among organizational units coexist. Accordingly, the MDM contains two OU-DSMs.

Here, the MDM is chosen for a number of reasons:

- The network structure of the process is modeled in all its facets. This way, no single relationship dominates over the others, i.e., the complexity of the process is captured more realistically. In fact, most process models can be converted into a MDM with little or no loss of information concerning their structure.
- Different models can be combined; this way, it is possible to check how well-aligned the different structures that are modeled actually are (e.g., a process and a team).
- Qualitative and quantitative models can be combined to some extent, if they each can be represented using MDM methodology (e.g., using weights for nodes or edges or by introducing attributes via additional matrices).
- The process can be analyzed either based on the native data, with regard to the impact of an analysis on other domains (e.g., by finding clusters in a task-DSM and then constituting teams in the OU-DSM in Figure 4-1), or using aggregate views (e.g., by computing how tasks are interrelated via documents, and then using the computed task-DSM).
- Structural characteristics become accessible. This allows systematic analysis using available algorithms for DSM, DMM and MDM analysis. Based on native or aggregate datasets, structural characteristics can span one or more domains.
- Common DSM-based analysis is applicable, e.g., tearing, banding, or clustering. While these algorithms can be used on other dependency models, as well, their effect is directly visible in a matrix representation.

Of course, MDM-modeling possesses a number of disadvantages, too:

- The matrices grow rapidly. While, theoretically, almost all information in a common process model can be converted into an MDM, this hardly makes sense. If, for example, an EPC model contains many attributes, e.g., starting- and end-times of every task, a small process chart will turn into a very large MDM. Therefore, it makes sense to convert only those parts of a process model that are of interest to an analysis.
- Reading a matrix is not very intuitive. While, in general, a matrix reflects an engineer's mindset, a MDM that contains several domains is nearly impossible to read manually and thus needs tool support. Modeling a process exclusively in MDM notation, therefore, rarely makes sense, as most users will be unable to understand the process model, and little transparency would be generated.
- There are different matrix notations. Generally, two different conventions about how a DSM or related matrix is written exist; either, an entry in the cell of a matrix is read as "row impacts column", or the other way round. Here, all matrices are designed as "row impacts column".

- The actual graphical structure of a flow chart is lost when turned into an MDM. This is a major shortcoming, as the "structuredness" and "style" of the flow chart layout are important to understand a process model [GRUHN & LAUE 2007b] as well as transmit part of the meaning of a process model.

## 4.3 The Structural Process Architecture model

As shown in section 2.2.3, models are commonly defined using a meta-model to describe the entities within the modeling language and their possible relations. This section makes use of this approach to describe a meta-model suitable to model multi-layered process networks as a Multiple-Domain Matrix.

### The need for a meta-model

There are many different process models available, each of which has a particular focus on the structure of a process. Thus, there is no need to develop any further modeling languages, as specific models exist to suit almost all possible needs [BROWNING 2009].

At the same time, however, choosing one modeling language will produce only a limited number of domains and relationship types. Therefore, developing structural metrics based on only one process modeling language is not practical, as it will limit the application of the metrics to this one kind of modeling language. Thus, the transferability of the metrics to another process model using a different modeling language would be more difficult.

This section is, therefore, intended to design an adaptor that allows plugging different process models into a common denominator to which to apply the structural metrics. The meta-model developed in the following is, therefore, meant as a "set of models" [HÖFFERER 2007], i.e., to create a modeling scheme capable of describing relevant aspects of structural modeling and a goal-oriented process analysis. This Structural Process Architecture is thus a consequent extension of the regrouped process models. It represents a meta$^2$-model approach, as previously outlined. Figure 4-2 visualizes the concept: For example, the domain task is tailored to represent different tasks or activities as found in common process models. The same applies to the relationship types, which are equally collected from relevant meta-models (level 2) for processes.

When used, the meta-model provides, in turn, orientation when modeling a structural process model, and it serves as a guide and an example when submitting a process model to structural analysis using metrics. The principal reason is that the meta-model systematizes and collects relevant domains and relationship types and puts these into a common framework.

This framework is necessary for the development of meaningful metrics. As defined, structure consists of a particular pattern of nodes and edges in a graph, but a structure only has meaning if it is related to a certain semantic context. This context is provided by the meta-model that describes the types of nodes and edges concerning their meaning in an industrial application.

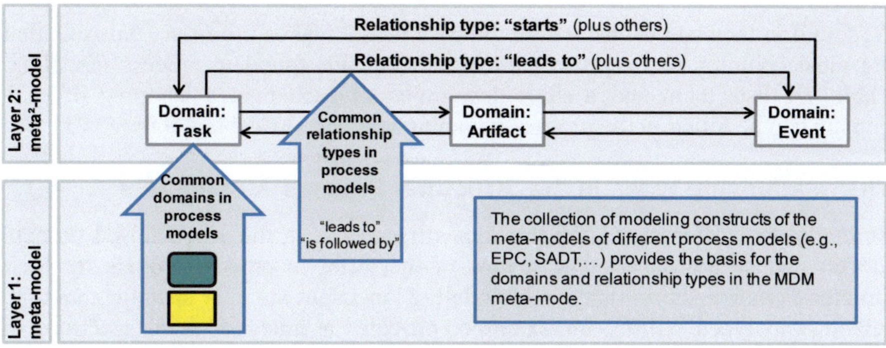


*Figure 4-2: Generation of the SPA (application of the approach shown in Figure 2-20)*

The meta-model consists of two views on a structure: The domains, describing what types of entities are common to process modeling, and the relationship types, describing how the domains are commonly related. This meta-model is, by no means, exhaustive; it makes use of the most common process models and integrates their concepts which, regarding common models, appear to be the principal aspects of process modeling. However, it may be necessary to refine the model for a new concept, using, for example, the principles shown in Figure 2-3.

*Table 4-1: Domains in the Structural Process Architecture*

| Domain | Description |
|---|---|
| Task | The task domain collects all entities that describe the execution of work done in the process; further terms are: Function, method, action, activity, unit of behavior, gateway, transition, and work package. |
| Artifact | The artifact domain regroups all objects that are intermediate and final input and output objects in the process. Some modeling languages differentiate between artifacts focused on value generation and on process control; however, here, both kinds are included. Further terms are: Input / output, object, product, data, parameter, and information. |
| Event | This domain addresses non-persistent occurrences in time that present a certain status or progress. Further terms are: Message, order, initial / final node, label, and place. |
| Org. Unit | The organizational unit domain contains all human resources in their respective ordering; further terms are: Staff, responsibility, team, pool, lane, actors, roles, and committee. |
| Resource | The resource domain is intended for all non-human resources necessary to enable the process execution, such as IT-systems, equipment or knowledge. Further terms are: Attribute, mechanism, method, pool, and lane. |
| Time | The time domain addresses persistent time issues such as the start time of a task or milestones in the process. Further terms are: Attribute, duration, starting time, end time, average time, milestone, and phase. |

## The domains in the Structural Process Architecture

As found in the state of the art, the process model makes use of six domains that are most common and represent the usual domains found in process modeling. Table 4-1 lists them and a short description of each one; a mapping of each domain can be found in the process modeling language in Table 2-9 (page 71).

## The relationship types in the Structural Process Architecture

Similarly, the different relationship types that occur in the 13 reviewed process models can be regrouped. The review of the different process models for their structural content shows that some modeling languages are very specific about the relationship types. EPC, for example, provides a finely detailed spectrum of different types, while others leave this refinement to the user. For this reason, relationship types were consequently collected, and those types that are common across the majority of models or that are at a higher level of abstraction were designated as **principal relationship types**. These can be found in bold print in the list of relationship types for each possible tuple of domains. Of the 36 possible relationship types between the six domains, only 28 actually occur in the reviewed models; while the other relationships can, of course, also be designated, they appear to be irrelevant.

Table 4-2 shows how many relationship types there are between any pair of domains. As can be seen, the most important focus in common process models is placed on the interplay of tasks and artifacts. Many of these relationship types are similar or even identical. They were, therefore, regrouped; all detailed relationship types are listed in Table 4-3.

*Table 4-2: Occurrence of relationships among domains of process modeling for all 13 reviewed methodologies (read as "row relates to column")*

| | Task | Artifact | Event | Org. unit | Resource | Time | Product attribute |
|---|---|---|---|---|---|---|---|
| Task | 7 | 12 | 6 | 1 | 2 | 2 | 0 |
| Artifact | 12 | 5 | 2 | 1 | 2 | 2 | 1 |
| Event | 5 | 2 | 1 | 1 | 0 | 0 | 0 |
| Org. unit | 4 | 1 | 0 | 2 | 1 | 1 | 0 |
| Resource | 8 | 3 | 0 | 1 | 3 | 1 | 0 |
| Time | 0 | 1 | 0 | 1 | 0 | 0 | 0 |
| Product attribute | 0 | 1 | 0 | 0 | 0 | 0 | 0 |

Looking, for example, at the relationship types from tasks to resources, two relationship types are common, as the table shows: "Task *requires* resource" and "task *processes* resource". Of the two, "requires" is the more general one, and it is, therefore, designated as the principal relationship type.

Besides the specific relationship types, general **decomposition** is relevant to each domain, as well. Some common process models do, in fact, explicitly allow decomposition to a finer granularity or the regrouping to larger entities, e.g., collecting tasks into value chains. This is why the following decompositional relationship types need to be considered in addition to any relationship type specific to process models to address the possible levels of granularity in a process model:

- Is part of
- Is a generalization of
- Consists of

The decompositional relationship types are not represented in the Structural Process Architecture, as it does not focus on the interdependencies among different levels of detail. However, they need to be considered when entities of the same super-domain (e.g., the domain's "phases" and "milestones" for the domain "points in time" from the meta-model) are combined in an overall process model.

### The complete Structural Process Architecture (SPA) model

The SPA model regroups the domains and relationships that were collected across the most common process models. This way it serves as a supermodel that recombines these models and serves as a consistent structure for the development and operation of the structural metrics. It provides a semantic background to both design the measures and to give them a meaning when interpreting them.

Table 4-3 shows the overall model. It is read as "row impacts column". For example, a "task has the output of an artifact" (row 1, column 2), using the principal relationship type. To be more specific, it is possible that this output is, for example, only a change of an existing artifact; in this case, a "task changes an artifact", using the refined relationship as shown. In the following, further aspects of using the meta-model are explained, and an example for the application is given. In section 4.4.1, the meta-model is complemented by an additional domain; therefore, a complete meta-model is only shown in appendix 10.4.

*Table 4-3: Structural Process Architecture with domains and relationship types suited for most modeling and analysis purposes*

| | Task | Artifact | Event | Organizational unit | Resource | Time |
|---|---|---|---|---|---|---|
| **Task** | • **Precedes temporally**<br>• Precedes logically | • **Has output of**<br>• Processes<br>• Controls<br>• Changes | • **Generates**<br>• Processes<br>• Ends in | • **Belongs to** | • **Requires**<br>• Processes | • **Is active at**<br>• Is finished at<br>• Has duration of<br>• Starts at |
| **Artifact** | • **Is input for**<br>• Supports<br>• Controls<br>• Starts | • **Transits into**<br>• Is used to create<br>• Is in conflict with | • **Sends**<br>• Is created at<br>• Is needed at | • **Belongs to** | • **Is processed by**<br>• Is stored in | • **Is created at**<br>• Is needed at |
| **Event** | • **Starts**<br>• Controls | • **Produces** | | • **Occurs in** | | |
| **Org. unit** | • **Is responsible for**<br>• Is necessary for<br>• Supports<br>• Is involved in<br>• Takes decision about<br>• Has to be informed about<br>• Processes data for | • **Has responsibility for**<br>• Uses<br>• Creates / deletes / updates<br>• Is authorized to read | | • **Communicates with**<br>• Belongs to<br>• Is formed by<br>• Is head of<br>• Is responsible for<br>• Takes role as | • **Uses** | • **Is responsible for**<br>• Assists reaching of<br>• Is finished at<br>• Starts at<br>• Is active during |
| **Resource** | • **Is required for**<br>• Supports<br>• Is occupied by | • **Processes**<br>• Creates<br>• Emits<br>• Stores<br>• Transmits | | • **Transmits message to** | • **Transmits message to** | • **Is available at**<br>• Is available during |
| **Time** | | • **Is attribute to** | | | | |

## 4.4 Specific aspects of modeling engineering design processes

While the Structural Process Architecture model originates from the domains and relationships encountered in common process modeling languages, there are a few more specifics that need to be accounted for, especially for process models:

- The link to the product architecture: As any engineering design process creates a product, the process setup is oriented to serve the architecture of the product.
- The occurrence of attributes to nodes and edges: Depending on the information needed in the process analysis, different aspects need to be modeled in the MDM besides nodes and edges. Specifically the information transfer, commonly modeled as edges, is often of interest and may need more detailing.
- The occurrence of decision points: During the generation of knowledge in the process, many decisions have to be made about how to proceed with the process, e.g., by iterating or continuing downstream. These decision points are often modeled as Boolean operators to represent decision logics.

To suit the set of modeling constructs to incorporate these aspects of process modeling, the following sections first extend the meta-model to product attributes. Then, MDM modeling techniques are extended to include attributes in a manner coherent with common MDM modeling, and, last, an extension to model decision points in MDM is proposed.

### 4.4.1 Alignment of the process structure with the product architecture

Process improvement based on the requirements and constraints set by the product architecture has generated a lot of interest in research in order to discover the interdependencies of a technical product and its design process [HENDERSON & CLARK 1990, p. 9]: "Architectural knowledge tends to become embedded in the structure and information processing procedures of established organizations"[59].

It is neither the specific focus of the approach presented in this research to propose a process structure that corresponds to the needs of the product architecture, nor to analyze how well a process is suited to efficiently creating a specific product. However, a framework to analyze a process in engineering design needs a specific "adapter" for the content processed in the process organization. As such, the alignment of process and product addresses the fact that the organizational

---

[59] There are many examples available: [KREIMEYER et al. 2007c] and [SOSA et al. 2004a], for example, align the process structure with the product architecture to facilitate communication across the overall process. DANILOVIC proposes to use a clustering of product attributes to define workgroups that work more efficiently because they are closely related for the development issues for which they are responsible [DANILOVIC & SANDKULL 2002]. PONN defines the need to align the process in response to the specific situation, i.e., the need to generate knowledge about certain aspects of the product [PONN & LINDEMANN 2005]. HERFELD provides a concept that allows close cooperation between design and simulation engineers based on team structures that originate from the dependencies of requirements and their embodiment in the product's components [HERFELD 2007].

dependencies of a process (which are the focus here) are only in place because they are meant to generate a certain content.

The process has as a specific environment for the product it is generating knowledge about. As the first-level interface in the SPA model presented in the previous section, these can be understood as attributes to entities of the process. If, for example, three tasks of an automotive design process are linked to developing the rearview mirror of a car, the rearview mirror could be the content attribute of these three tasks. As such, the three tasks are linked across the common attribute. Additionally, four other tasks of the same process might be linked to designing the sheet metal parts of the driver's door. As the mirror is attached to the door, there is a relationship between the two attributes, which possibly indicates a relationship between the tasks. Figure 4-3 visualizes this example and shows how the additional domain "product attributes" are introduced (1) to link different tasks that are connected to a certain product attribute, and (2) how these attributes can be linked to each other (3).

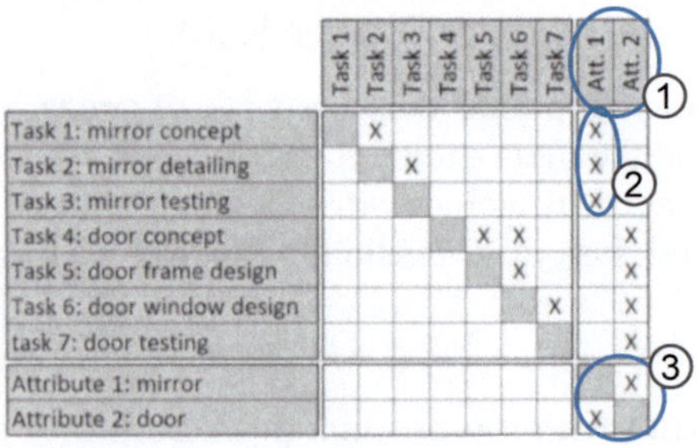


*Figure 4-3: Use of product attributes in process model*

Product attributes can be of different kinds, and they can be related in different ways to the process. Basically, a product attribute is a "characteristic" of the product at a given level of abstraction and perspective to the product [PAHL & BEITZ 2007, p. 220]. Each level of abstraction has a different scope, as shown by the Munich Model of Product Concretization [PONN & LINDEMANN 2008, p. 21]: requirements, functions, principles of operation, geometry, concept. Each of these can be further detailed using different perspectives of Design for X, e.g., design for assembly, design for maintenance, and more.

Relations among product attributes are manifold and are not the focus here[60]. The relation of product attributes to the process is only vaguely described, as many authors addressing the topic often do not clarify their understanding of structure any further; generally, "product attribute is implemented by process" [EPPINGER 2001] or "product attribute connects to process / product attribute has affiliation to

---

[60] [PIMMLER & EPPINGER 1994] suggest four basic kinds: Spatial, energy, information, and material. [JARRATT 2004, p. 125] extends these to mechanical steady state, mechanical dynamic, spatial, thermal steady state, thermal dynamic, electrical signal, electrical earth, electrical dynamic. However, many other relationship types are possible.

process" [SOSA 2008] is what is provided. However, there appears to be no detailed descriptions of how the different domains of a process relate to a product architecture [O'DONNELL & DUFFY 2005, pp. 12-14].

The goal of introducing product attributes is the coherence of the process necessary for a high degree of efficiency of the process [O'DONNELL & DUFFY 2005, pp. 12-14], i.e., the "alignment" of the structures of a product and its design process including all supporting domains [SOSA et al. 2004b].

Table 4-4 shows how product attributes can be introduced into the Structural Process Architecture from Table 4-3. It proposes basic relationship types that represent a general view of the product for the best generalization possible.

*Table 4-4: Extension of the Structure Process Architecture to include product attributes*

| | Task | Artifact | Event | Org. unit | Res. | Time | Product attribute |
|---|---|---|---|---|---|---|---|
| **Task** | (Compare Table 4-3 for relationship types of the process meta-model) | | | | | | • **Processes** |
| **Artifact** | | | | | | | • **Represents**<br>• Describes (part of) |
| **Event** | | | | | | | |
| **Org. unit** | | | | | | | • **Generates**<br>• Is responsible for<br>• Consults about<br>• Has knowledge about |
| **Resource** | | | | | | | • **Processes** |
| **Time** | | | | | | | |
| **Product attribute** | | • **Starts process by sending** | | | | • **Is defined at**<br>• Is needed at | |

## 4.4.2 Inclusion of attributes to nodes and edges

To allow the comprehensive modeling of a process, either from scratch or by converting one or more existing models into an overall structural process model, the modeling methodology needs to be as extensive as possible to include all possible modeling constructs that are necessary to represent a given process. Besides a principal flow of a process, there may be additional attributes that complete the process description and that form part of the overall structure. For example, the IT systems that are used by individual tasks are, in fact, attributes of the domain "IT-system" that are related to the elements in the domain "tasks".

Basic MDM notation is able to include entities of different domains, being related within one domain as a DSM or across two domains as a DMM. It is, therefore, able to represent attributes of any element; the attributes can simply be mapped to the entities to which they belong. However, MDM notation is unable to include attributes (i.e., entities or nodes) of the relations between different nodes (i.e., to the edges). Figure 4-4 illustrates these two kinds of attributes.

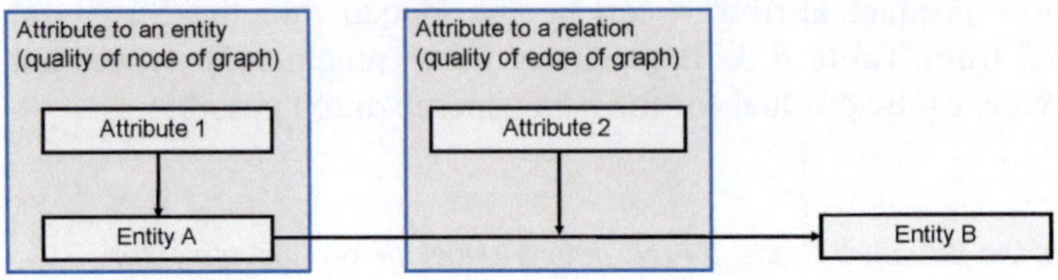


*Figure 4-4: Attributes to nodes and edges*

Figure 4-5 illustrates an example: Business object c is generated by task B and transferred using IT system 3, attributed to edge γ as the transferring system (as opposed to another system the business object originates from). Several different process models use such constructs. Another integration is given by [BRAUN & LINDEMANN 2007], for example, who link a process layer to a basic product architecture and, at the same time, to a resource layer to estimate the expected expenditures in a development process.

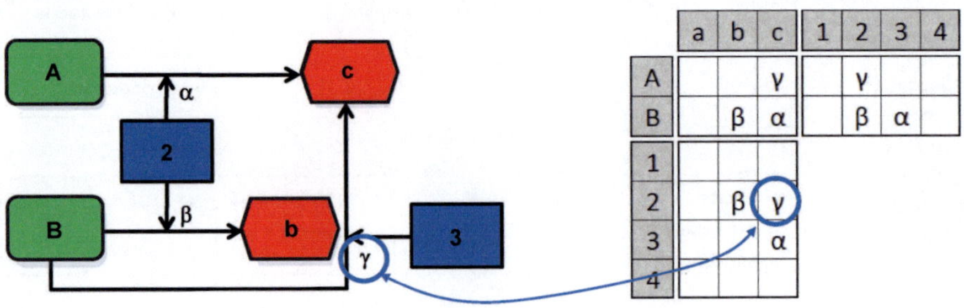


*Figure 4-5: Three individual chains of relations (α, β, γ) in graphical and matrix notation using three DMMs*

A basic approach would be to model relationship γ using three matrices, linking two domains at a time as shown in Figure 4-5. However, such a model can turn out to be ambiguous in certain cases. In fact, if the chains of relations (α, β, γ) are not explicitly named in the matrix-based notation, another unwanted chain of relation occurs. Figure 4-6 uses the common notation with an "x" indicating the existence of a relation. It shows how task B produces business object c via IT system 2, which is actually not the case in the modeled process. The model is, therefore, not unambiguous and thus insufficient.

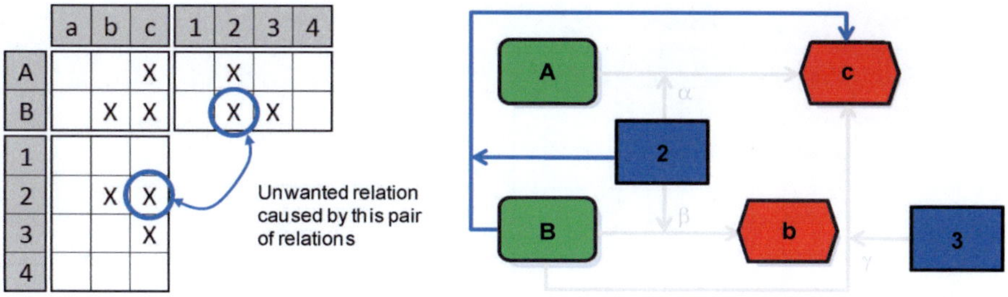


*Figure 4-6: Unwanted relation in the example from Figure 4-5 because of ambiguous modeling*

The reason for such unwanted edges is the fact that the individual denomination of a chain of relation is lost when replacing the chain with the simple existence of a relationship. Figure 4-7 illustrates this phenomenon for the previous example: There, the chains of relations a-A-1, b-B-1, and a-B-1 are modeled as DMMs. However, as the individual chains of relations are not differentiated, a fourth unintended chain of relations occurs that draws one edge from each of the intended cases. In fact, the more tightly a structure is coupled, the higher the probability that such unwanted edges occur. Therefore, an explicit denomination needs to be introduced which indicates each chain of relations in MDM notation.

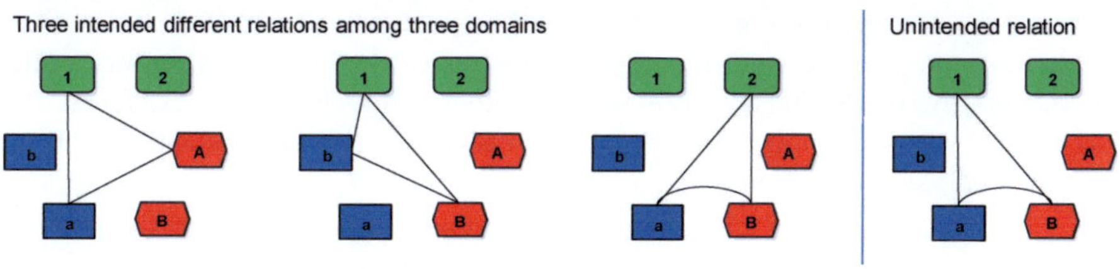


*Figure 4-7: Occurrence of unwanted relations using three DMMs*

In fact, another domain needs to be introduced to identify each chain of relations. Even though this approach is complex and makes the generation of the matrices quite large, it is the only way to ensure unambiguity. In Figure 4-8, the new domain is included in the MDM. The empty matrices constituting the whole MDM are visualized, too, although only six basic matrices are needed. As can be seen, business object c is produced via chains α and γ (1). However, only chain γ is possible with task B (2). As chain γ only contains IT system 3, business object c is produced only via IT system 3 (3) and not via system 2, as originally indicated by the matrices depicted in Figure 4-6.

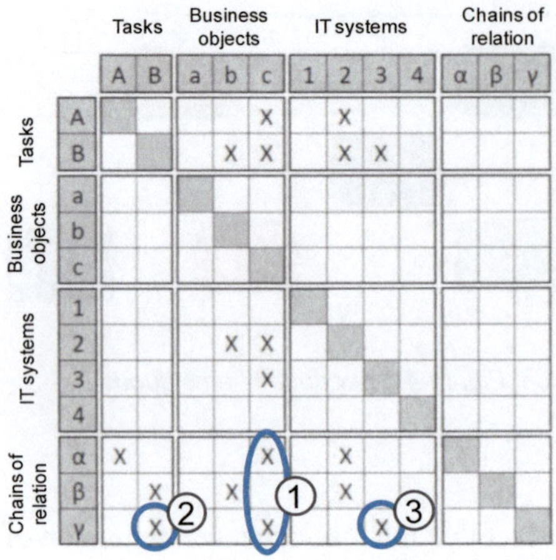


*Figure 4-8: Unambiguous MDM-based description of attributes to edges*

The approach is equally applicable to DSMs, as a DSM can be understood as a DMM linking two identical domains. As the same ambiguity, as in the DMM case, can occur for DSMs, a denomination of the chain of relations is necessary.

The approach serves as a basis for extending matrix methodology to facilitate the completeness of modeling constructs. It is especially valuable to complete matrix methodology for handling large systems with many elements; in such cases, it is not just the compactness of a matrix that is of interest but also the strict modeling scheme and the possibility to represent multiple relations in their coexistence. As it is in line with common MDM notation, it allows the effortless application of common analysis methodology.

### 4.4.3 Decision points modeled as Boolean operators

Almost all available process models can model the split or merge of the process flow using logic operators. These operators represent decision points that, according to the results of the process up to this point, take it along one path or another. To make a process model containing such operators accessible to a systematic analysis using structural metrics, a conversion into an MDM is necessary. The sub-section below describes different ways of doing this.

#### Basic logic operators and possible conversions

Commonly, logic operators are modeled as Boolean operators to represent the choices that are possible. Boolean algebra provides three basic operators AND, OR, and NOT which can be used to model all other more complex operators such as XOR, NAND, or NOR [Pahl & Beitz 2007, p. 47]. Table 4-5 shows the basic Boolean operators and explains their behavior.

*Table 4-5: Basic Boolean operators (according to PAHL & BEITZ 2007, p. 47)*

| Logic operator | AND (conjunction) | OR (disjunction) | NOT (negation) |
|---|---|---|---|
| Symbol according to DIN EN 60617-12 | 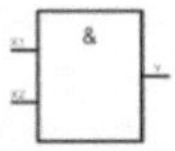 | 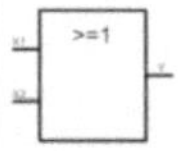 | 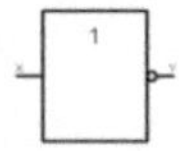 |
| Truth table | X1: 0 1 0 1<br>X2: 0 0 1 1<br>Y: 0 0 0 1 | X1: 0 1 0 1<br>X2: 0 0 1 1<br>Y: 0 1 1 1 | X: 0 1<br>Y: 1 0 |
| Boolean algebra | $Y = X_1 \wedge X_2$ | $Y = X_1 \vee X_2$ | $Y = \overline{X}$ |
| Symbol according to IDEF-3 |  |  | (not available) |

To model the control flow of a process (see section 2.2.3), typically only AND, OR, and XOR are used; for example, in IDEF-3, EPC, or IUM models, these operators are commonly used. The NOT operator is, in fact, not used in any common process modeling methodology. For this reason, the logic modeling for processes explained in this section focuses only on AND, OR, and XOR.

*Table 4-6: Process flow for the common split and join operators*

| | AND | OR | XOR |
|---|---|---|---|
| Behavior of the process for a join | All incident edges are required to continue | Activated by at least one incident edge | One incident edge is allowed at a time |
| Behavior of the process for a split | All outgoing edges have to be followed | One or more outgoing edges are to be followed | Exactly one outgoing edge has to be followed at a time |

Whereas no decision is made when using the AND connector, and the process simply splits or joins, both OR and XOR are based on a decision that influences the process behavior. In most process models, the actual decision is made in the entity before the operator, i.e., a task that is followed by a split-operator will have different possible outcomes which influence any further procedure. As such, both OR and XOR cause a non-deterministic behavior of the process and influence the structure of a process (see Table 4-6).

As Table 4-6 shows, the process model can take different shapes if it is reduced to a structure that does not involve any logic operators. In fact, the operators in common models such as IDEF-3 and others represent different variants of the possible behavior of the process in one single model. If the process model is unfolded to reveal the relations of its entities across the connectors, the entities can interact differently, depending on the operators involved. [BELHE & KUSIAK 1996] illustrate this with an example of a simple process: The conversion of a process with only seven tasks and three Boolean operators will turn into six different process flows (see appendix 10.2.1 for the complete example and algorithms).

*Table 4-7: Conversions of a process model with logical operators into matrix-based notation*

| Rule | Concept | Advantages | Disadvantages | Application |
|---|---|---|---|---|
| 1 | • Creating alternate process matrices | • Consistent notation with simple matrices<br>• Characteristics of structure are kept | • Many different networks are generated<br>• Complex conversion | • Simple models<br>• Models with few logic operators<br>• High relevance of logic operators |
| 2 | • Neglecting the operators | • Structure easy to analyze<br>• Sufficient for many applications<br>• Consistent with all DSM notations | • Neglects dynamic information<br>• Structural analysis impossible for decision points | • Big and complex models<br>• Dynamic information less important |
| 3 | • Decisions as probabilities for related paths | • Detailed description of model quality<br>• Model is good for simulation | • Needs run-time information<br>• Results highly susceptible to changes in data | • Numerical optimization of decisions<br>• Limited focus on general structure |
| 4 | • Logic operators as additional entity in DSM | • No algorithmic conversion necessary | • Limited analysis of logic operators possible | • High number of entities in matrix<br>• Analysis difficult |
| 5 | • Logic operators modeled as additional domain with "type of" attribute | • Conversion back to flow-chart possible<br>• Model for basic analysis (without operators) and extended analysis (with operators) | • Complex MDM is generated | • Large processes<br>• Extensive use of logical operators<br>• Combination of several process models into one<br>• Impact analysis of decision points |

Overall, five ways of converting a process flow involving logic operators into an MDM are possible. The five rules are ordered according to their degree of completeness of converting the structure of a process. Appendix 10.2.5 shows the algorithms and resulting matrices for each conversion rule in detail. Table 4-7 sums up all conversion rules with their advantages, disadvantages, and recommended adaptations. Depending on the application case, any method is suitable to produce a valid MDM.

**Rule 1: Resolve all logical connections [BELHE & KUSIAK 1996]**

Logical operators are eliminated by creating different graphs and matrices for each alternate process given by each decision in the process. Because of the large number of different matrices eventually obtained, this rule is only of theoretical interest, while its application is of little practical use. The number $n$ of all possible graphs amounts to $n = 3^k * 2^m$ with $k$ the number of binary OR-operators and $m$ the number of binary XORs.

**Rule 2: Neglect the operators [KREIMEYER et al. 2007d]**

By dropping all decision points and turning their connections into simple edges, only the basic structure of the process remains. This way, flow characteristics can be analyzed, while a critical path across different decision points (Critical Path Method), for example, cannot be observed. Thus, only analyses based on structural characteristics, i.e., those that do not rely on decision points, are possible.

**Rule 3: Translate operators into probabilities [GÄRTNER et al. 2008] [CHO & EPPINGER 2001]**

By resolving all possible paths into or after a decision point as numerical values that correlate to the probability for taking each path, it is possible to evaluate the sequence of decisions that take place numerically. As such, the decision points are basically modeled like a Bayesian network. However, the appropriate numerical data (e.g., as a numerical DSM) is necessary, which often is not the case.

**Rule 4: Logical operators as additional entities in the process domain**

The operators are kept as an additional entity, losing information about the type of operator. The operators lose their meaning, and only the pure existence of a relationship is transferred, as if all operators were AND operators. This approach extends the simple disregard of the operators, as in rule 3, and integrates an additional number of entities into the network that can be analyzed using a common methodology. A process network with $n$ entities will, therefore, grow to a network with $n + k$ entities, with $k$ as the number of distinct logic operators. The approach is mainly useful if the process model only consists of a single DSM and if decision points are of little importance.

**Rule 5: Carry along the logical operators and their characteristics**

Extending rule 4, this approach (explained in the following) extends the process MDM by a new domain that models the existence and connectivity of connectors (i.e., connectors are modeled as nodes of a new domain "connector") and that uses another additional domain to model the type of the connector (i.e., each connector node is attributed with its type using a DMM). Although somewhat complex in both execution and result, the resulting matrices can transfer the structure of any process model without loss of information bi-directionally.

### MDM-based modeling of logic operators

With MDM as a chosen common basis to model the relationships in any given process notation, not only is their structural impact relevant, but also their modeling in MDM in order to assemble a complete, correct, and consistent process model that is then submitted to analysis. Therefore, logical operators need to be unambiguously represented in the model. As rule 5 is the most complete, it is the one that is chosen for the representation in the following. However, for simpler cases, the other rules can also be applied to generate a process model. In this subsection, the EPC notation is used to show that the rule is applicable even for complex process models.

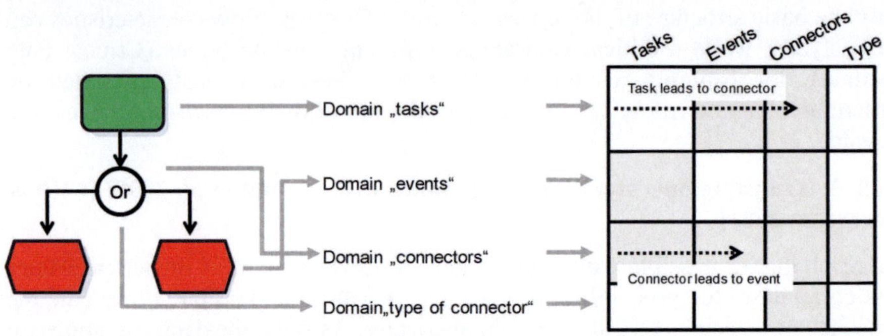


*Figure 4-9: MDM with appropriate domains to translate the primary flow of an EPC model (domains that are potentially not empty are shaded)*

To allow the conversion of logic operators, an MDM can be used that handles decision points as an additional domain; the nodes of this domain serve as an intermediate connection between the tasks and events that alternate in the process flow of an EPC model. To characterize the type of each of these connectors, each node that represents a connector is attributed with its type (AND, OR, XOR) using an additional DMM (the characteristic domain). Figure 4-9 shows how a basic EPC process element (left-hand side) necessitates that four domains be converted.

Theoretically, all 16 matrices within the emerging MDM could contain entries representing dependencies. If, however, the EPC model that is used as a basis is semantically correct, only the shaded matrices in the MDM are needed, while the others remain empty. Yet, it is often possible that process models in practice do not fully adhere to all rules that are set in the process meta-model. Therefore, dependencies can also occur in the empty matrices, for example, if tasks are directly linked (especially for notations other than EPC or for "dirty" models).

For EPC, ten elementary combinations of tasks, events, and logic connectors are possible, as shown in Table 4-8. In fact, it is impossible that any entity of a process not executing a task can make a decision to change the primary flow of the process. Therefore, events being only static states, the process cannot lead to different tasks. However, real-life modeling of process does not always adhere fully to the process modeling notation; therefore, the MDM conversion is also able to cope with the impossible cases of EPC.

*Table 4-8: Logics in EPC notation [SCHEER 1999]*

| | **Several events** | | **Several tasks** | |
|---|---|---|---|---|
| | Join based on more than one event | Split generating more than one event | Join based on more than one task | Split generating more than one task |
| | 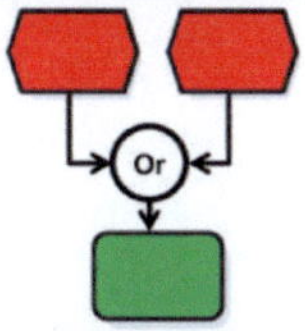 | 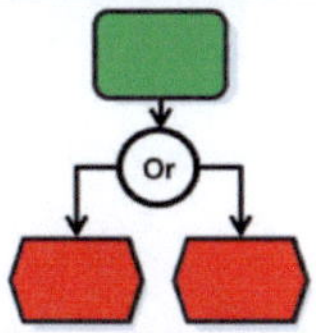 | 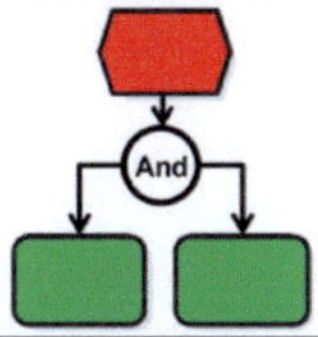 | 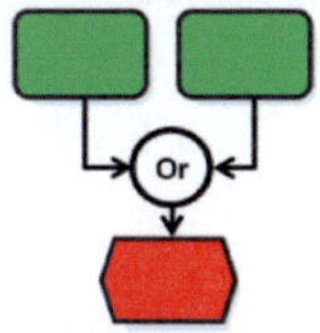 |
| **AND** | All events necessary for continuation | All events are generated by task | Event can lead to multiple tasks | All tasks lead to the same event |
| **OR** | At least one event necessary for continuation | One or more events are generated by task | **Impossible**: event cannot make decision | One or more tasks lead to one event |
| **XOR** | Exactly one event necessary for continuation | Only one of the possible events is generated | **Impossible**: event cannot make decision | Only one out of several tasks leads to one event at a time |

To accommodate all possible constellations of logical operators in a process model, those combinations shown in Table 4-8 are possible. Each can be understood as the smallest possible building block of a process model, as from these blocks, all possible processes can be assembled. These Elementary Building Blocks (EBB) thus embody the smallest units of a process. Interlinking them results in the entire process. Representing the elementary connection types, they consist of two process elements (either tasks or events), and one logical connector. Altogether, there are ten possible EBBs: six for tasks with AND, OR, XOR for splits and joins (compare Table 4-8). Figure 4-10 illustrates two EBBs: an XOR-join EBB for events and an AND-split EBB for tasks. All other EBBs are formed in the same way.

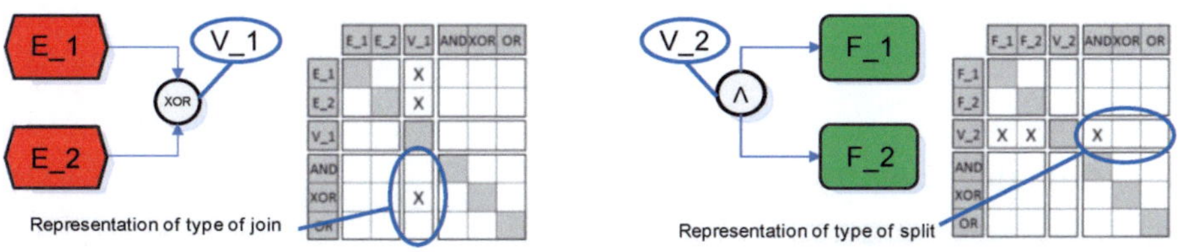


*Figure 4-10: Example for EBBs of XOR-join (left-hand side) and AND-split (right-hand side)*

The example in Figure 4-11 provides a possible recombination of EBBs constituting a complete process. It combines the two EBBs from the previous example, linking the two connectors in a DSM. Except for this entry, the overall MDM is the superposition of the two previous smaller MDMs.

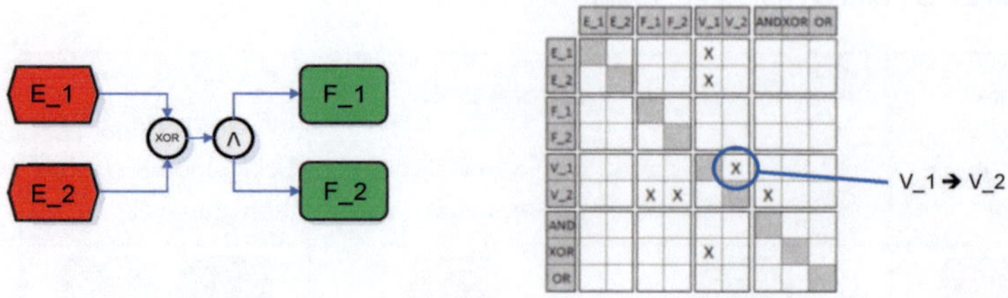


*Figure 4-11: Recombination of the two EBBs from Figure 4-10*

Thus, the approach of converting a process model into an MDM using rule 5 has a number of advantages: Above all, it is compliant to the "classic" matrix-based description of systems, thus extending the modeling base to describe and analyze the structure of a process.

While MDM and EPC are somewhat different forms of notation, the proposed mode of conversion shows that their content is compatible in terms of the structure of the process, i.e., the interaction and dependencies of the various entities that prevail in the process. In fact, a direct inter-conversion between both notations is possible (omitting the fact that, of course, the actual design of the graphical process model is lost). As such, it supports a consolidation in the form of a clearly defined interface between the modeling methods.

As EPC was only used as an example in this sub-section, the proposed method is applicable for any other graphical notation involving logical operators, e.g., IUM or IDEF-3 (an example is found in appendix 10.2). The definiteness of the approach allows modeling iterations and other structural characteristics common to processes. It is suitable for representing a large process unambiguously, as long as all entities are uniquely named (logical operators, too).

As the outcome of the conversion is an MDM, all structural characteristics and metrics applicable to an MDM are relevant. This holds true for the interpretation, as well. However, when evaluating a structural characteristic or metric for a decision point other than AND, the underlying structure does not necessarily represent a network that is present at all times but that *can be* present. As such, interpretations need to be made more carefully, considering the fact that the relationships within the structure represent possibilities and are not permanent.

## 4.5 Building the process model

The meta-model presented as the Structural Process Architecture lays the foundations of modeling a process. Its application has different facets, which are explained in this section. Above all, the SPA model is used to generate a process model that can later be submitted to structural analysis. To this end, the process MDM can be analyzed either using the native data from the process model, or the domains and relationship types can be recombined to generate aggregate views.

### 4.5.1 Generating a process model

Two modes of generating a process model are possible; either, the model is generated as an MDM from scratch, or one or more existing process models are imported. Which of the two possibilities is chosen to create a model depends on the actual context of the process improvement and the availability of other models: If existing process models are available (compare the models in appendix 10.1), these can be processed directly. If other models are available, these can be turned into an MDM by identifying their domains and relationships. If no models are available, a process MDM has to be created from scratch.

To **create an MDM from scratch**, a procedure proposed independently by [MAURER 2007, p. 69] and [DONG 2002] can be used, based on best practices: Using workshops, interviews, existing documentation, questionnaires, or web-based forms [SABBAGHIAN et al. 1998], the system is delimited by collecting relevant entities and relations. Classifying these, a list of domains and relationship types is generated that forms the meta-model for the successive model-building. In a second step, using the same methods of information acquisition, the domains are then refined to their individual elements, and these are reviewed in a pair-wise manner to collect the existence of all relevant relationships. Figure 4-12 integrates this procedure into the overall context of structural analysis.

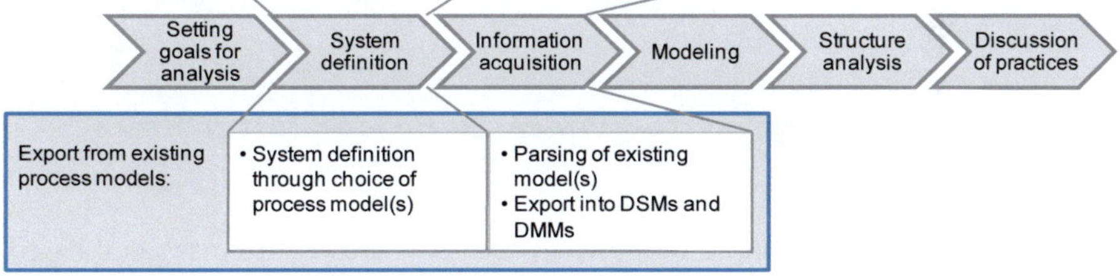


*Figure 4-12: Two approaches to setting up a structural process model*

The Structural Process Architecture is thus a reference to define domains and relationship types. If the analysis follows a certain goal (see section 6.2.1), the meta-model furthermore helps by including domains that are relevant to answer questions related to the chosen goals.

The following questions should guide setting up the system definition. They operationalize the principles of a system description as shown on page 38.

- Which domains are available in the existing process organization?
- Which domains are needed to answer specific questions or to get a complete picture?
- How can the domains of the meta-model be refined to generate a meta-MDM for the analysis in question?
- How are these domains related?
- Which of these relationship types are relevant for the analysis in question?
- How can the relationship types of the meta-model be refined to describe the domains of the analysis in question?

| | Committee | Role | Department | Product attribute | Phase |
|---|---|---|---|---|---|
| Committee | Committee is superior to committee | --- | Committee is of relevance to department | Committee decides on product attribute | Committee is active during phase |
| Role | 1) Role is part of committee 2) Role makes agenda | Role informs role | Role is part of department | Role processes product attribute | Role is active during phase |
| Department | Department has responsibility for committee | --- | Department is part of department | --- | --- |
| Product attribute | --- | --- | --- | --- | --- |
| Phase | --- | --- | --- | --- | --- |

*Figure 4-13: Example of refinement of a domain into three more detailed sub-domains*

Figure 4-13 shows an example of an MDM that was created from scratch, based on the SPA model. Here, the communication structure among the 45 committees relevant for a development process was modeled. For this problem, it was necessary to detail the committees that regroup personnel from the company across departments and specific roles. Therefore, the domain "organizational units" from the meta-model was further decomposed into departments, committees, and roles. Equally, the relationship types have been refined from the basic propositions of the meta-model to fit the refined domains; some pairs of domains are connected by two different relationship types paralleling each other. For example, roles are, on the one hand, connected to committees as being part of them and, on the other hand, as being responsible for creating the agenda for a committee's meeting. To adapt both domains and relationship types, the principles laid out on page 38 were applied.

As a result, a specific meta-MDM[61] for the process analysis was generated, as shown. Then, the meta-MDM was instantiated, i.e., its relevant sub-matrices were filled, as proposed above. In the example, 12 matrices were needed and generated through workshops and from document analysis to build the complete MDM.

To **generate an MDM based on existing process models** — for example, an EPC chart, as shown in the introductory case study — a different procedure applies, as here the meta-model plays a slightly different role. Basically, either one model can be parsed and converted, or several models can be combined into one MDM. In all cases, the existing process models need to be exported from their native systems and converted into appropriate matrices; the model shown in the introductory case study was, for example, generated in a modified version of the ARIS Toolset, exported into a spreadsheet format, and then the individual export files were assembled as an MDM. However, this procedure is different for every modeling tool and not regarded here, as it is mostly an application development specific to the modeling system.

If one model is converted, the Structural Process Architecture model generally supports the goal-oriented analysis and the selection of the appropriate domains within the existing process model, which need to be converted into the MDM.

If two or more models are combined into an MDM, the Structural Process Architecture serves as a frame of reference to collect models for all domains and relationship types necessary; furthermore, it helps to combine possible different levels of granularity in a process model [WYNN et al. 2009]. While, of course, it is not possible to combine two process models with different levels of abstraction into one homogenous model, it is possible to connect models that only differ slightly in their level of detail. To this end, it is often possible to combine several sub-models into an overall model by finding the correct abstract terminology that bridges the elements of the sub-models. If several models are combined this way, it is important to introduce a common naming scheme for all entities involved, as these form the "docking points" among the models. Experience has shown that an enumeration of all entities is a good basis to do so. See the validating case study in chapter 7 for an example; here, 99 different models were assembled into one MDM to generate a coherent model.

### 4.5.2 Aggregate views recombining domains and relationship types

As shown on page 39, a process model will commonly consist of more than two coupled domains [GUILLAUME & LATAPY 2004]. Yet, when analyzing the model and comparing the entities of these domains among each other, it is necessary to compare only entities of one kind with each other. To do so, it is often necessary to incorporate indirect relationships among these entities, which only exist via an intermediate domain. To do so, an aggregate view can be used that only contains entities of one domain and their (computed) relations among each other. This is especially necessary for analyzing structural characteristics [MAURER 2007, p. 82] or structural metrics based on intra-domain networks (i.e., DSMs).

61 Also called a system-graph [LINDEMANN ET AL. 2009]: It shows, what domains and relationship types are used and how they relate. As such, it is similar to an Entity-Relationship-Diagram at a meta-level.

As Figure 4-14 shows, five basic patterns of relationship types are possible at the meta-level, either as inter-domain relations, connecting two domains, or as reflexive, intra-domain relations, relating the elements of a domain to other elements of the same domain. These relationship types can be directed or not.

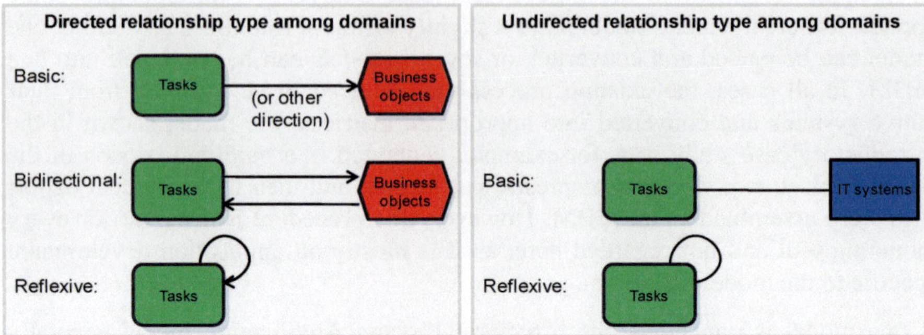


*Figure 4-14: Basic patterns of aggregation of domains*

Based on these basic patterns, three forms of aggregation are possible, and each can lead to either an intra-domain aggregation (producing a DSM) or an inter-domain aggregation (producing a DMM). Table 4-9 shows all six kinds: Aggregation is possible via different paths at the meta-level, via a common entity of reference, or via superposition. As the table shows, the aggregation via path searching produces a directed relation (dashed) between two business objects, the second reached via an intermediate task. If the two tasks are connected to the same task, their aggregate relation is not directed as in path searching. If two models are superposed, the directedness of the resulting aggregate relation finally depends on the available input data.

If an aggregate view is produced via **path searching**, the possible paths within the meta-MDM are followed and computed, as suggested by [BIEDERMANN & LINDEMANN 2008]. In such cases, directed networks are the result of directed native data: if the native data is an undirected network, of course, the aggregate view is undirected, as well; in mixed native networks, containing directed and undirected relationship types, each aggregation has to be checked individually for possible directedness.

Aggregate views can, furthermore, be created by evaluating the concurrent **attribution to a domain of reference**, as shown in the center row of Table 4-9. Here, two entities in the domain "business objects" both relate to an entity in a second domain, and thus share a common partner in that domain. The aggregate views are always undirected.

*Table 4-9: Different forms of aggregation*

| | Aggregation into intra-domain network (aggregate DSM) | Aggregation into inter-domain network (aggregate DMM) |
|---|---|---|
| Aggregation via path searching | 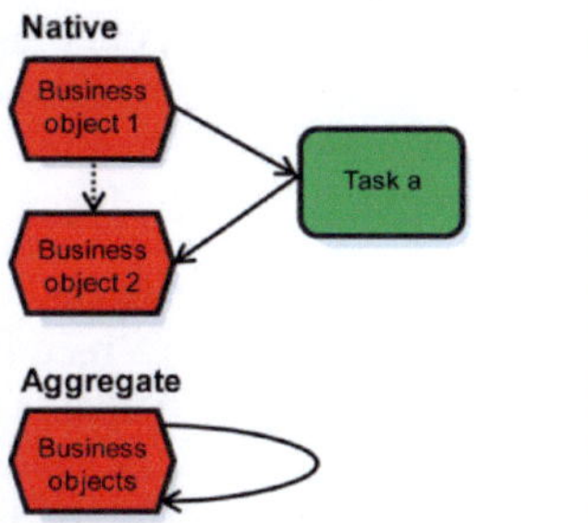 | 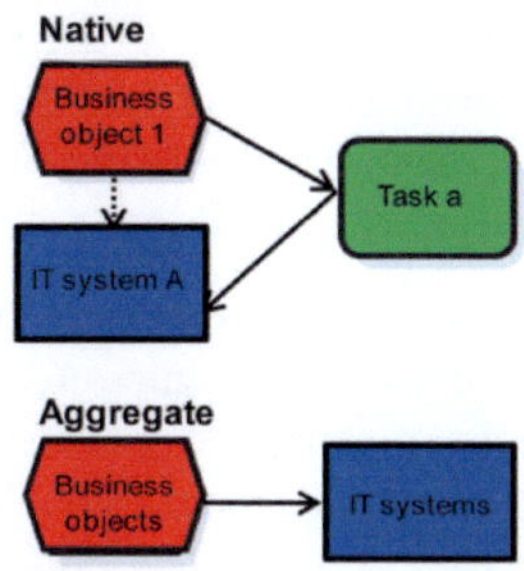 |
| Aggregation via attribution to reference | 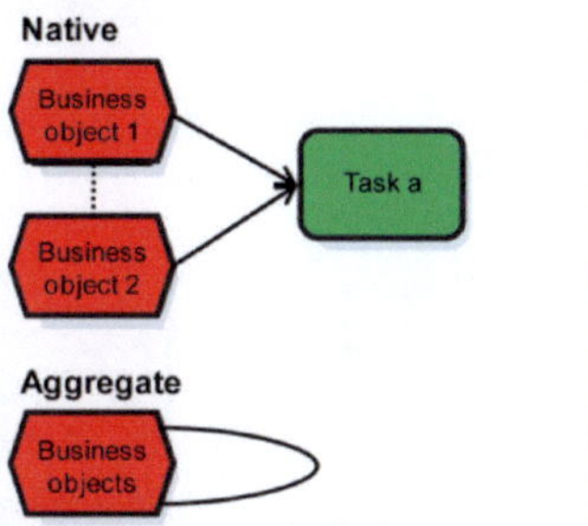 | 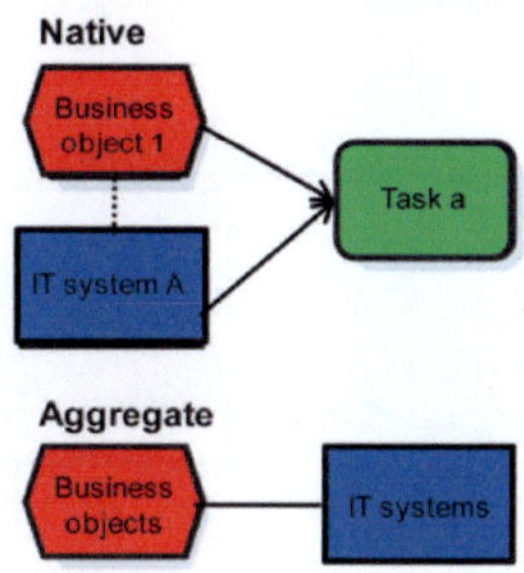 |
| Aggregation via superposition | 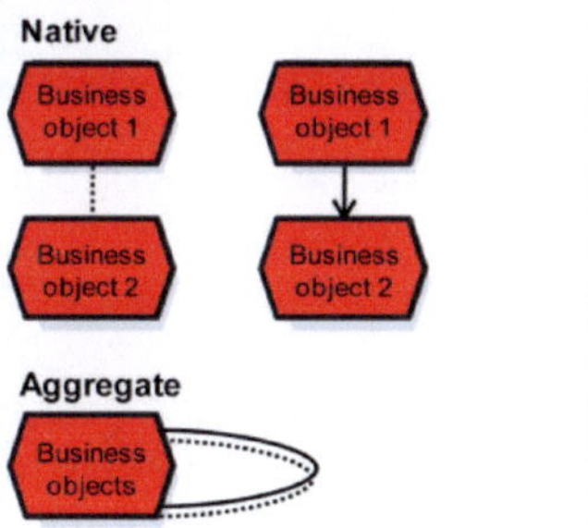 | 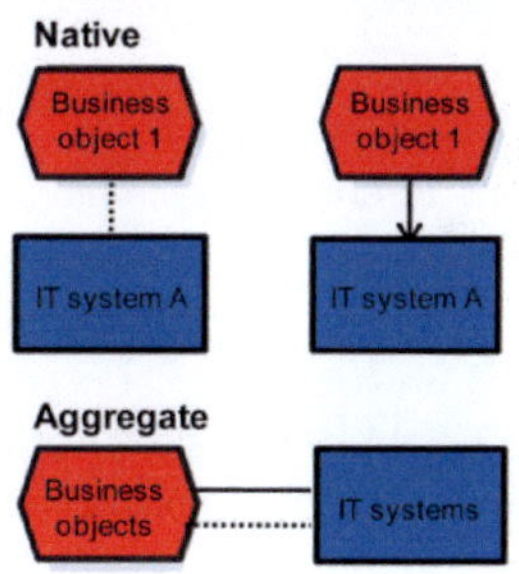 |

Thirdly, it is possible to overlay two networks that are of the same domain ("**superposition**"), i.e., the same kind of entities at the same level of granularity, but that have possible different relationship types. Commonly, such an aggregation only makes sense if the two networks are very similar, e.g., if either partial models are combined or if the relationship types within the models to be combined are a decomposition of a higher-level relationship type that will then govern the aggregate view. Of course, an aggregation of directed native data will provide a directed, aggregate view, while undirected models will naturally bring forth an undirected model. Combined models, integrating directed and undirected relationship types, are not advisable, as the outcome will then be mixed within one aggregate view, making it impossible to differentiate this directedness in further analyses. This third principle is, in fact, used when two process models are combined to generate a process model with a wider system border.

However, for the purpose of process analysis, only directed relationship types are considered in the following. To this end, undirected relationship types are treated as bidirectional directed relationship types, which is sufficient to calculate aggregate domains and structural metrics. There is an ongoing discussion whether this is permissible and whether DMMs especially are directed matrices or how; however, from a pragmatic point of view, treating undirected relationship types as bidirectional relationships is sufficient for structural analysis

Using this convention and the three remaining patterns of relationships, the example shown in Figure 4-15 can be assembled by recombining the different patterns to constitute a process MDM including its domains (four of them in the example: organizational units, business objects, tasks, IT systems) and relationship types[62] (four, again). The undirected basic relationship type between tasks and IT systems is resolved as bidirectional-directed.

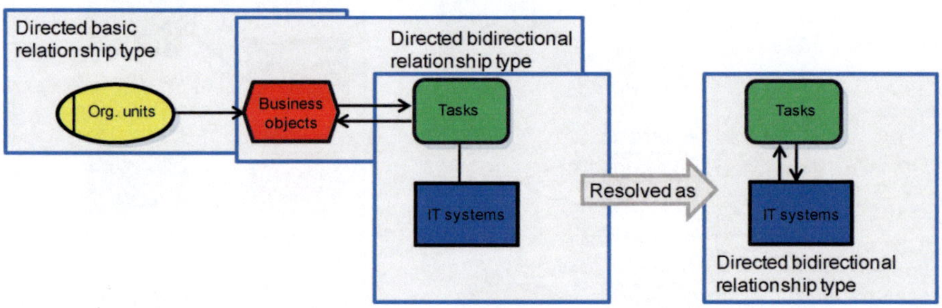


*Figure 4-15: Example of recombination of domains and resolving of undirected basic relationship*

In the example in Figure 4-15, different aggregate views can now be calculated. If, for example, the business objects are in focus, two possible intra-domain networks can be computed to generate a business object-DSM: Either, the business objects can be related via just the tasks, or they can be related via the tasks across the IT systems and back via the tasks. In the first case, one intermediate domain is used; in the other case, the view is aggregated via two intermediate domains.

The computations of path searching can, therefore, be brought down to the computation of **reachability within the MDM** (compare Table 2-2 on page 50), following all possible paths starting from a domain of reference (the business objects in the example above) back to itself. Every individual circuit (i.e., a directed cycle) though the MDM will generate a new aggregate view for the domain of reference that is – at the same time – the start and end domain for this circuit. In the same manner, all paths in the MDM running between two different

[62] The figure only serves as an example; whether organizational units actually use a directed relationship type to connect to a business object or not is not the focus in this section.

domains can be brought back to an aggregate DMM; this, however, is not further regarded here, as no aggregate DMMs are needed for the structural metrics.

Figure 4-16 lists a few examples of native data (left-hand side) and the way these can be aggregated towards a domain of reference (red hexagon) following all possible circuits with the domain of reference as the start and end domain. For example, in the first case, no closed loop is found in the graph of domains and relationship types. Therefore, no aggregate view can be computed. In the second row, one circuit from the domain of reference (the red hexagon, serving as a reference towards which the network is to be aggregated) to the green rectangle and back exists. Here, an aggregate view for the domain of reference via this intermediate domain is possible. For the last row, for example, four different aggregate views of the domain of reference are possible. Of course, this principle of aggregation can also be used for more domains than shown in the figure.

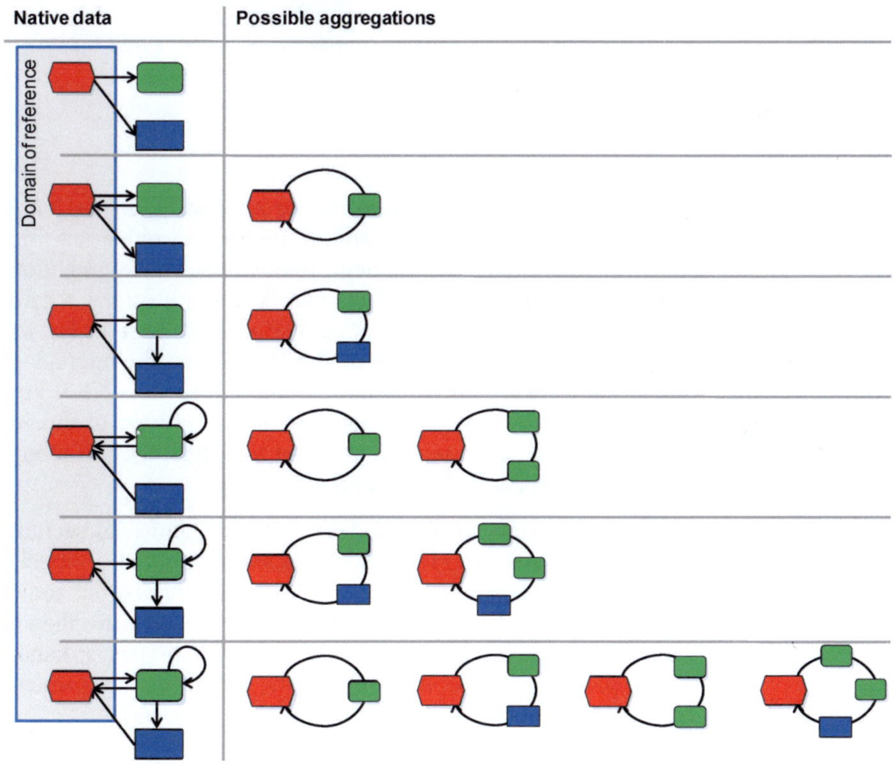


*Figure 4-16: Possible aggregations of up to three domains (using path searching)*

When creating aggregate views, domains and relationship types are condensed into a single domain, changing the relationship type of the new, aggregate view. As such, the relationship types and intermediate objects are chained up to create the **aggregate relationship type**. The example in Figure 4-17 explains the process of aggregation, first collecting the initial relationship type, then the intermediate domain, then the second relationship type. Unfortunately, there is no other systematic way to condense the aggregate relationship type any further. In many cases, however, a higher level of abstraction can be found; for the example, in Figure 4-17, "task delivers information task" can be applied, which is reasonably close to the original aggregate relationship type. However, these simplifications always include a loss of precision and, therefore, should be considered with care.

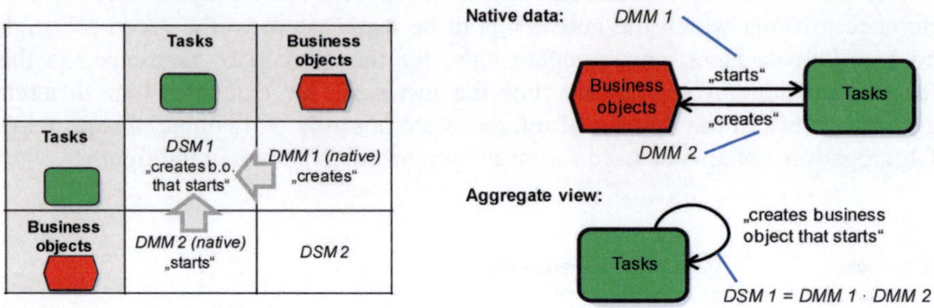


*Figure 4-17: Aggregate relationship type for the example from Figure 4-15*

To **compute aggregate views**, the rules shown in Figure 2-9 and Figure 2-10 are applied. For each path, the matrices along the path are multiplied, either starting from the beginning of the path or the end of the path. For example, an aggregation path via one other domain, therefore, consists of two matrix-multiplications, a path across two other domains of three multiplications, and so on. Figure 4-17 shows the matrices that embody the relationship types of the flow chart; the aggregate DSM is calculated according to case 3 of the rules presented in section 2.1.3. For larger systems, in particular, involving many matrices, it has proven useful in practice to use an ID for each individual matrix, and first to collect the IDs along each aggregation path, and then to compute the aggregate views.

When using logic operators according to the modeling scheme shown in section 4.4.3, aggregation is more complex, as Figure 4-18 shows. Commonly, models involving logic operators do not connect all entities via these operators, but some are directly connected. Thus, on the one hand, aggregate views exist, as shown above via the relationships entered in DSMs and DMMs. On the other hand, aggregate views exit in parallel via the logic operators, which can be nested among themselves (e.g., an OR can be connected to an XOR, and so on).

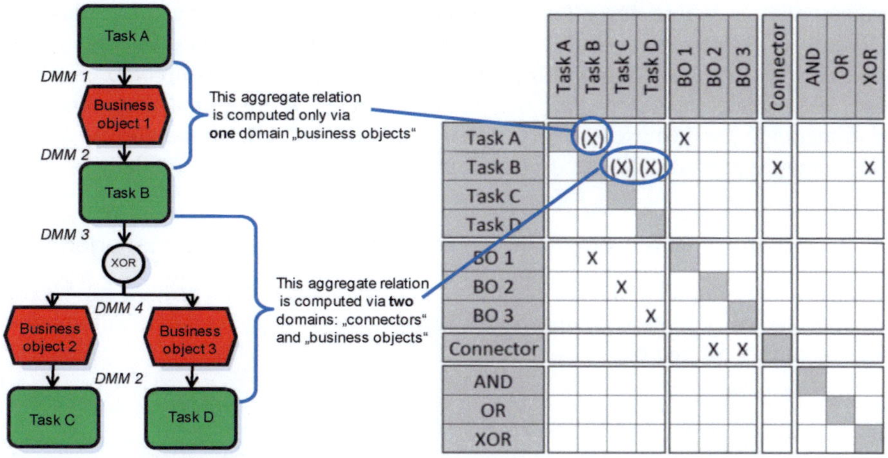


*Figure 4-18: Combined aggregation with logic operators*

Figure 4-19 shows how the aggregate view for the domain "tasks" is computed. DMMs 1 and 2 are multiplied to generate the intermediate DSM of relationships that do not use any logic connector between them; then, a second intermediate DSM including the logic operators is calculated. As a third step, both intermediate matrices are added. This aggregate view thus represents the minimum set of relations among the tasks.

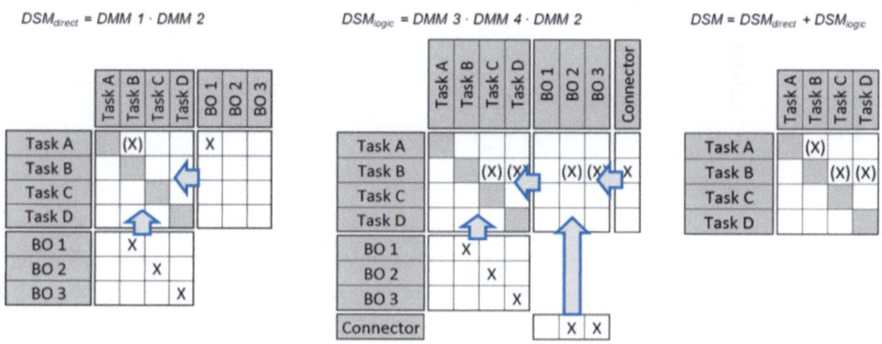


*Figure 4-19: Computation of aggregate DSM from MDM with logic operators (computed relations in brackets)*

Regarding interpretation, the summing up of two (or more) matrices means that the full set of relationships between two domains is spread over two (or more) other domains; in this case, the domain "connectors" only contains one entity, the XOR connector, and, therefore, only contains the relationships between those entities, whose behavior is governed by this logic operator. In a larger context, however, this principle of **aggregation via superposition** is another possible strategy that is viable for other contexts, too, e.g. the data exchange between different IT systems via different interfaces and intermediary systems. Here, however, it is only used to combine models of the same relationship type, as is the case with logic operators, as the aggregation of similar but not identical relationship types requires a very detailed review of the combinability of the relationship types to ensure a purposeful analysis of the aggregate model.

### 4.5.3 Example of a process model for engineering release management

Using the meta-model and its principles of application shown, a process model for the analysis of a release management process at a large premium automotive manufacturer was created.

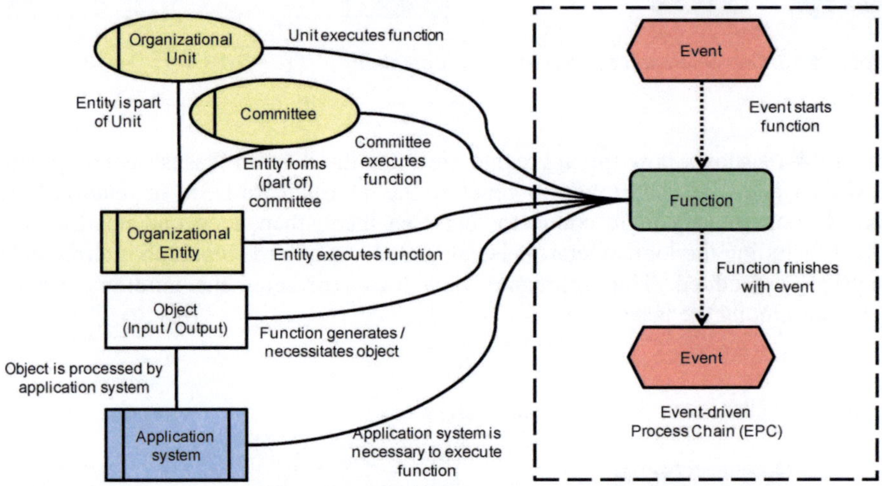


*Figure 4-20: EPC meta-model of the process analysis*

The goal of the analysis was to identify improvement measures for the communication among the different organizational bodies along the principal process flow for a new product line based on the experience from other projects. Specifically, the following aspects were targeted to be improved:

- Alignment of committees and organizational entities with process chains
- Improvement of the assignment of tasks to organizational entities by reshuffling the work distribution based on the process sequence, by the targeted deduction of interdependencies between organizational entities, and through the generation of suggestions for an improved organizational setup
- Improvement of the composition of committees by re-ordering the committees based on the process sequence and by deducing communication channels between committees

The process model was first assembled as an EPC model and then exported into an MDM to be further analyzed. Figure 4-20 shows the input process model as an EPC meta-model that was then translated into a meta-MDM (Figure 4-21). Based on this meta-MDM, the actual process model was then imported and analyzed.

From this native data, aggregate views for the organizational entities and for the committees were computed to analyze the use of resources and to identify necessary communication channels. Equally, aggregate views for the documents and the related IT systems were computed to better analyze the data flows between these domains. The following analyses were computed in detail:

- Analysis of aggregate organizational entities-organizational entities DSM (via functions) and aggregate committees-committees DSM (via functions)
- Analysis of aggregate objects-objects DSM (via functions)

| | F | E | OU | C | OE | I/O | AS |
|---|---|---|---|---|---|---|---|
| Functions F | | E takes place after F | | | | F generates I/O | |
| Events E | E starts F | | | | | | |
| Organizational Units OU | OU executes F | | | | | | |
| Committees C | C executes F | | | | | | |
| Organizational Entities OE | OU executes F | | OE is part of OU | S forms C | | | |
| Objects I/O | I/O is necessary input for F | | | | | | I/O is processed by AS |
| Application Systems AS | AS is necessary to execute F | | | | | | |

*Figure 4-21: Meta-MDM adapted to the analysis of the EPC model used for the process analysis*

Based on this model, the structural characteristics of the aggregate DSMs were examined (e.g., clusters, bottlenecks, start-/end-nodes) to gain a better understanding of the character of the overall system, and the native organizational structure was compared to the aggregate organizational entities–organizational entities DSM to see how well aligned the existing organization was in comparison to the needs exercised by the principal process flow.

As core findings, a potentially lean process chain was eventually proposed, reducing media breaks through better integration of the information flow and through the elimination of redundant communication efforts by the definition of specific communication efforts.

## 4.6 Conclusion: MDM-based process modeling

Common process models are mostly focused on the flow of information through the process. Into this sequence, they integrate boundary conditions, for example, that are set by other domains in the process organization, such as organization units or points in time. However, all of these different domains impact the process organization concurrently, and thus different network structures coexist in a process.

Multiple-domain matrices (MDM) enable a more balanced modeling of these different networks, as they are able to clearly differentiate various domains and relationship types in a large dependency model, integrating all into one coherent model. This model can be extended as needed by adding one matrix at a time, making it possible to import data from various sources to combine them into one overall process model. While losing to a time-oriented representation of the flow of time, leaps back in time or loops in the principal process flow, as well as in the supporting domains, can be represented at the same time.

Furthermore, an MDM allows condensed aggregate views to be created that take into account the relations via any number of domains, reducing them into a single matrix that enables running specific analyses of a domain of reference without losing sight of the implications that originate from the interplay with other domains. As such, a process model that consists of several domains becomes accessible for more straightforward analysis without reducing its complexity. By doing so, one-sided improvements can be avoided during the analysis of a process model. Furthermore, existing algorithms and metrics that are only suitable to work with one domain, as commonly established, can be applied to more complex process networks. Therefore, an MDM is advantageous for a more comprehensive analysis.

Yet, MDMs are difficult to use. In fact, large matrices are not intuitive to read, and they are not meant to replace the graphical modeling of a process. In the context of this research, MDMs are, therefore, used essentially to serve as a common adapter, representing the structural content of different process modeling languages.

# 5. Complexity Metrics for Design Processes

This section lays out available complexity metrics in order to assess the structure of engineering design processes to discover indications about their behavior. These metrics receive, as an input, a structural model (preferably based on the Structural Process Architecture, as shown in the previous chapter, although any other graph or dependency model can be used) of only the entities and relationships of the system. The metrics are, therefore, suited to work with qualitative models, as commonly found in process models set up from "boxes and arrows". These metrics and their description are referred to as the Structural Measurement System (SMS). They support the following purposes as discussed in chapter 3:

- Provide a comprehensive toolbox to analyze a process chart for the occurrence of all relevant patterns among its entities and relationships
- Analyze patterns to describe their occurrence as metrics for every single entity, group or network (depending on the scope of the pattern)
- Connect the patterns to their structural significance for all domains and relationship types as defined by the Structural Process Architecture from the previous chapter
- Provide empirical evidence for the metrics available
- Ensure that the metrics are compliant with measurement foundation (representation, uniqueness, meaningfulness) and Weyuker's Properties
- Rank the results of the analysis by their degree of distinctiveness for the process by identifying results that "stick out"
- Describe the metrics in an intuitive and understandable manner
- Present the metrics in a way that their computation can be automated

The set of metrics was developed to assess design processes in a comprehensive way; thus, the resulting metrics should fill in the solution space as completely as possible. The solution space for structural metrics with a focus on engineering design processes is spanned by three axes, as shown in Figure 5-1.

Principally, process analysis is guided by common *goals* or, more generally, concepts as already shown in Table 2-7, namely, planning, resource consumption, quality, flexibility, organizational decomposition, interfaces, and transparency of process. These concepts guide the use of different process models that assemble the relevant entities and relationships in a process, which can be regrouped under the Structural Process Architecture's *domains and relationship types*, consisting mainly of tasks, artifacts, events, organizational units, resources, and time, as well as the appropriate relationship types (Table 2-9 and Table 4-3). The system of entities and relationships of a process then creates different *structural characteristics* that dictate the behavior of the process, a fact that is referred to as emergence.

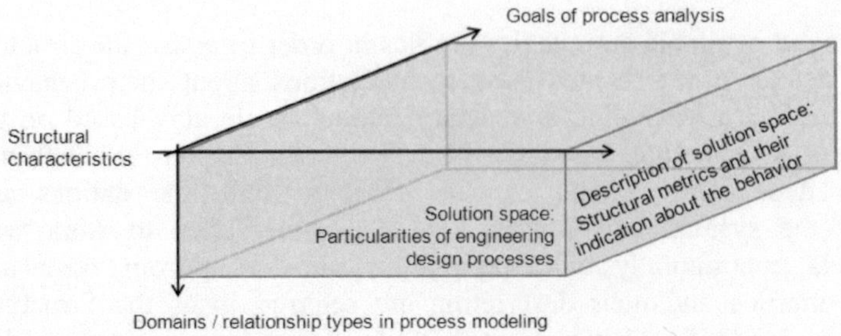


*Figure 5-1: Solution space for the development of structural metrics for engineering design processes*

These three axes form the solution space to the behavior of a process from a structural point of view. Generally, the solution space is filled with the **particularities of engineering design processes** that, in part, relate to the structure of a process. Such particularities are, mostly, the dynamics of a process, its creative nature, iterations in the process (loops, leaps), the fact that results are not predictable, the driving influence of continuously appearing changes, the work based on imperfect definitions, the uncertainty inherent in the process and the accompanying risks, the growing maturity of artifacts, the fact that a definite process path is commonly not determinable, and the involvement of many stakeholders (see section 2.2.1).

These particularities relate, at least to some extent, to the structure of a process, and they can thus be related to structural characteristics. Yet these structural characteristics are, at an abstract level, independent of the semantics of process analysis, as a structure is a constellation of nodes and edges that gains its meaning by the semantics and purpose transported by the model (for a definition of structural characteristics see page 49).

Therefore, possible **constellations of nodes and edges as basic constituents of structural characteristics** are shown first to develop structural metrics independently of domains and relationship types. In a second step, the metrics are then combined with common domains and relationship types to evaluate the particularities of engineering design processes and, thereby, give the structural metrics a process-focused meaning as a Structural Measurement System. The development of structural metrics independent of their application allows, at the same time, the development of a more general concept that can be adapted to different needs of analysis, possibly not only for engineering design processes but, for example, product architectures.

## 5.1 Assessing structural characteristics using metrics

To develop the complexity metrics, a two-stage process is followed, which is commonly used in such cases [GEIGER 2000, p. 95] [MUTSCHELLER 1996, pp. 63-83]. At the first stage, basic structural characteristics are developed that serve as elementary constituents of any network structure; from these elementary components, the combined structural characteristics can be assembled which many different disciplines, e.g., Network theory, have generated.

In the second stage, **basic structural metrics** can be used to embody these structural characteristics in basic measures; recombining and refining these measures then generates **combined and special structural metrics** to assess different structural aspects of engineering design processes. This last step is, in fact, a tailoring of existing metrics to the needs of the empirical object. This procedure is in line with the common reasoning that led to a classification of two kinds of metrics: basic and combined structural metrics, similar to fundamental and derived measures [ZUSE 1998, p. 95]. Figure 5-2 visualizes the procedure.

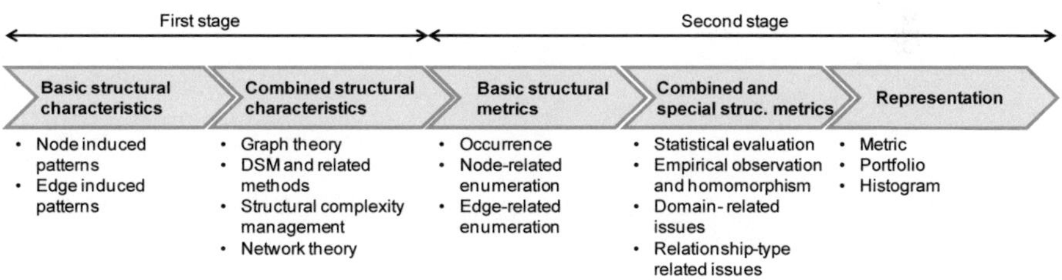


*Figure 5-2: Procedure to set up structural metrics*

### 5.1.1 Basic and combined structural characteristics

All structural characteristics are constructed from basic patterns[63] of nodes and edges that are the constituents of any graph. Such basic structural characteristics can be node-induced or edge-induced, as Figure 5-3 shows. Although the focus in this research is essentially on directed graphs, node-induced basic structural characteristics are differentiated according to the directedness of the network, as the underlying concepts in graph theory[64] vary accordingly.

**Node-induced** basic structural characteristics relate to the connectivity of a graph. If a path from any node to any other node of a graph can be found, the graph is

[63] Patterns or particular patterns are the constellation of nodes and edges of a structural characteristic; however, this constellation does not relate to any semantics of the model and, therefore, is just a pattern of entities, whereas a structural characteristic only gains its right to exist from the pattern AND its meaning. Yet, since the two terms are very closely related, the terms are not further differentiated and used synonymously.

[64] See section 2.1.2 for the basics from graph theory that are used here.

called a connected graph; otherwise, the graph is disconnected. A block is either a maximally 2-connected graph (i.e., a graph that remains connected if one edge is removed), a bridge (including its nodes), or an isolated node. A 2- or biconnected component, therefore, is also a block. A directed graph is said to be strongly connected if every node can be reached from every other node. A clique is a completely connected graph with a relational density of *1*.

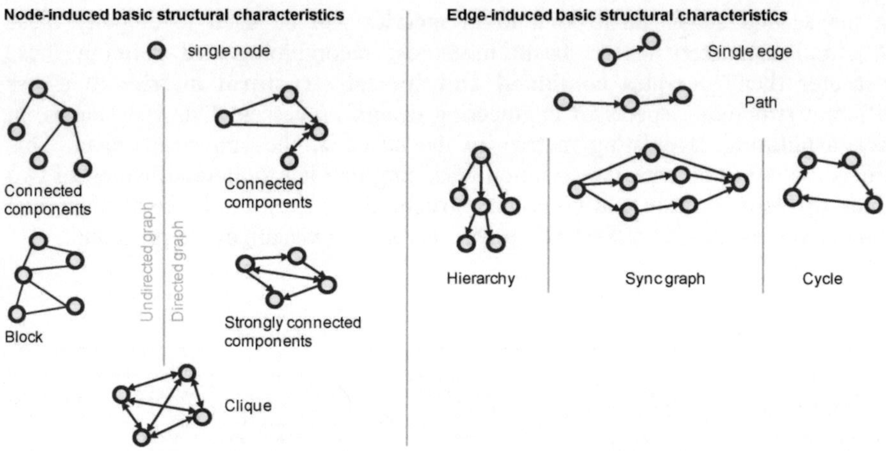


*Figure 5-3: Basic node-induced and edge-induced structural characteristics*

**Edge-induced** basic structural characteristics take shape as different kinds of paths. Hierarchies are a special set of paths to all reachable nodes that can be attained from a root node (also called arborescence). Sync graphs are graphs consisting of a set of parallel paths connecting two nodes. A cycle is a path starting and ending at the same node (called a circuit if a direction is given; a cycle with only bidirectional edges has two circuits).

From these basic structural characteristics, combined structural characteristics can be set up; different disciplines of research have provided a set of structural characteristics that can be considered more or less complete and computable, as was shown in Table 2-2. Table 5-1 summarizes the structural characteristics currently available. Some disciplines claim, in fact, basic structural characteristics as specific to their domain; however, in such cases, these structural characteristics already bear a specific meaning and are no basic constituents.

Based on these structural characteristics, common structures can be described in a comprehensive manner. Each of these structural characteristics serves, in the next step, as a basis for a measurement that produces structural metrics.

*Table 5-1: Combined structural characteristics (according to relevant disciplines)*

| Graph theory | Design Structure Matrix (DSM) | Structural complexity management | Network theory |
|---|---|---|---|
| • Size<br>• Adjacency<br>• Connectivity<br>• N-partite-ness<br>• Paths<br>• Cycles<br>• Reachability<br>• Planarity | • Sequencing<br>• Tearing<br>• Banding<br>• Clustering | • Isolated node<br>• Leaf<br>• Transit edge / node<br>• Bridge node<br>• Splits/joins / adjacency<br>• Bus<br>• Hierarchy<br>• Similarity<br>• Biconnected comp.<br>• Bloc<br>• Spanning tree | • Size<br>• Small world effect<br>• Transitivity<br>• Resilience / connectivity<br>• Degree distribution<br>• Mixing patterns<br>• Community struct. / clustering<br>• Navigation<br>• Centrality<br>• Motifs |

### 5.1.2 Solution principles for structural metrics

Table 5-1 provided a list of currently available structural characteristics. Classifying them at a higher level of aggregation, several solution principles can be deduced that serve, together with the structural characteristics shown in Figure 5-3, as basic aspects that are detailed using different structural metrics. In the following, each solution principle is explained.

Metrics assessing the **size and density** of a network are basic counters to characterize a process by the occurrence of domains, entities, and relationships; as such, they mostly serve as a scaling reference to other metrics. Isolated nodes and leafs (as start- or end-nodes) are also part of this group of metrics.

**Adjacency** addresses the fact that a node is connected to neighboring elements; metrics based on adjacency as a feature of a network, therefore, only regard the relationships of a node within its immediate environment, while secondary impacts or even those farther on are not regarded. As such, the direct impact among nodes and the distribution within the overall network are focused on. Also, splits and joins, and more generally, structural busses are relevant to adjacency as a feature of a structure; likewise, the number of independent sets addresses the number of bands obtained in banding a DSM.

Looking at the propagation and long-range impact of a node across the whole process, **attainability** (also referred to as reachability) extends the metrics relevant to adjacency to the whole network. Thus, nodes are regarded in terms of their embedding in the overall network.

Metrics using **closeness** do not simply look at the overall embedding, but refine this feature by how closely related any two nodes are within the network. Although actually related to paths, centrality uses a count of path lengths to attribute to a node its centrality within a network, thus indicating whether it is well integrated into the network or whether it has a position on its border.

By evaluating **paths** a process is characterized by how it can be navigated, and it is analyzed by which paths across the network are relevant to this navigation. Furthermore, each individual path can have special properties, as it constitutes, in fact, a dependent subset of the overall network; therefore, specific metrics assessing relevant features of a path are part of this group, too.

**Clustering** is an important feature. Here, metrics assessing clusters, i.e., densely or completely connected groups of entities, are counted; equally, transitivity, i.e., potentially existing clusters, are regarded. Ultimately, modules as pre-defined groups of entities that may form a cluster are of interest, especially with regard to whether their border is purposefully drawn.

Metrics that are part of the group of **connectivity** are oriented towards the resilience of a network, i.e., its robustness against individual entities and relationships dropping out.

**Cycles** are another important feature of complex networks, especially if they represent engineering design processes that are almost always subject to iterations. The metrics within this group are tailored to characterize cycles in general, the involvement of different entities and relationships in the cycles, and possible decision points that initiate or re-start iterations within a process.

Metrics that involve **several domains** address the fact that not all process networks are set up with only one domain and that aggregation towards one single domain is not always practical. These metrics thus make use of the ideas behind the features n-partite-ness and mixing patterns. Overall, these metrics strive to assess the degree of alignment between a set of domains from different angles.

Metrics on **cognition** in structural analysis are still at an early stage of research; the basic concept is to evaluate the human capability to actually grasp or understand a network structure using empirical concepts or planarity.

Ultimately, metrics involving **Boolean Operators** can be used for structures that are modeled using decision points, as shown in section 4.4.3. However, the metrics are good for any other dependency model as well, as long as decisions are explicitly modeled.

### 5.1.3 Evaluation of structural characteristics using structural metrics

There are different measurement philosophies, which have lead to different kind of metrics (nominal, ordinal, interval, and ratio scales), as provided by measurement foundation (section 2.3.1), and discussion is still ongoing about what measure is a good measure. The very existence of this discussion indicates that the question may have no fundamental answer, but rather that a selection of the type of measurement needs to be made according to the nature of the empirical object under observation. Generally, two major kinds of measurement philosophies can be differentiated: comparative measures – comparing two or more relative measures to each other – and absolute measures, requiring a reference or scale.

It is scarcely possible to measure a characteristic of a structure in an absolute manner, as there is no structure of reference that could be used. Therefore, complexity metrics will typically be comparative measures. To identify structural outliers, such measures are fully sufficient.

Furthermore, the quality of the measurement needs to be related to the model, and again the quality of the model needs to be related to the empirical object; across both of these stages, information is lost due to increasing abstraction. Errors may be introduced at both stages, as a complete aggregation of an empirical object into a model is not possible, and neither is it possible to completely represent such a model in a (set of) metric(s). In other words, a strict reductionism is not possible, and the basic structural characteristics, therefore, cannot be understood as epiphenomena[65] in a strict sense. Thus, they cannot serve as an absolute scale for a possible absolute metric.

However, the application of structural complexity metrics does not call for a precise measurement, but for the identification of possible weak spots that indicate parts of a process (e.g., an entity, a cluster, a relationship, a domain) which need further attention and which, when improved, may potentially render the overall process more efficient. Thus, a metric is not suited to be a stand-alone means of process improvement, but rather it supports systematic analysis and improvement by prioritizing certain structural characteristics in a process over others and by indicating how structural changes of a process impact its behavior.

*Table 5-2: Available means to set up basic structural metrics to identify structural outliers*

| Basic metrics | Characterization |
|---|---|
| Occurrence of a structural characteristic | The number of times a structural characteristic appears in a system; all basic counters, e.g., the number of nodes, use occurrence as a basic metric, but also more complex metrics as, e.g., the tree-robustness that assesses the penetration of the structure by hierarchies. |
| Node-related enumeration of a structural characteristic | Some structural characteristics have a particular impact on one or more nodes of the structural characteristic; to assess, e.g., the cycles in a system, counting the nodes that occur in these cycles provides deeper insight into the drivers for the cycles. |
| Edge-related enumeration of a structural characteristic | Similarly to node-related structural characteristics, the edge-related occurrence is interesting for, e.g., the edges that keep re-appearing across all cycles in a system, thus dominating, e.g., communication. |

Therefore, the complexity metrics should mainly be used to identify **structural outliers** (see next section), i.e., such instances of a structural characteristic that significantly stand out from the rest of the system. Of course, statistical significance cannot be reached for the analysis of most process models, as common process models only have a limited number of nodes, and, therefore, the population of the analysis will be, from a statistical point of view, too limited to

[65] An epiphenomenon is a phenomenon at a higher level of abstraction that, as a constituent, is able to explain a phenomenon at a lower level of abstraction; the paradigm of strict reductionism postulates that one lower level phenomenon can be completely explained using more fundamental epiphenomena [ANDERSON 1972].

obtain a mathematically sound significance level or p-value. Rather, a structural outlier can be identified using the Pareto principle [REED 2001].

To identify outliers, a distribution is necessary, and thus **basic structural metrics** use the occurrence of a structural characteristic, node-related enumeration, or edge-related enumeration (Table 5-2).

Statistics provide further measures, for example, mean values, variance, regression analysis, correlation, factor analysis, and others. However, due to the limits of the dataset (i.e., the process model being analyzed generally possessing only a small number of entities and relations), these cannot be employed in most cases and, thus, are not part of the basic metrics.

In fact, for the structural analyses shown here, basic metrics are generally sufficient; to refine them, however, **combined and special structural metrics** can be generated, as shown in Table 5-3. These metrics make use of basic metrics but take a more focused perspective.

*Table 5-3: Available means to set up combined and special structural metrics to identify structural outliers*

| Combined and special metrics | Characterization |
|---|---|
| Statistical evaluation | Depending on the network and dataset, different statistical methods can be applied, e.g., the mean path length of all paths across the network to compute the size of the network, or the correlation of degrees of all nodes of the process network to see whether the coherence and information transfer of the process depends on a few highly connected nodes or whether the connectivity of the nodes is approximately evenly distributed. |
| Empirical observation and homomorphism | In some cases, observations from the empirical object relate directly to modifications in the model to create the appropriate homomorphism; e.g., the cluster-coefficient translates into the probability that two nodes that are connected to a root-node are probably connected among each other, too. |
| Domain-related and relationship-type issues | To connect, for example, two domains and the structure of their relationship, metrics such as bipartite density, e.g., assess the networks in two domains at a time, comparing them via a mapping; similarly for domain-related issues, relationship type-related issues address the recombination and combined occurrence of different relationship types within one domain (e.g., comparing a native to an aggregate domain using one and the same metric). |

In the last step, the results of the measurement need to be visualized using an appropriate form of representation.

Table 5-4 lists the different forms of representation. The mode of representation[66] is chosen according to the type and quantity of results of the metric. The quantity

[66] A good overview of different modes or representation, their dependencies, and their strengths and weaknesses can be found in [TUFTE 1992]

of the results of the metric depends on whether a metric delivers a single value for an entity of the process or the overall process. In the former case, histograms are used; in the latter the metrics are given for each reference (a sub-process, a domain, etc.). If a metric is used separately for outgoing and incident edges (see section 5.2.4), portfolios can be used to relate the results of a metric to the relevant two or three axes. Likewise, portfolios or tables can be used if a reference of the measurement is used, for example, when comparing one domain to another.

*Table 5-4: Different forms of representation*

| *Representation* | *Description* |
|---|---|
| 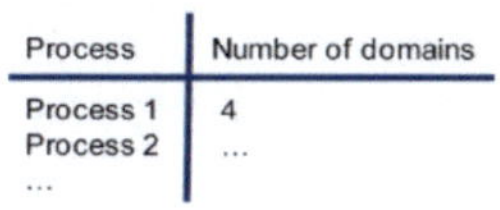 | Metric per domain<br><br>e.g., Number of domains |
| 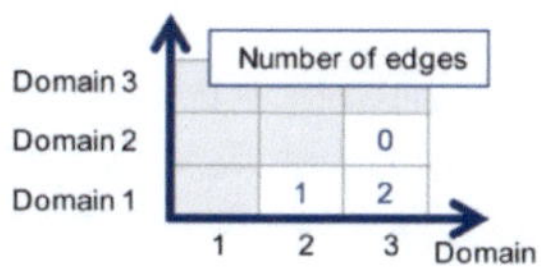 | Table of metric per pair of domains<br><br>e.g., Number of interfaces between domains |
| 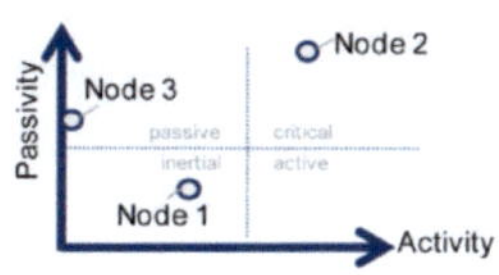 | Two-dimensional portfolio for all nodes<br><br>e.g., Activity / Passivity |
| 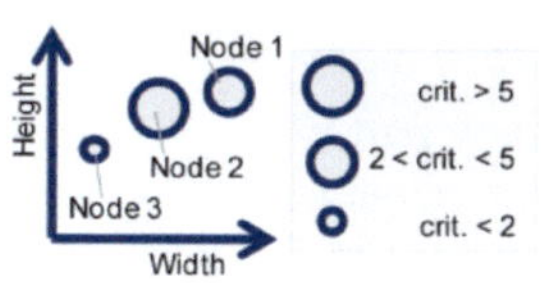 | Three-dimensional portfolio for all nodes<br><br>e.g., Tree criticality<br><br>Also possible as correlation plot of occurrence of a metric for each pair of reference values (e.g., nodes, degrees)<br><br>e.g., Degree correlation (based on nodes) |
| 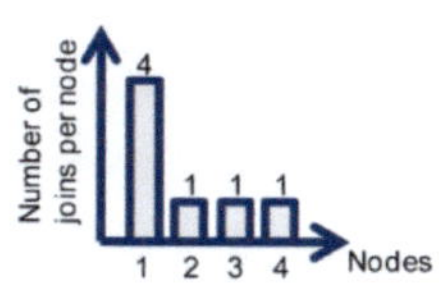 | Histogram of metric for all nodes, possibly as Pareto distribution<br><br>e.g., for Impact of synchronization-points<br><br>Also possible as continuous (interpolated) curve to represent distribution functions<br><br>e.g., for Degree distribution |

### 5.1.4 Structural outliers

As the metrics provide a highly condensed picture of the process, they do not provide detailed information about the process's behavior; however, their main focus is to identify structural outliers that characterize the process's structure. These outliers are the entities and relationships of the actual process that "stick out" the most and therefore are the most interesting for process improvement. A structural outlier is defined as follows:

> Outliers are such instances are of a network structure that particularly stand out with regard to their involvement in a structural characteristic. The identification of outliers makes it possible to pinpoint entities that are of extremely high or low impact to the system represented as the network, thus significantly driving a pattern of entities [HAWKINS 1980]. Outliers are, therefore, those results that are "numerically distant" from the main population of results are understood as outliers, and they commonly show up in histograms or other distributions [BARNETT & LEWIS 1998, p. 16]. While, of course, a process has a limited number of entities that is often too small to obtain statistically significant results, the concept of the outliers essentially embodies the Pareto principle[67] [REED 2001] by highlighting the core entities of a system.

Different modes of identification of structural outliers are possible. All necessitate the existence of a histogram or a distribution that presents the results of a metric per node, per edge, per process-module or for any other reference that is part of the structure being focused on.

- The most intuitive outliers are **upper-bound outliers** that appear at the top of a distribution. Most commonly, the Pareto principle will be applied to spot the top five or top ten outliers that drive and govern the process. Typically, these entities will be the most relevant ones for the process.
- At the other end of the scale, **lower-bound outliers** are of interest, too; commonly, these will be such entities that are almost not integrated into the process, therefore contributing very little to the process.
- Structural outliers can, furthermore, appear within any part of a distribution, e.g., as a characteristic spike or an abrupt drop in the range of results. These **abrupt-drop outliers** will generally appear as a particular footprint of the process.
- Lastly, the comparison of two datasets (either a native and an aggregate one, or two aggregate ones) will allow the identification of **cross-aggregation outliers** that will not show up as one of the three kinds above but only become visible if two distributions are compared based on identical abscissa.

Figure 5-4 visualizes their appearance in the distribution[68] of the metric "Relative centrality" for the process models shown in the later case studies. In the left aggregate view (points in time via documents), node 24 is the most interesting

[67] Also called the 80/20 rule.

[68] This example is taken from the second case study, as shown in section 7.2.

outlier, as it clearly sticks out above the rest in the process. This point in time is, therefore, the one that drives the process the most, as most paths within the overall process run across this point in time. It will, therefore, be most important in driving the timeliness of the process. Furthermore, nodes 49, 33, 44, 22, and 43 appear at the lower bound. They appear little integrated and may be of little importance to the process. Their analysis could, therefore, provide possible cost-saving potential. Thirdly, a small drop appears from the plateau after node 67. This drop points to two plateaus, in fact, the one with a centrality of approximately 12 and the one with a centrality lower than five. These two groups could point to, for example, different levels of importance of the points in time in question. Lastly, node 33 appears as an outlier if the left-hand aggregate view is compared to a second aggregation of points in time via tasks. While node 33 does not appear as highly relevant in that distribution, it still shows a very different characteristic if compared to the initial position in the Pareto distribution. This could indicate that the document structure and the task structure may vary considerably in their integration of node 33.

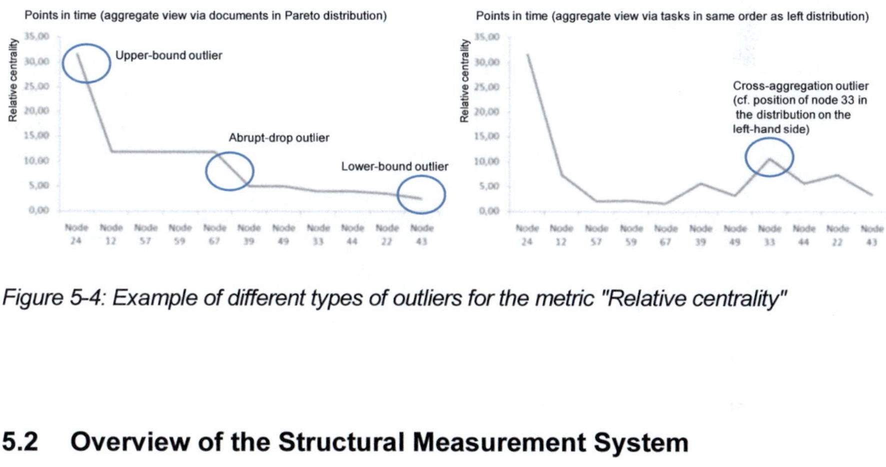


*Figure 5-4: Example of different types of outliers for the metric "Relative centrality"*

## 5.2 Overview of the Structural Measurement System

To represent a comprehensive overview of available complexity metrics for the structure of engineering design processes, the Structural Measurement System (SMS) regroups available complexity metrics. For each, the general concept of the metrics, i.e., the different aspects of their description and the different perspectives that can be taken for a metric are described. While many basic metrics are not explained in detail, those metrics that use a more complex rationale are illustrated in this section. All metrics are detailed in appendix 10.4.

### 5.2.1 A comprehensive set of complexity metrics

Overall, 52 metrics were assembled to represent a comprehensive toolset for the structural analysis of processes. Table 5-5 provides an overview of these metrics, grouped by basic solution principles of a structure that govern each metric.

*Table 5-5: Metrics that are part of the Structural Measurement System*

| **Size and density** |
|---|
| Number of domains |
| Number of nodes |
| Number of edges |
| Number of classes |
| Number of interfaces between domains |
| Number of edges per node |
| Relational density |
| Number of unconnected nodes |
| **Adjacency** |
| Activity / Passivity |
| Degree correlation (nodes) |
| Degree correlation (edges) |
| Degree distribution |
| Fan criticality |
| Synchronization points / distribution points |
| Number of independent sets |
| **Attainability** |
| Number of reachable nodes |
| Reachability of a node |
| **Closeness** |
| Proximity |
| Relative centrality (based on between-ness) |
| **Connectivity** |
| Node connectivity |
| Edge connectivity |
| **Paths** |
| Number of paths |
| Path length |
| Weight of an edge |
| Centrality of path (based on centrality) |
| Centrality of path (based on degree) |
| Degree of progressive oscillation |

| **Hierarchies** |
|---|
| Height of hierarchy |
| Width of hierarchy |
| Tree criticality |
| Snowball-factor |
| Forerun-factor |
| Tree-robustness |
| Maximum nesting depth |
| **Clustering** |
| Number of cliques |
| Cluster-coefficient (local) |
| Cluster-coefficient (global) |
| Module quality 1 (flow of information) |
| Module quality 2 (compactness) |
| **Cycles** |
| Number of cycles |
| Number of cycles per node |
| Number of cycles per edge |
| Number of feedbacks |
| Activation of cycle |
| Number of starting points for iterations |
| Iterative oscillation |
| **Several domains** |
| Bipartite density |
| Number of organizational interfaces |
| **Cognition** |
| Cognitive weight |
| Degree of non-planarity |
| **Boolean Operators** |
| McCabe Cyclomatic Number |
| Control-Flow Complexity |

All metrics were developed starting from the solution principles shown. Each solution principle was, to this end, reviewed for its different aspects. Each aspect of a solution principle was then translated into a structural metric. This process was done from two sides to collect a list of metrics as complete as possible. First, existing metrics were attributed to the solution principles bottom-up, i.e., existing metrics (as shown in Table 2-11, 2-12, and 2-13) were attributed to the solution principles to gain an overview of available metrics and the completeness of existing solutions. This attribution was shown in Table 2-14. Most of these metrics are **basic structural metrics**, as they are directly derived from structural characteristics. Second, those solution principles (and their aspects) that are relevant to engineering design processes were broken down into their structural content to fill the gaps in the list of existing structural metrics (shaded in Table 2-14). These metrics are explained in detail in section 5.2.3. This second group of metrics represents mostly **combined and special structural metrics** that are based on more specific evaluations.

During the design of the metrics, each metric was defined and its structural significance and representation described. Table 5-6 provides an example.

*Table 5-6: Description of the metric "Reachability of a node"*

| | |
|---|---|
| **Definition** | • Number of possible starting nodes to reach a designated node<br>• Can be normalized to the total number of nodes within the graph |
| **Structural significance** | • Influence of the overall process for a node<br>• Influence is not weighted according to distance (as opposed to hierarchies) to distance (as opposed to hierarchies) |
| **Representation** | • Pareto-distribution of nodes ordered by the number of passively reachable nodes (see metric "Number of reachable nodes")<br>• Also possible as a portfolio of reachabilities and number of reachable nodes |
| **References** | [ALBERT et al. 2000], [DANEVA et al. 1996], [ROLÓN et al. 2006b], [NEWMAN, 2003a], [MAURER 2007, p. 202] |

Generally, each metric is visualized as a graph representation to show one example of a constellation of nodes and edges. This *graph* serves as an example for the description of the metric where possible. The *definition* explains the algorithm that is followed to calculate the metric; where applicable, it also integrates possible modes of normalization. Each metric has a basic *structural significance* which is, however, very generic if no domain and no relationship type serve as a reference for a possible interpretation. For this reason, only the basic aspects of the meaning transported by the metric are explained. The principal *representation* is then detailed, as shown in Figure 5-5. Again, the example shown in the graph representation serves as a depiction. For some metrics, more than one representation is possible, for example, a histogram and a portfolio differentiating active and passive metrics (i.e., outgoing edges and incident edges for the "Number of reachable nodes" that can actively be reached and the "Reachability of a node" as passively reachable). In such cases, a reference refers to further depictions, and the description of the metric may be split into two descriptions, one for the active and one for the passive metric – compare Table 5-4 for common forms of representation. Ultimately, available *references* explain algorithms and empirical evidence relevant to the metric.

Figure 5-5 presents an example of the metric "Reachability of a node", taken from page 326 in the appendix. As can be seen in the example, node 1 has a high impact on the overall network, being simultaneously the start-node, generating a number of reachable nodes as 6, i.e., all available nodes in the network outside this node. It is, however, impossible to reach the node from any other point in the network, thus resulting in a reachability of zero. It, therefore, influences the network in an important manner; however, the degree of impact on any downstream node is not further assessed (for example, using weights or other means).

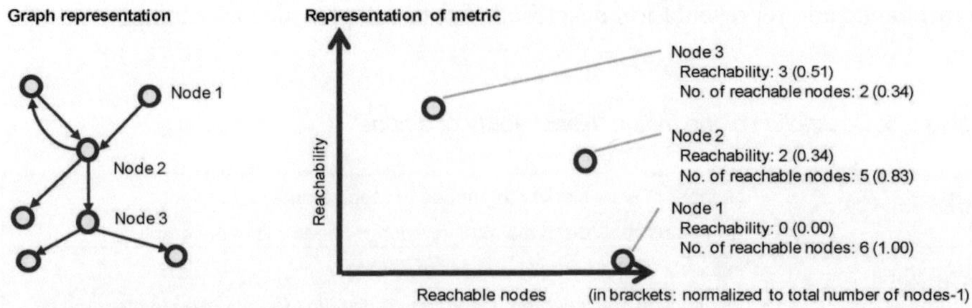


*Figure 5-5: Visualization of the metric "Reachability of a node"*

Using the Structural Process Architecture, the meaning of the metric is further detailed in a second table, as shown in the example in Table 5-7, to outline its detailed domain-specific significance. For each domain, the relevant meaning and informational value transmitted by the metric are explained. The interpretation is, furthermore, based on the principal relationship, as shown in this meta-model in Table 4-3, as these represent common intra-domain relationship types.

If the network shown in Figure 5-5 represents, for example, a network of tasks, it can be deduced that node 3 depends highly on previously compiled information that is transferred across the process; thus, if there were any errors in the previous process, there is a high risk that the wrong input information may impact the task, as the task serves as an information sink for the three tasks previously executed.

*Table 5-7: Domain-specific significance of metric "Reachability of a node"*

| | |
|---|---|
| **Task** | • Degree to which a task serves as an information sink within the overall process<br>• Level of risk of a task to be based on wrong assumptions<br>• Degree to which a task depends on previously compiled information |
| **Artifact** | • Extent of necessary consistency with previous artifacts<br>• Level of impact of the overall, previous process on an artifact<br>• Level of risk of an artifact to work as an information sink and include errors from the subsequent process |
| **Org. unit** | • Visibility of an organizational unit within the organizational setup<br>• Extent to which information is likely to arrive at an organizational unit |
| **Time** | • Degree of control of the schedule previous to a point in time over its timeliness<br>• Level of importance of a point in time for its integration into the overall schedule<br>• Degree to which a point in time serves as a buffer for the overall, previous process |
| **Event** | • Level of relevance of an event for all previous events in the overall process<br>• Degree of control of the process previous to a event over its timeliness<br>• Degree to which an event serves as a buffer for the overall, previous process |
| **Resource** | • Extent of potential of a resource to serve as an information sink that receives and compiles information and knowledge from the overall process<br>• Level of potential for consistency among all artifacts in the process |

### 5.2.2 Relevance and limits of basic structural metrics

Before the complexity metrics are further detailed, the foundations to consider whether the proposed metrics embody the foundation of a good measurement are laid out. As shown in section 2.3.1, the representation, uniqueness, and meaningfulness are of relevance for any measurement [STEVENS 1946]. The meaningfulness can furthermore be broken down into the content, the criterion, and the construct [MENDLING 2008, p. 106].

All metrics necessitate certain semantics (i.e., their content) to validate the construct. All metrics are, therefore, examined with a focus on the domains and principal relationship types provided by the Structural Process Architecture shown in chapter 4: tasks, artifacts, events, organizational units, resources, and time. Only the domain "Product attributes" was not considered more closely, as this domain only serves as an adaptor for the product architecture.

The results of the examination of each metric provided, in part, indications of the structural significance of each metric, as shown in appendix 10.5. The rationale for the development of these indications is given in the following.

A good **representation**[69] demands that the scale of a metric be proportional to the phenomenon in the empirical observation. Most of the proposed metrics are counters, thus producing a result proportional to the counted phenomenon, e.g., edges or nodes. Only a few metrics use other concepts. The following section 5.2.3 looks deeper into those metrics that escape reasoning based on counters.

As most metrics are listed as original counters, normalization is optional, particularly with a regard to only comparing measures among themselves; as the main goal of this research is to spot outliers, normalization is of limited interest if only a single process is under review. If, however, two or more processes are compared for certain structural characteristics, the base for which each metric is normalized needs to be carefully chosen. In most cases, the number of nodes (or edges, respectively) in the graph will be the base. In more complicated cases, e.g., measures of centrality, the normalization is explained as part of the description of the definition of the metric.

The **uniqueness** of each metric refers to the invariance of a metric in mathematical operations. In simple terms, it refers to the fact that the result of the metric can be obtained in one and only one way. As this proof necessitates extensive mathematical background, it is not provided for any of the given metrics, and there is no dedicated literature available that reviews even the common metrics, such the Activity metric or McCabe's Cyclomatic number.

The **meaningfulness** of the metrics is argued in detailed description in the appendix. The *content* of each metric is designed to involve the full scale of possibilities, including the non-existence of a structural characteristic, thus producing ratio scales (see section 2.3.1). This fact is highly relevant for the development of metrics that are meant to discover structural outliers, as outliers that show up on the top of the scale as well as outliers that range on "the long tail"

[69] This term is not to be confused with the term "representation" as used for the visualization of the results of the metric; here, representation refers to the proposed criterion as cited in [STEVENS 1946].

are relevant [ANDERSON 2007]. The *criterion* is, at each time, based on empirical evidence where available, describing the domain-specific significance of each metric in detail in the appendix (see Table 5-7 for an example). Paired with the description of the graph representation of the metric's focus, the criterion is, therefore, shown for each metric. Ultimately, the *construct* behind each metric refers to the theoretical reasoning, and it is thus closely related to the definition and interpretation of a metric. The theoretic reasoning is, furthermore, provided in the given references for each metric

An important part of the meaningfulness is the granularity of the model, which impacts the results of the metrics. If one and the same process were modeled at three different levels of detail, an intuitive expectation would be that, if there were one important improvement potential in the process, analyzing the process using structural metrics would yield this problem at any level of detail. Of course, the metrics are, as such, tailored to consistently analyze a given process model at a given level of detail. Comparable results across different levels of detail of a process are, therefore, mainly an issue of the appropriate process model. However, different levels of detail do not imply an increase or decrease in the number of nodes and edges in the associated graphs in proportion to the level of detail. As the metrics are mainly conceptualized as counters, they will, therefore, not necessarily yield results that remain comparable among the levels of details.

Figure 5-6 illustrates this example: While – at the task level – the process only contains one cycle, at a higher level one further cycle occurs. Furthermore, at a work package level, the two tasks are of different sizes, task two involving six work packages and eight relations, but task four only three work packages and three relations.

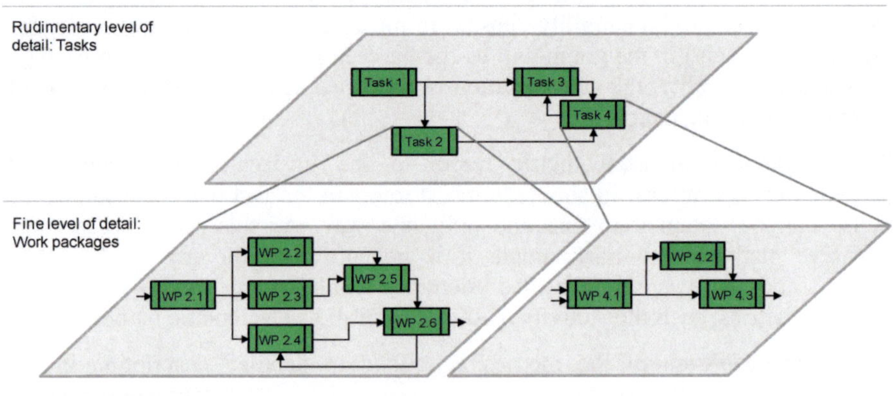


*Figure 5-6: Disproportional increase of number of nodes and edges of process graph at different levels of detail*

To further discuss the validity of the metrics, Weyuker's properties form an important basis. Particularly, but not exclusively, in software quality assurance, the following nine properties are commonly considered relevant for the development of metrics to assess a structure [WEYUKER 1988]. The compliance of a metric with these properties commonly indicates the conceptual soundness of a metric [CARDOSO 2005a]:

1. A metric cannot measure all software programs as being equally complex.
2. There are only a finite number of programs of the same complexity.
3. Even a program perceiving a different complexity may map into the same complexity measurement value.
4. The complexity of a program depends on its implementation, and even if two programs solve the same problem, they can have different complexities.
5. The complexity of two programs joined together is greater than the complexity of either program considered separately.
6. A program of a given complexity when joined to two other programs does not necessarily imply that the resulting program will be of equal complexity, even if the two added programs are of equal complexity.
7. A permuted version of a program can have a different complexity; thus the order of statements matters.
8. If a program is a straight renaming of another program, its complexity should be the same as the original program.
9. The complexity of two programs joined together may be greater than the sum of their individual complexities.

Intuitively, different processes have different levels of complexity; the same is true for the results of the metrics proposed in this research, as counting different processes that are set up from different numbers of nodes and edges, *property 1* is fulfilled. Yet, to be able to compare processes at all in terms of their levels of complexity, they need to be modeled with the same scope (domains and relationship types) at a comparable level of detail. This is also true for *property 2*, which is equally fulfilled by the metrics as proposed. While there are a great number of processes in any company, they are typically of different size and complexity. The case studies in chapter 7 will clearly illustrate this fact. *Property 3* is fulfilled as well; in fact, it is possible for every metric to construct different graphs that yield the same result for a given metric. However, it is not possible that two different graphs have the same results for all metrics, as at least one metric will always deviate. *Property 4* is only partly relevant, as it originally addresses different programming languages and paradigms that will make one and the same program take different shapes (and thus vary in complexity) if it is written in different languages; of course, a process that is modeled in two different process modeling languages can result in different results for the metric, as well. Yet, if the Structural Process Architecture is used as suggested to condense the structural content of the different process models to a comparable level, i.e., the same domains and relationship types, this property is not relevant. The fulfillment of *property 5* is based, again, on the enumerative nature of the metrics: the more elements, the higher the results of the metrics. This extends also to *property 6*, which will generally lead to a greater complexity if two processes are joined. As the order of tasks in a process is essential to the purpose of a process, swapping the order of tasks is only of limited interest to process management, and therefore *property 7* is of limited relevance. However, metrics (e.g., the number of feedbacks) that relate to an ideal sequence of a process from a structural point of view use this property to point to processes with more feedbacks (thus having a less ideal sequence) as being more complex. *Property 8* is intuitive; if a process model is assessed twice using the same metrics, the results will be identical. However, in some cases heuristic algorithms are used for the computation of a metric; in such cases the actual outcomes of the computations can differ, even though they should be identical from a theoretical point of view. Ultimately, *property 9* states that joining two processes does not mean that the results of the metrics for each initial process can be summed to obtain the results for the finalized process. Of course, the metrics are not simply added, as most of them (except for the "number of" metrics) involve other mathematical operations than just the enumeration of nodes and edges. In fact, in some cases the resulting complexity may even be lower than the sum of the initial results, for example, if two processes are joined via a common node. In this case, the resulting process will have one node less than the sum of the numbers of nodes of the initial processes. However, as WEYUKER states, this property is not a very strict one.

Ultimately, automation is of relevance in addition to the above criteria [KERNLER 1996, pp. 35-38]. Unfortunately, not all metrics are applicable in practice due to computational limitations, even though algorithmic support is – in theory – available. Appendix 10.6 gives an overview of those metrics that have proven computable with reasonable computing time at a desktop workstation.

### 5.2.3 Relevance and limits of combined and specific structural metrics

A few metrics use more extensive computation schemes and, therefore, cannot be as simply argued as shown in the previous section. Additionally, some of these metrics use extended reasoning to suit the needs of process analysis and are, therefore, explained in this section. In particular, Boolean operators (activity / passivity, McCabe's Cyclomatic number, Control-flow complexity), module quality, hierarchies (tree criticality, snowball factor, forerun factor, tree-robustness), cycles and feedbacks (number of feedbacks, activation of cycle, number of starting points for iterations), cognitive weight, and degree of non-planarity are reviewed more closely. Every metric is first described at a conceptual level, then a small example is used, and finally the entire metric is discussed.

The metric **activity/passivity**, in its common form, evaluates how "active" or "passive" an element is, i.e., how much it impacts other nodes directly or how much it is impacted by them [DAENZER & HUBER 2002, p. 558]. To calculate the activity or passivity, the number of outgoing or incident edges is counted for each node. If, however, a logical connector appears between a set of nodes, it is not possible to calculate the activity or passivity.

In the case of AND and XOR, this is a simple problem to solve, as either all nodes or simply one node are connected. If, however, an OR is in between two nodes, neither the maximum number of points (= *n*) that may eventuate nor the minimum number (= *1*) is correct. Rather, a mean value in between is relevant from a structural point of view, to show the impact weight of logically connected entities. Thus, from a structural point of view, there is a need for a "mean impact value". Table 5-8 shows the basic idea: If a join-OR has two incident edges, it can have three possible structural constellations. Cases 1 and 2 are single edges each; cases 1 and 2 correspond to two edges. This way, the mean impact equals 1.3333, supposing a uniform distribution of all possible cases.

*Table 5-8: Example of an OR and possible structural constellations*

| Behavior | Structural relevance | Passivity of case | Partial passivity |
|---|---|---|---|
| 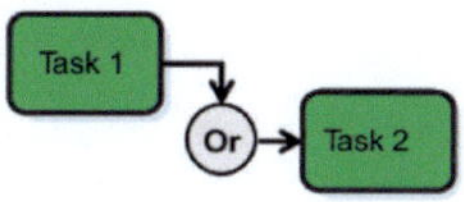 | $r = 1/3$ | $P(task\ 1) = 1$ | $P_{task\ 1} = 1/3 * 1$ |
| 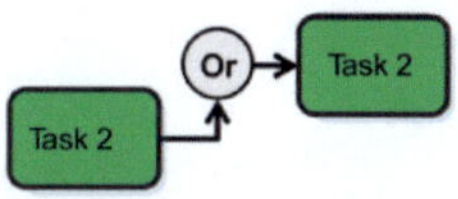 | $r = 1/3$ | $P(task\ 2) = 1$ | $P_{task\ 2} = 1/3 * 1$ |
| 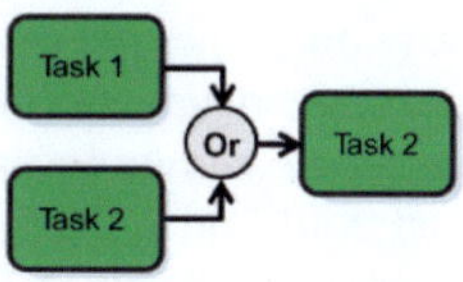 | $r = 1/3$ | $P(task\ 1\ and\ 2) = 2$ | $P_{task\ 1\&2} = 1/3 * 2$ |
| | | Total passivity | $P_{total} = 4/3$ |

More generally, the weight of the impacting connector can be calculated using binomial coefficients. Each time, the case to which no edge is connected is omitted, as this would mean the process is interrupted. Table 5-9 provides the necessary formulas. Instead of the direct calculation of the activity and passivity, weights are proposed that are necessary for the calculation of nested operators, as shown below.

*Table 5-9: Calculation of structural weights of different logic operators in processes*

| Weight of AND connector | Weight of XOR connector | Weight of OR connector |
|---|---|---|
| $weight_{AND} = n$ | $weight_{XOR} = 1$ | $weight_{OR} = \frac{1}{2^n - 1} \sum_{i=1}^{n} \binom{n}{i} \cdot i$ |

A special case arises for nested operators. In such a case, the operators can either be recombined, or the respective weights have to be split (Figure 5-7). In practice, a fusion of similar connectors has proven a very adequate approximation, as, on the one hand, from a structural point of view there is commonly no information on the factual structural relevance of each case, and, on the other hand, often in process modeling, logical operators are nested for reasons of visualization and not because of a nested dependency structure among the operators. In general, nested operators are calculated successively, starting at the deepest level of nesting (i.e., from the left to the right for the example in Figure 5-7).

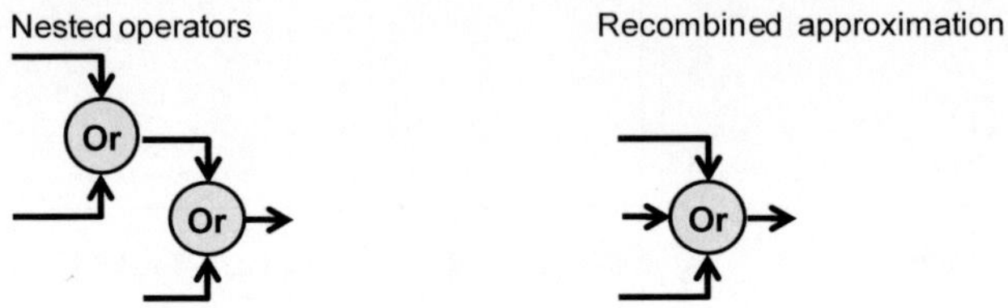


*Figure 5-7: Nested operators and approximation through recombination*

For nested operators, the weights are calculated for each individual operator and then summed for each entity in the process model. Figure 5-8 visualizes the calculation scheme, consisting of the following steps (an example is shown in the appendix): First, splits and joins are separated, then for each split and join, the appropriate weights are calculated according to the formulas from Table 5-9. Then, if no nesting of operators is to be resolved, the weights are directly translated into corresponding activity or passivity; if nested operators exist, a further differentiation of preceding and succeeding operators is made, and finally, the activity and passivity are calculated along these paths of nested operators.

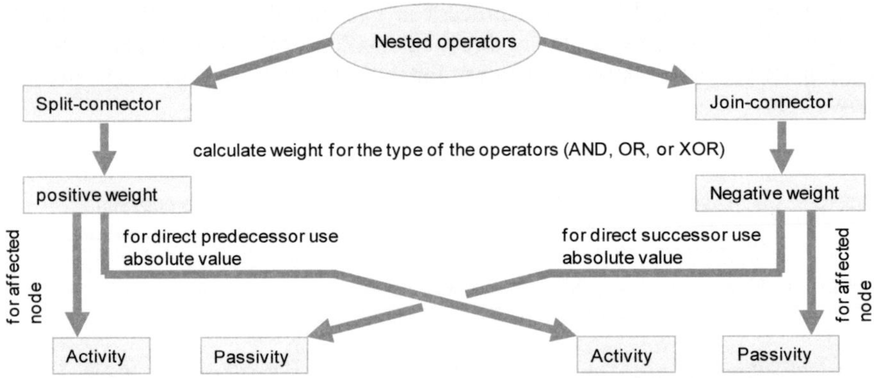


*Figure 5-8: Calculation scheme for using weights to calculate the activity and passivity including nested Boolean operators between the concerned entities of the process*

If Boolean operators are part of the process model, not only the activity and passivity as shown above can be calculated to better understand decision points in the process, but also McCabe's Cyclomatic number and the Control-flow Complexity can be calculated.

**McCabe's Cyclomatic Number** (MCC) [McCabe 1976] counts the number of linearly independent paths through a process, enumerating edges, nodes, and connected components in the control flow. In principle, the number is, therefore, a direct measure of the number of binary decisions[70] in a process. Thus, the lower the number, the fewer options there are in the process. As many process models are of a bipartite nature, this calculation can be extended to assess this bipartite network directly, i.e., without first aggregating the two domains into an aggregate view. This way, EPC models, for example, can be assessed directly.

*Table 5-10: McCabe's Cyclomatic number for unipartite and bipartite process graph*

| MCC for one domain [McCabe 1976] | MCC for two domains |
|---|---|
| $MCC = e - n + 2 \cdot p$ | $MCC = e - (F + E + V) + 2 - n(AND_{split})$ |

The original MCC is valid for unipartite graphs, calculated for edges (e), nodes (n) and connected components (p), with p being the probability of a decision at a connector. If, for example, an EPC model is transferred into an MDM-based process model, as shown in section 4.4.3, the bipartite version of the MCC is calculated as shown in Table 5-10 using the matrices spanned by E, F, and V; there, F, E, V are the powers of the respective matrices for functions, events, and connectors. While this matrix-representation contains additional edges in both the basic and the MDM representation (because connectors are modeled as nodes), this does not change the MCC, as each additional edge leading to a connector ($e \rightarrow e + 1$) is compensated by an additional node ($n \rightarrow n + 1$), i.e., a connector. This is also true for "dirty" models that do not fully adhere to the modeling scheme (which often happens with pragmatic models), as long as a connector only acts as a split or a join of the process. Therefore, with F the number of functions, E the number of events, and V the number of connectors, the MCC can simply be summed up from the matrix; e is the total number of Xs within the matrices p equals 1, as only Boolean decisions (i.e., no non-binary decisions) are possible. The number of the AND splits have to be deduced to make sure these are not counted as decision points (unless this is intended).

The **Control-flow complexity** (CFC), proposed by [Woodward et al. 1979] and extended to workflows by [Cardoso 2005], expands the idea of counting binary decisions to counting the number of states a process can take, i.e., how many different pathways there may be considering how many paths the process can split

[70] The term "binary" refers to the existence of only two possible options for each decision, i.e., "yes" and "no".

(or re-join) at each connector. The CFC has additive properties and can be added up from the individual CFCs of each connector. Each time, n is the number of outgoing (or incident, respectively) edges at the connector. For AND, the CFC is *1*, as no decision is taken [CARDOSO 2005]. For XOR, the process splits into (joins out of) *n* different options, therefore, the CFC is *n*. For an OR, there are *2n* options possible. However, this includes the option that no edge is followed, which would stop the process; therefore, this option is taken out of the equation (Table 5-11).

*Table 5-11: Formulas to calculate the control-flow complexity (CFC)*

| CFC for AND | CFC for XOR | CFC for OR | Total CFC |
|---|---|---|---|
| $CFC_{AND} = 1$ | $CFC_{XOR} = n$ | $CFC_{OR} = 2^n - 1$ | $CFC_{process} = \sum_i CFC_i$ |

The evaluation of **hierarchies**[71] is difficult, as a combined assessment of the reach of a hierarchy (i.e., the longest path in the graph of reachable nodes, starting from a dedicated root node) and of the width of each level of the hierarchy always demands a trade-off, prioritizing either one or the other aspect (Figure 5-9).

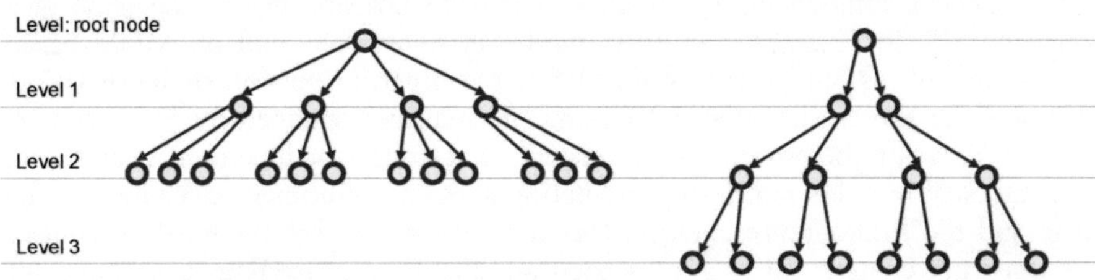


*Figure 5-9: Short wide hierarchy on the left, long narrow hierarchy on the right-hand side*

As hierarchies can have different meanings (see Table 5-12), different trade-offs may be necessary; hence, a general metric “Tree criticality” is used with any weighing function. The more specialized “Forerun factor” and “Snowball factor” use the reciprocal value of the level (i.e., the shortest path from the root node to the level) as weight for each level to favor the impact of levels closer to the root node over those that are further away, as in most cases those levels closer to the root node will be more important to it.

[71] “Hierarchies” are also referred to as “trees”; here, the term hierarchy is chosen, as the algorithmic concept is also used to compute the nesting depth of a model, which is often understood as an extension to different levels of the hierarchy of a model [GRUHN & LAUE 2006a]; for this reason, the start node of the hierarchy is called the root node.

The metric "Tree-robustness" makes use, again, of the assessment of all existing hierarchies in the process of calculating the degree to which the process is permeated by different hierarchies. Therefore, the appropriate weighing has to be chosen if the "Forerun factor" and the "Snowball Factor" are not applicable.

*Table 5-12: Different possible meanings of hierarchies in a structure*

| Meaning of relationship type | Meaning of hierarchy | Meaning of root node |
|---|---|---|
| Flow of information (communication, transfer of documents, input-output relationships) | Information from the root node is transmitted to all nodes of the hierarchy | Errors can spread rapidly; the root node is therefore a highly critical node to the information transfer |
| Logic sequence (sequence of tasks, consistency among documents) | Automatic execution of sequence once the root node is active | Root node is initial event to start sequence |
| General causal relationship (logic operators, dependencies among decisions, sequence of milestones) | Nodes subsequent to root node can only be active after root node has been active | Activity of root node can be deduced if any element of the hierarchy is known to be active |

**Cycles** are evaluated by counting the number of cycles, involved nodes, and edges [MAURER 2007]. However, in a process, where an iteration starts is particularly of interest (i.e., where the cycle is initially entered for the first time), and where the decision is taken to iterate (i.e., re-run the cycle after an initial first run) and how many edges actually cause the feedbacks (several iterations may share one common communication channel that leads back in the process flow) is also of interest. Therefore, the metrics in Table 5-13 detail the assessment of cycles.

*Table 5-13: Metrics to detail cycles in a process structure*

| Metric | Description |
|---|---|
| Number of feedbacks | Number of relationships in an idealized process sequence that lead back to the flow (upstream) and thus cause an iteration to re-run |
| Activation of cycle | Number of nodes that are at the start of a cycle in an idealized process sequence before the cycle is run |
| Number of starting points for iterations | Number of nodes that are at the end of a cycle in an idealized process sequence and that lead to a feedback (as in the metric "number of feedbacks"); in these nodes, the decision is taken to re-run the cycle or not |

These three metrics are not straightforward counters, as they are based on the triangularization of the DSM representing the process in focus to first obtain an idealized sequence before cycles are sought; this triangularization cannot always be calculated exactly. Triangularization is commonly done using an algorithm that removes one edge after another for the cycles in a DSM until no more feedbacks exist [KUSIAK & WANG 1993]. This is done in the order of importance – at first, feedbacks causing the largest cycles are removed, then the list is pared down to the shortest cycles. In some cases, it is possible that different edges could be removed for cycles of equal length, where removing one edge or the other will yield different results for the overall triangularization. So far, no algorithm is available to overcome this ambiguity. Thus, the result is often only an estimate.

Figure 5-10 shows a small process graph and the reciprocal DSM. There, node 1 is the activation of cycle 1, which only alternates between node 1 and node 2. It is identified in the DSM on the left, which is already in a triangularized format. There, node 1 is the uppermost node in the cycle among nodes 1 and 2. Node 1 is, at the same time, the activation node for cycle 2. Cycle 2 is only rerun if the feedback as marked is used. Overall, there are thus three feedbacks that can be identified. Ultimately, node 4, for example, is the starting point for the iteration marked as cycle 2. This iteration is, however, only executed in some cases, which cannot be deduced from the structure of the process; yet, node 4 is of particular interest, as here the process either continues further downstream or back upstream.

| | Node 1 | Node 2 | Node 3 | Node 4 | Node 5 |
|---|---|---|---|---|---|
| Node 1 | | X | | | |
| Node 2 | X | | X | | |
| Node 3 | | | | X | |
| Node 4 | X | | | | X |
| Node 5 | | | | X | |

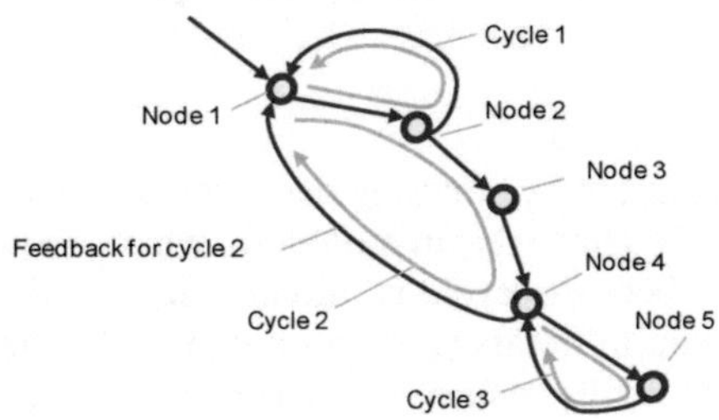


*Figure 5-10: Example for rationale behind the three metrics shown in Table 5-13*

In contrast, the **cognitive weight** makes use of empirically-founded basic constituents of a structure that each have an assigned numerical value of the degree of difficulty to understand such a constellation in a structural context. These weights have been proven useful both in software science [SHAO & WANG 2003] [WANG 2006] [MCQUAID 1997] and in process management; in the latter application, however, the cognitive weight has not fully matured [GRUHN & LAUE 2007a]. A basic overview of possible weights is shown in Table 5-14

Ultimately, the degree of non-planarity uses an average of the count of all edges that need to be removed and thus is not fully proportional to the concept that is being measured. While common empirical evidence exists which indicates that networks with many relations that cross each other are considered more complex [HENRY et al. 1981] [RICHTER 2007, p. 140], there is no evidence available that the measurement by the degree of non-planarity as proposed is relevantly scaled. However, as experiments on generic networks show [KORTLER et al. 2009], the measure is clearly able to differentiate between graphs of different complexity.

All of these metrics still lack detailed empirical evidence of their application in process analysis and only present a conceptual level that is, each time, based on general empirical observation and existing publications. Yet, the basic reasoning of the structure of a process is similar among them, i.e., the design of the metrics of assessment of structural characteristics that are accepted in literature.

*Table 5-14: Empirically researched cognitive weights of structural characteristics*

| Category | Software structure | Structural characteristic | Cognitive Weight | | |
|---|---|---|---|---|---|
| | | | [WANG 2006] | [SHAO & WANG 2003] | [MCQUAID 1997] |
| Sequence | Sequence | Transit node, transit edge, bridge node | 1 | 1 | 1 |
| Branch | If...then, case | Split, join | 2 to 3 | 3 to 4 | 3 |
| Iteration | For...do, repeat...until, while...do | Cycle | 3 | 7 to 8 | 3 |
| Call of component | Function call, nesting level | OR and XOR split, repeated node | 2 to 3 | 7 to 11 | |
| Concurrency | Concurrency | AND split, hierarchy | 4 | 15 to 22 | |

At the same time, these metrics close the gap between the available structural characteristics and the existing body of structural metrics that could be adapted from other fields of science. Thus, their right to exist is obvious. Their use is, therefore, shown in the case study to provide basic evidence of their suitability.

## 5.2.4 Classification of available metrics

To better characterize the metrics and to make them accessible for different analysis without using the framework shown in the next chapter, a classification of the metrics shown is provided in this section to illustrate different views of a structure.

While grouping metrics of the principal features of a structure already represents a certain classification, metrics can be further regrouped in different ways. Table 5-15 provides an overview of the relevant taxonomies that will be outlined in the following. The 52 metrics of the Structural Measurement System can be regrouped differently for each classification; these groups are listed in appendix 10.4. In the following, only the categories are explained to detail the multi-facetedness of the structural metrics.

Classifying the metrics in terms of their **scope of analysis** yields groups of metrics, which access the process at different levels of granularity. Single entities focus on, for example, individual tasks or relationships and how they interact with the process. Modules are pre-defined groups of elements such as teams or phases; for these, coherence within the module and interaction with the surrounding system are the primary focus of the analysis. Metrics that regard the alignment of entities recombine two domains at a time to see whether the structures of both domains fit together. The overall process, i.e., the complete network, can be characterized in two ways. First, it can be characterized as a single metric which supports the comparison of several such networks, for example, comparing them by the number of nodes and the relational density. Second, it can be characterized as a histogram, where the characteristic distribution of values of metrics for individual entities provides a picture of each process. In fact, many of these distributions, for example, for activity/passivity or for the degrees of all nodes, can be approximated as a set of coefficients of a distribution function. However, this concept is not further regarded here, as it does not provide information about the identification of structural outliers.

*Table 5-15: Available classifications of the structural metrics*

| | Description | Categories |
|---|---|---|
| **Scope of analysis** | Granularity of the analysis | • Single entity<br>• Module<br>• Alignment<br>• Overall process as histogram<br>• Overall process as single metric |
| **Particularities of the model** | Metrics with particular regard for specific aspects of modeling | • Decision points (AND, OR, XOR)<br>• Comprehensibility of the model<br>• Quality of modeling |
| **Focus of analysis** | Basic structural concept in analysis | • Robustness<br>• Propagation<br>• Grouping<br>• Extent |
| **Direction of impact** | Directedness of relationship type | • Active (influencing)<br>• Passive (influenced)<br>• Related (not directed) |
| **Type of entity** | Type of entity | • Nodes<br>• Edges |
| **Type of network** | Number of domains | • Intra-domain<br>• Inter-domain<br>• Multiple-Domain |

The categorization of metrics as to their suitability to analyze **particularities of the model** excludes those metrics that only have a general structural focus. The group of metrics that are capable of analyzing decision points (i.e., AND-, OR-, and XOR-gates) is limited to three basic metrics. The comprehensibility of the model assesses the basic human capability to grasp and understand the network represented by the model. Ultimately, some metrics generate information about possible modeling errors, for example, when nodes are not connected to the mode. This kind of information can be extracted from metrics regarding the quality of modeling.

The **focus of analysis** intends to regroup metrics into primary categories from which a structure can be analyzed. Metrics on robustness focus on the cohesion and connectivity of the overall network. Metrics on propagation regard the speed of information transmission and the reach and impact of information and errors, both in terms of direct impact among neighboring entities and across the overall network. Grouping refers to the buildup of sub-graphs that stand out within the process, such as complete clusters that may be the foundation of the constitution of a team, for example. Metrics which are concerned with the extent of a network describe its size and the quality of its crosslinking.

The **direction of impact** regroups metrics into those that only consider active metrics, i.e., such metrics that assess the outgoing edges of a node or a module, whereas passive metrics assess only incident edges. In some cases, the direction is not of interest, and metrics, therefore, are classed as related. The Snowball factor, for example, is the active metric assessing the reachability in a network as an outgoing hierarchy, and the Forerun factor assesses only incident edges in the same manner. Although some metrics can be split into an active and a passive metric, they are not always listed as separate metrics, e.g., the degree distribution. Although not in all cases, a separate metric for each direction may provide substantial additional informational value.

Equally, metrics can be regrouped by the **type of entity** they consider, either the nodes or edges of the network. Like the basic structural characteristics (see section 5.1.1), they assess either nodes (i.e., node-induced structural characteristics) or edges (i.e., edge-induced structural characteristics).

Ultimately, the **type of network** characterizes the input information that is necessary for a metric. Intra-domain networks are DSMs consisting of only one domain and one relationship type. Inter-domain networks are DMMs that relate two domains at a time via one relationship type. Multiple-Domain networks are MDMs that consist of at least two domains and at least two relationship-types.

## 5.3 An example application of the Structural Measurement System

As the Structural Measurement System presented up to here is the main focus of this book, a short case study[72] is introduced to provide a better picture of how the metrics work and what results they are able to deliver. Two further example applications are shown in chapter 7. The case study here uses an existing, published process model in order to be able to compare newly calculated and existing results. Therefore, first, the process model is introduced, then the different metrics are computed and the results are discussed in detail.

### 5.3.1 The process in focus

The process used for this case study was already described and analyzed in [Braha & Bar-Yam 2004]. As it is available online[73], it was used here as a basis to illustrate the complexity metrics shown up to here.

---

[72] Adapted from Kreimeyer, M., Bradford, N., Lindemann, U., Process Analysis using Structural Metrics, Design 2010, with friendly permission by the Design Society.

[73] available at http://necsi.org/projects/braha/largescaleengineering (last checked on 24 October 2010)

The process model represents the basic vehicle development process at General Motors. It lays out the 26 weeks of General Motors' automotive development separated into phases: Expert opinion phase, quick study phase and integrated vehicle concept model and o.d. deliverables phase. Within the process, 120 tasks are linked in a Design Structure Matrix (DSM) containing 417 immediate directed relations. The tasks are accomplished by 19 organizational units. Each set of tasks belonging to an organizational unit is referred to as a module.

The model was built from interviews with engineers and from documentation. For each task, it was asked 'Where do the inputs for the task come from?' and 'Where do the outputs generated by the task go to?'. The answers were used to construct the network of information flows. In the following, the tasks shown in Table 5-16 will occur repeatedly and are therefore listed here as T1 to T120.

*Table 5-16: The most important tasks in the process [Braha & Bar-Yam 2004]*

| ID | Task | ID | Task |
|---|---|---|---|
| T1 | Develop Nine Box Summary | T88 | Run Updated Workload Model |
| T2 | Provide Preliminary 2-D Sketches of Vehicle | T89 | Update Financial Assessment and Finalize Business Case |
| T3 | Identify Target Architectures | T90 | Update Decoupled Development Plan |
| T6 | Provide Key Volume Drivers Chart | T91 | Review Quick Study Deliverables |
| T10 | Develop Critical Product Characteristics | T98 | Finalize Body BOM |
| T11 | Set Engineering Target Parameters (Concept Technical Descriptors) | T111 | Create Physical/Virtual Models |
| T33 | Run Initial Workload Model | T112 | Assess Risks in Performance Req's |
| T37 | Recommend Final Architecture | T114 | Develop Manufacturing Program Timing Plan |
| T72 | Track Total Vehicle Issues | T115 | Provide Final Volume Forecast |
| T73 | Maintain Vehicle Mainstream Chart and Update Engineering Product Content Sheet | T116 | Develop Final Integrated Concept |
| | | T117 | Develop Option Plan for NASB Review |
| T83 | Generate Surface Limits for Bodyside | T118 | Review Integrated Vehicle Concept Model and O.D. Deliverables |
| T85 | Conduct Marketing Clinics | T119 | Provide Transition to Vehicle Program |
| T86 | Provide Customer's Perspective to Option Team | T120 | Provide Transition to Program Quality Manager |

In an earlier analysis of the process [Braha & Bar-Yam 2004] the following main results were elicited. They concern especially the small-world properties (i.e. most nodes are not adjacent but reachable via a short average path length) and the degree-related properties (i.e. direct coupling among immediate neighbors) of the involved tasks:

BRAHA & BAR-YAM consider the process to exhibit clear small-world properties. Accordingly, the task network's entities have a relatively high cluster coefficient, whereas the characteristic path length is relatively short and approximately equal to a characteristic path length of a random graph having the same number of nodes and edges. A modular organization (defined by a higher degree of internal information exchange than across the borders of modules) was found to be a consequence of high cluster-coefficients and small word properties.

They furthermore identify an imbalance concerning the relation of out-degrees and in-degrees (activity and passivity). Most tasks were found to have relatively low in- and out-degrees, whereas few have high degrees. Those few having a high out-degree, or respectively passive (high in-degree) tasks are characterized as information generators (or information consumers, respectively). In turn, tasks with a high in-degree have a low out-degree and vice versa. The process is dominated by a small number of such tasks.

BRAHA & BAR-YAM declare their results as typical for product development processes, with the following consequences: The most effective way of improving the overall process is to improve the central, dominating tasks, similar to the concept of structural outliers. Secondly, they conclude that a failure of those tasks is likely to impede the correct function of the overall process.

### 5.3.2 Overview of the analyses using structural complexity metrics

Only the core metrics are shown here, some metrics, such as the Number of Organizational Interfaces could not be calculated for the dataset at hand, and others, such as the Degree of Planarity or the different metrics towards the cognitive weight of a network could not be computed due to a lack of sufficient algorithmic support. Table 5-17 lists the 34 structural metrics that were used. For each, the relevant dataset is listed in order to generate meaningful results. The metrics are arranged by categories related to the underlying structural patterns.

*Table 5-17: Overview over the applied structural metrics*

| | Overall process | Tasks | Modules |
|---|---|---|---|
| **Size and density** | | | |
| Number of domains | x | | |
| Number of nodes | x | | |
| Number of edges | x | | |
| Number of classes | x | | |
| Number of interfaces between domains | x | | |
| Number of edges per node | x | | |
| Relational density | x | | |
| Number of unconnected nodes | x | | |
| **Adjacency** | | | |
| Activity / Passivity | | x | |
| Degree correlation (nodes) | | x | |
| Degree correlation (edges) | | x | |
| Degree distribution | | x | |
| Fan criticality | | | x |
| Synchronization points / distribution points | | x | |
| Number of independent sets | x | | |
| **Cycles** | | | |
| Number of feedbacks | x | | |
| Number of cycles per node | | x | |
| Number of clycles per edge | | x | |
| **Attainability** | | | |
| Reachability of a node | | x | |
| Closeness | | x | |
| Proximity | | x | |
| Relative centrality (based on between-ness) | | x | |
| **Paths** | | | |
| Number of paths | x | | |
| Path length | x | | |
| **Hierarchies** | | | |
| Height of hierarchy | x | | |
| Width of hierarchy | x | | |
| Snowball-factor | | x | |
| Forerun-factor | | x | |
| Tree-robustness | | | x |
| **Clustering** | | | |
| Number of cliques | x | | |
| Cluster-coefficient (local) | | x | |
| Cluster-coefficient (global) | | x | x |
| Module quality 1 (flow of information) | | | x |
| Module quality 2 (compactness) | | | x |

The results of the metrics generate distributions, within which the individual values can be compared to identify relevant outliers. Three views can be identified that each relate to a distinct dataset: The overall process model as a whole, the tasks individually, and the modules as formed by the tasks belonging to the different organizational units. Thus, the results are organized accordingly.

### 5.3.3 Analyses using complexity metrics for the overall process model

At first, a basic analysis of processes is the calculation of metrics concerning size and density as well as metrics delivering characteristic values for overall networks. The number of domains in the MDM is two: Tasks and organizational units. The number of nodes is 139: 120 tasks and 19 organizational units. The number of edges is 537, of which 417 edges are entries of the task-task DSM. The number of classes (i.e. number of different kinds of nodes) is 139, equally, as no node is instantiated more than once in the actual process model, which sometimes happens when one task is instantiated several times for easier modeling or to show how results are transferred. The number of interfaces between domains is 120, i.e. each task is executed by exactly one organizational unit. The number of edges per node is 3.457 for the task-task network and 3.891 for the overall network (including the organizational units). Respectively, the relational density is 0.029 and 0.028. Both values show that a rather low part of all possible connections is exhausted and the process is, possibly, rather linear. This concurs with the initial process model that can be triangularized easily, i.e. the task sequence can be put into an ideal order without severe conflicts. The number of unconnected nodes, which could reveal possible mistakes in the process model, is zero, pointing to the fact that, at least, no errors were made by forgetting or not connecting a node. However, such missing links can, of course, also be intentional and do not necessarily always point to errors. The number of independent sets (i.e. the number of sets of tasks accomplished concurrently and independently from each other, as found when banding the respective DSM) is eight, i.e. the process can be broken down into eight phases. The number of paths across the overall process is especially useful for estimating the importance of root nodes. From root node T1 36 paths lead to the process' five end nodes, whereas 33 paths start from root node T2, leaving both starting tasks relatively equal in their impact. The average path length of these shortest paths is 3.6 between T1 and the end nodes and 5.6 between T2 and the end nodes, showing that information spreads faster throughout the process starting at root node T1. This metric, although it describes connection between tasks, concerns the overall process as properties of start and end nodes.

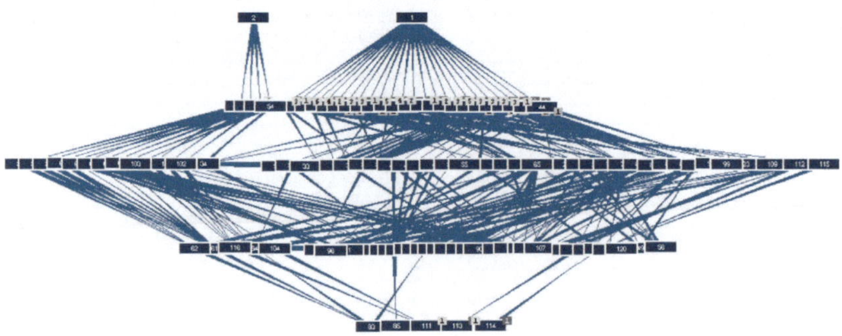

*Figure 5-11: The five levels of the process*

Figure 5-11 visualizes the different paths as a graph: The maximum height of the process' hierarchy, i.e. the number of levels from start to end nodes, is 4. The width of the process' levels (the number of nodes per level) is 2 on first level (i.e. the start nodes), 28 on second level, 50 on third level, 35 on fourth level and 5 on fifth level (i.e. the end nodes).

The process has two start nodes, namely T1 and T2, and five end nodes: T85, T111, T117, T119 and T120. The maximum nesting depth, i.e. the number of splits retraceable to a root node, is 100 for root node T1 and 33 for root node T2, showing again a higher influence of root node T1, as the process bifurcates noticeably more from this task. The number of cliques, i.e. the number of complete clusters within the network, is zero, i.e. no groups of tasks that are completely mutually connected exist within the model. The global cluster-coefficient (quotient of the sum of all cluster-coefficients per node and the number of information distributors) is 0.27, indicating that many tasks are likely to be coupled more intensely than the number of cliques shows; this potential for coupling relies on the concept that two tasks connected to a third task are likely to be interrelated because they are coupled to a third task in the same way. The number of feedbacks within the process, i.e. the number tears in a triangularized DSM, is 24, a rather low percentage (5.75% of all 417 connections), showing that the overall iterative nature of the process is broken down rather well into only few intended relations.

### 5.3.4 Analyses using complexity metrics for each task

The majority of structural metrics is applied to compare the entities of one domain to each other, i.e., the tasks of the task-task-DSM. As shown in Table 5-17, adjacency and attainability are the categories concerning most metrics applicable on the behavior of the correlations between tasks. Metrics referring to adjacency and attainability are predestinated for measuring the importance of single entities for the function of the complete network.

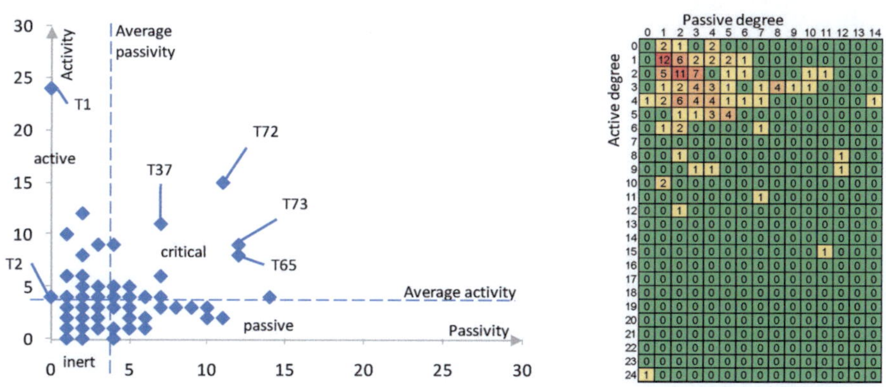


*Figure 5-12: Activity and passivity per task, left, and degree correlation (nodes), right*

The most basic metrics are activity and passivity (also referred to as out- and in-degree) of a node. Figure 5-12 (left hand side) shows a concurrent plot of both metrics per node. Four tasks (T72, T73, T65 and T37) stand out most, accordingly being highly active and highly passive at the same time. The four fields inert, active, passive and critical are defined by the average values for both axes. The start nodes T1 and T2 are positioned on the activity-axis with a value of 0 for passivity, as they only deliver information. However, T1's out-degree is nearly five times higher than the value for T2, indicating a higher initial impact of T1 onto the overall network.

The degree correlation can be based on edges as well as on nodes. The representation of the correlation based on nodes as in Figure 5-12 (right hand side) reveals a high number of connections between nodes with values of one or two for activity or passivity. Accordingly, most nodes within the process have relatively low in- and out-degrees at the same time; at the same time, the correlation plot shows that many nodes are connected with similar in- and out-degrees, as the diagonal axis of the plot contains many non-zero entries. Similarly, the representation of the correlation based on edges (not shown) indicates that 64% of the edges link two nodes both having more than one incident as well as more than one outgoing edge, i.e. most information transfers between two tasks will be based on several inputs into the first task and generate more than one output at the second task. The conclusion of both correlations is that a major part of the network consists of connections between nodes with low degrees, of which most have a degree larger than *1* (i.e. each task being coupled to more than one other task). Nonetheless, there are twelve highly important edges that are the only connection between the nodes (i.e. directed forwarding between two tasks).

The degree distribution reveals the occurrence of similar in- and out-degrees within the process. The plot underlines the occurrence of low degrees in the process, pointing to a hub-and-spoke-like structure of the overall process (i.e. the network is a scale-free network, like many typical collaboration structures [Newman 2002]). High degrees appear rarely, as Figure 5-13 shows.

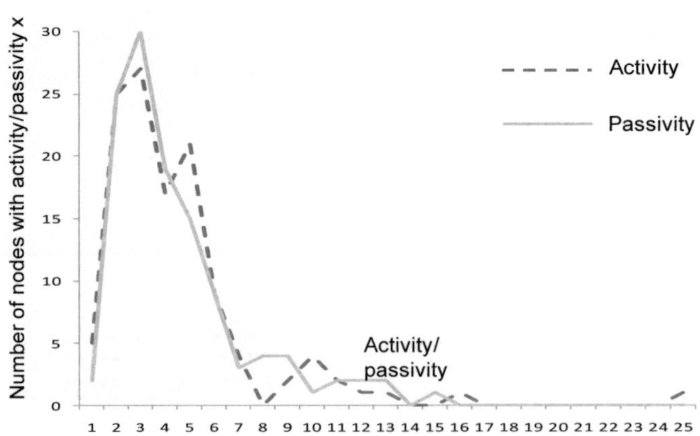


*Figure 5-13: Degree distribution*

The representation of the synchronization and distribution points produces no new results as all entries within the DSM have the value *1*. In fact, these metrics are especially useful to be used with weighted edges in a process graph. Accordingly the representation is identical to the activity/passivity plots. In other cases, if e.g. a weighted DSM is used, these metrics would be able to underline the importance of a task not just based on the degree but also on its coupling strength.

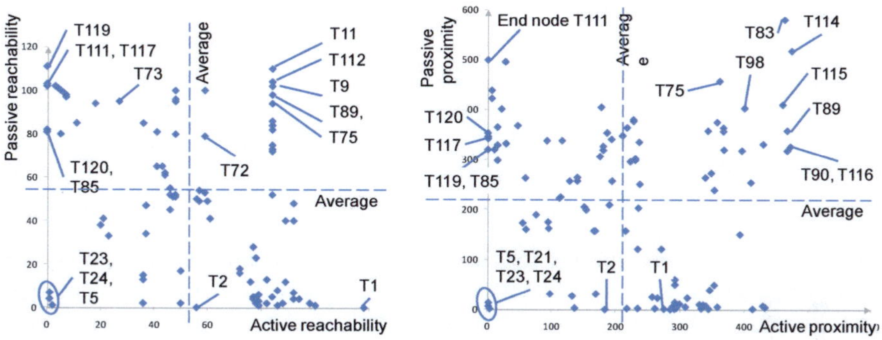


*Figure 5-14: Active and passive reachability, left, and active and passive proximity, right*

The active and passive reachability describe the number of nodes a designated node is able to reach (or the number of nodes that can reach this designated node, respectively). In a process analysis, these metrics are very important, as they show the propagation of information (and errors) across the overall process, thus estimating the impact (and impactedness) thereof for each task. They thereby extend the picture generated by the degree across not just adjacent tasks but across all tasks. The plot of both metrics per node (Figure 5-14, left) a few highly actively and passively reachable tasks: T118, T112, T89, T90, T91and T75. These tasks are therefore highly integrated into the flow of information through the process and play an important role in the supply of information of all other tasks.

The active and passive proximity are calculated by summating the rows (columns, respectively) of the distance matrix (listing the shortest path between any pair of tasks, zero if not reachable), i.e. describing the distance of one task to all others. As outliers for the active and passive proximity the following tasks appear: T75, T83, T89, T90, T98, T114, T155 and T116 (Figure 5-14, right). Once again, tasks T75, T89 and T90 (compare the results for reachability in the previous paragraph) seem to be of higher importance for the function of the overall network, a relatively high average path length represents high impact, as a high number of nodes positioned on according paths are involved into the incident and outgoing flow of information.

In a comparable manner, the relative centrality counts the number of shortest paths between any two nodes that cross a designated node: The higher the value, the more information flows go via that designated node. The following tasks stand out: T39, T39, T4, T72 and T73 (Figure 5-15). These nodes are therefore the central information brokers across the overall process.

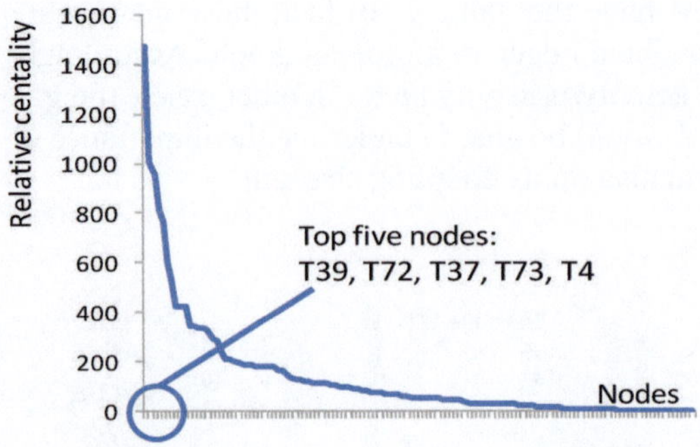


*Figure 5-15: Relative centrality*

Another pair of meaningful metrics for the estimation of influence of entities are the snowball factor and the forerun factor. They assess the outgoing (incoming, respectively) hierarchy of reachable nodes with decreasing impact for nodes that are farther away: They calculate as the sum of the products of width and height of the level in the hierarchy, weighted by to the inverse of the shortest path length to the root node. They thereby relativize the active and passive reachability, as nodes that can be reached but that are far away and have little impact are not counted as importantly. Figure 5-16 shows the plot for both metrics per node. Here, tasks T36, T37, T65, T72, T91 and T112 show up, having high values for both metrics. The distribution of the values shows that, each time, only a few nodes have high influence onto the process. Those particular nodes are one start node, T1, as well as the tasks T3, T10, T11 and T37. Start node T2 only has the 44th position in this ranking, which underlines the much higher importance of root node T1. The plot of all values for forerun factors, however, shows a more linear distribution, indicating that few tasks dominate the spread of information, while the tasks rely more homogeneously on the information intake from other tasks.

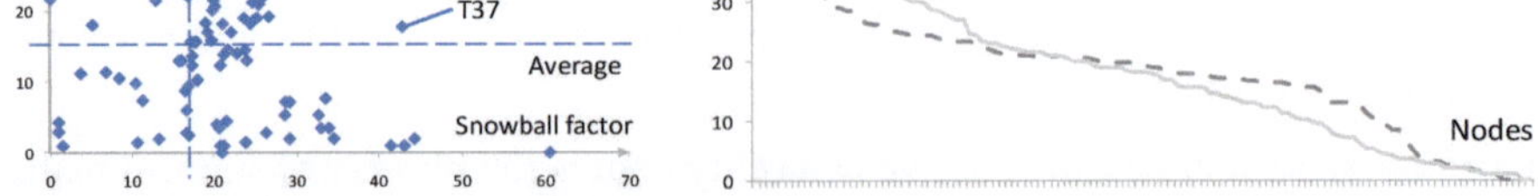


*Figure 5-16: Snowball and forerun factor per task (portfolio on the right hand side and individual distributions)*

The local cluster-coefficient shows how a task is likely to drive the clustering of tasks in the process. It is calculated as the quotient of existing edges to adjacent neighbors and the number of possible edges. Apart from five outliers with a maximum coefficient of one (T6, T33, T86, T88 and T111), the distribution shows a relatively linear behavior. Accordingly, those five tasks are connected to each possible neighbor, and close workgroups are likely to be necessary at this part of the process.

To assess iterations and uncertainty in the process, the metric number of cycles per node can provide insight. The more cycles take path via a node, the more this task will receive and distribute information from and to the overall network, and will therefore be of high influence. Figure 5-17 shows the distribution and the top values, also for how edges are involved in these cycles, pointing to important channels of communication. Both metrics describe task T4 as most influential; while this coincides with e.g. the relevance as detected e.g. through the degree distribution, this is incidental.

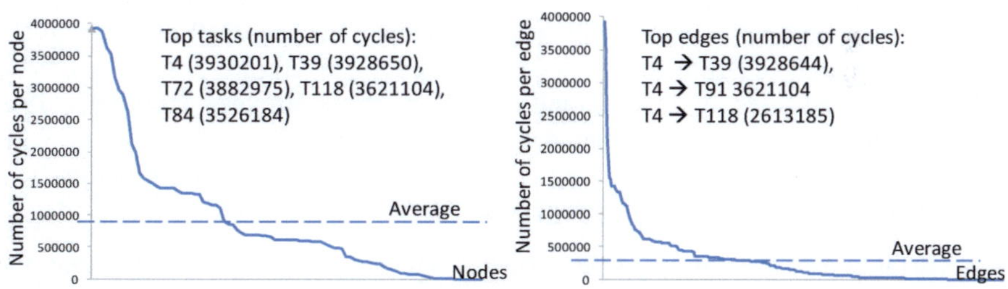


*Figure 5-17: Number of cycles per node (left) and per edge (right)*

## 5.3.5 Analyses using complexity metrics for each module

On a different level, the use of structural metrics delivers meaningful results concerning the properties of modules involved into development processes as well as interdependencies among the different modules. Modules are predefined groups of entities, in contrast to clusters that develop during the process' progress caused by their intense interaction. In case of the analyzed process, 19 organizational units are each taken as the designated modules of tasks they are responsible for, i.e., one organizational unit is responsible for one process module.

The fan-criticality (i.e. the number of outgoing and incident cross-border relations per module) allows comparing the out- and in-degrees of modules. The plot in Figure 5-18 (left hand side) shows module 17 stands out the most concerning both in- and out-degree, accordingly being most influenced as well as being the most influential module, being concerned with integration of a large set of components.

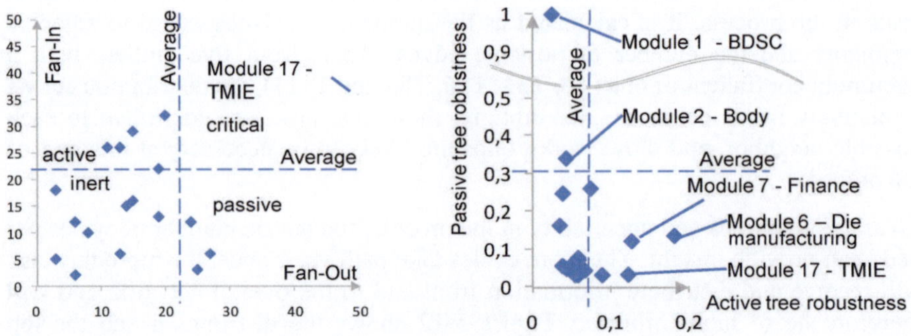


*Figure 5-18: Fan-out and fan-in per module, left, and tree-robustness, right*

The metric tree-robustness is applicable either on complete domains or on modules. The portfolio of the values for active and passive tree-robustness (i.e. the quotient of the number of nodes with a nonzero value for snowball or respectively forerun factor and the sum of the according factors) in Figure 5-18 (right hand side) shows modules 1, 2, 6, 7 and 17 as the most important outliers. This indicates that no module is dominated by both incoming and outgoing hierarchies of information flow, but that they either collect incoming information (all integration modules (6, 7, and 17) or generate it.

The distribution of the values of the global cluster-coefficient per module, i.e. the quotient of the sum of all local cluster coefficients and the number of nodes with an activity higher than one in a designated module, shows two outliers (modules 1 and 12). These modules are, therefore, the most likely to cause information transfers among the other modules.

The metrics delivering most information concerning the relation of internal and cross boarder flow of information of modules are module quality 1 and module quality 2. The metric module-quality 1 computes as the product of the number of edges that cross the border of the module and the number of edges within the module; module quality is calculated as the respective quotient. The first metric describes the flow of information through modules, while the second one describes the compactness of a module. For both modules, module 18 (TVIE – total vehicle engineering) can be identified as the most remarkable outlier.

Through these metrics the modules 1, 17, and 18 are determined as the most influential ones. In spite of the three more influencing modules, the process' organizational units are interconnected quite evenly. A rather low fraction of 122 (36,72%) of the 417 edges connect nodes within equal modules, characterizing the flow of information through the process as rather integrated among different modules. There is even one module without any internal connections (Module 15), for this reason it is positioned at the last positions for module quality 1 and 2 with value of zero. The importance of module 18 is rather logical because it contains most nodes per module. The importance of module 1 is a consequence of both of the process' start nodes being contained in this module, having high values for influence-describing metrics.

In summary, from a modules' point of view, the network can be described as well-balanced, with the modules 1, 17 and 18 having a higher importance because of their position in the process' progress at the beginning of the first phase or, respectively, at the end of the last phase.

### 5.3.6 Conclusions for the regarded process

For a better comparison, Table 5-18 lists the core results. There, the influence measuring metrics are sectioned into active and passive ones. Active ones determine the distribution of information, the passive ones describe information sinks. For all metrics, the top-ten upper-bound and lower-bound structural outliers are listed in the table. Start and end nodes are printed in bold.

Concerning the structure of the analyzed development process, one result approved by every metric is the difference concerning the importance of the networks' two start nodes. Start node T1 is much more influential onto the overall process than T2 is, which is reasonable given that the design sketch generated in T2 only impacts tasks that are related to the exterior design of the car. Therefore, T1 ranks first three times, whereas T2 is not even once among the top ten of the active-influence measuring metrics, describing the inequality concerning their importance. This points, however, to the fact that the development process seems to be little design driven.

In general, among the top ten positions of the active influence-measuring metrics many different tasks occur, showing a quite evenly distribution of importance. Not even do the process' start nodes occur among all top ten rankings, underlining the evenly distribution among the involved tasks. This result is consistent with the flow of information between the process' organizational units, which is likewise determined as homogeneous.

Some overall properties of the process can be deduced. Several indications categorize the development process as organized in a well-balanced and even manner: The two start nodes do have, logically, a high importance. But throughout the process, importance and influence onto the overall network is distributed among several tasks, a fact underlined by the high number of 65 different tasks appearing among the top and lowest ten positions depicted in Table 5-18.

*Table 5-18: Top ten and lowest ten outliers for selected structural metrics*

| Metric | | Upper-bound outliers, ranked (top 10) | | | | | | | | | |
|---|---|---|---|---|---|---|---|---|---|---|---|
| | | 1. | 2. | 3. | 4. | 5. | 6. | 7. | 8. | 9. | 10. |
| Active | Activity | **T1** | T72 | T11 | T37 | T3 | T10 | T17 | T27 | T61 | T73 |
| Active | Reachability | **T1** | T3 | T10 | T11 | T18 | T25 | T30 | T37 | T39 | T29 |
| Active | Active proximity | T114 | T90 | T89 | T116 | T83 | T115 | T12 | T15 | T87 | T6 |
| | Snowball factor | **T1** | T11 | T10 | T37 | T3 | T39 | T16 | T25 | T27 | T18 |
| Passive | Passivity | T84 | T65 | T73 | T72 | T91 | T92 | T112 | T96 | T93 | T94 |
| Passive | Reachability | **T119** | T118 | T112 | **T117** | T91 | **T111** | T113 | T114 | T116 | T104 |
| Passive | Proximity | T83 | T114 | **T111** | T82 | T75 | T104 | T113 | T115 | T99 | T98 |
| Passive | Forerun factor | T112 | T91 | T118 | T73 | T84 | T96 | **T119** | T94 | T95 | T93 |
| Relative centrality | | T39 | T72 | T37 | T73 | T4 | T65 | T91 | T84 | T112 | **T118** |
| No of cycles per node | | T 4 | T 39 | T 72 | T 118 | T 84 | T 65 | T 112 | T 73 | T 37 | T 106 |

| Metric | | Lower-bound outliers, ranked (bottom 10) | | | | | | | | | |
|---|---|---|---|---|---|---|---|---|---|---|---|
| | | 111. | 112. | 113. | 114. | 115. | 116. | 117. | 118. | 119. | 120. |
| Active | Activity | T108 | T109 | T110 | T113 | T114 | **T85** | **T111** | **T117** | **T119** | **T120** |
| Active | Active reachability | T113 | T5 | T21 | T23 | T24 | **T85** | **T111** | **T117** | **T119** | **T120** |
| Active | Active proximity | T113 | T5 | T21 | T23 | T24 | **T85** | **T111** | **T117** | **T119** | **T120** |
| Active | Snowball factor | T113 | T5 | T21 | T23 | T24 | **T85** | **T111** | **T117** | **T119** | **T120** |
| Passive | Passivity | T79 | **T85** | T86 | T88 | T105 | T113 | T114 | **T120** | **T2** | **T1** |
| Passive | Reachability | T22 | T36 | T3 | T5 | T6 | T8 | T10 | T21 | **T1** | **T2** |
| Passive | Proximity | T16 | T22 | T3 | T5 | T6 | T8 | T10 | T21 | **T1** | **T2** |
| Passive | Forerun factor | T9 | T36 | T3 | T5 | T6 | T8 | T10 | T21 | **T1** | **T2** |
| Relative centrality | | T6 | T33 | **T85** | T86 | T88 | **T111** | **T117** | **T119** | **T120** | **T1** |
| No of cycles per node | | T 82 | T 35 | T 20 | T 33 | T 29 | T 13 | T 14 | T 23 | T 24 | T 31 |

Another structural indication is the fact that within the entire process only two tasks with only one incident and one outgoing node succeed each other twice. All remaining connections (415 of 417) are edges connecting nodes with higher in- and out-degrees and are therefore far less critical. The number of edges being the only connection between two nodes (12 occurrences), is similarly low. The equally distributed interdependencies between the process' organizational units confirm the general properties from another point of view.

The small percentage (5.575%) of feedbacks among all connections also describes the flow of information as straight and evenly. However, there are about 4 million cycles that are, in particular, driven by T4; this coincides not only with its importance based on the degree but furthermore with its centrality (second outlier), confirming that especially here, decisions are taken and systems architecting is taking place.

The results are partially consistent with the results of an earlier analysis [Braha & Bar-Yam 2004] speaking of few nodes being of high importance for the overall process. For single metrics, assessing single views onto the structure, this may be right. For example, task T39 ranks at first position of values for the metric relative centrality, appearing to be of outstanding importance. But, regarding all active influence measuring metrics simultaneously, it ranks at position 15, representing a rather high but not significantly outstanding importance.

Concerning another result of the earlier analysis, the newly calculated results are identical. Most nodes do have quite low degrees most connections within the network link entities with small activity and passivity. The metrics degree correlation based on edges and on nodes as well as degree distribution confirm this result unequivocally.

## 5.4 Conclusion: Structural metrics

The review of the state of the art (section 2.3) showed that good foundations are available to measure the complexity of, in particular, flow-oriented systems such as processes or workflows. Yet, no comprehensive set of metrics could be identified. In this chapter, a Structural Measurement System was generated to fill this gap, based on existing evidence. As methodical support, measurement foundation provides the requirements for metrics design.

The metrics shown were developed based on a generic procedure for metrics design. The procedure is based on classifying different existing structural characteristics that can be considered as complete as possible. However, these structural characteristics are still abstract and necessitate more research into their significance for processes to support the development of refined metrics. As an intermediate option, therefore, solution principles for the assessment of structures were used as input for both basic and combined and special structural metrics. The solution principles are, in fact, comparable to different classes of problems that can occur in a network. As a consequence, they relate to different problems that are currently regarded in process management and serve as a viable basis for metrics design.

In conclusion, the metrics, therefore, meet the demands set by process management in general. The set of metrics is as comprehensive as the solution principles and the underlying set of structural characteristics permits; yet, many metrics remain conceptual, e.g., the Cognitive Weight or path-based metrics that are still too complex to compute. At the same time, the metrics numerically evaluate all relevant patterns as provided by the solution principles and the available structural characteristics, which are currently the main means of assessing the completeness of the solution.

All structural metrics are connected to their structural significance, as listed in appendix 10.5. Here, the meaning of each metric is based on the meta-model as provided by the Structural Process Architecture shown in the previous chapter. However, the indications of the structural significance of the metrics remain abstract and need more expertise to be applied. As no generalization is possible without losing, to some degree, the concreteness of a solution, this is due to the checklist-like character of how the metrics are presented. A compromise between the degree of abstraction and the level of detail is practical as general applicability and transferability to other cases are not lost.

All structural metrics are based on measurement foundation, as reviewed in section 2.3.1. As detailed, the enumerative nature of most structural metrics is in line with these requirements. All other metrics were reviewed in detail to explain the rationale for them and their empirical foundations in more depth. Yet, more empirical evidence is needed in the long run to extend their applicability. To this end, more case studies and experience derived from their application are necessary and should be collected in ongoing projects. This is especially important as not all empirical evidence is centered on metrics per se but on structures in a more general way.

Therefore, some metrics are more viable than others. Case studies show that the metrics for size and density, activity/passivity, degree distribution, attainability and closeness, the snowball factor and forerun factor, the number of cliques, and the number of cycles (including related counters) are the most practicable metrics as they bear sound empirical evidence, good computability (limited only for cycles), and good structural significance across all domains and relationships defined in the meta-model. All other metrics lack either the detailed algorithms, which make them impracticable to use, or too little evidence is available to fully understand the structural significance of the metrics in detail.

So far, the Structural Measurement System covers common cases of application as found in process management. However, the interdependencies of the metrics are not reviewed yet. For example, the correlation of metrics between their meaning and mathematical foundation is needed to generate more proof of their significance. Furthermore, a more formal description of all metrics needs to be developed to point to further missing metrics.

Overall, the complexity metrics provide a much more condensed overview of a structure than structural characteristics themselves. Therefore, a comprehensive analysis at a rudimentary but global level of process management is possible without examining every detail. This way, the structure and the behavior that is coupled to it becomes accessible. Possible examples are the hierarchies in a process: It is almost impossible to assess all outgoing hierarchies for every task to judge the impact of each task on the process. The metric “number of reachable nodes” and “forerun factor” makes it possible to easily spot the tasks of interest. In a second step, a more detailed analysis of the structure can then be undertaken and the actual structural characteristics can be examined individually.

# 6. The S-GQM framework to select metrics

To enable the goal-oriented analysis of a process using complexity metrics as e.g. provided by the Structural Measurement System, a framework is necessary to select the metrics and to guide their application. The solution presented here is based on the GQM approach and therefore called Structural Goal Question Metric framework (S-GQM). However, the framework is not meant as the *only* means of systematically applying metrics to a process analysis. Nevertheless, the framework is designed to help address the most common elements of a process analysis, as shown in the requirements outlined in chapter 3.

- Enable a simplified access to the analysis by providing goals common to process management as primary points of entry to a process analysis.
- Connect the goals of the analysis to an operational layer and to relevant metrics and the necessary semantics (domains and relationship types).

As Figure 3-3 showed, different elements are relevant for the analysis of a process from a structural point of view. Globally, such an undertaking is guided by goals and, more generally, concepts[74]. To concretize these goals, questions can be asked to examine the intent behind a goal and to demand answers be delivered by the analysis. In fact, structural characteristics can then be used to analyze an existing process using these questions; in the context of this research, structural characteristics are embodied in structural metrics that allow a condensed analysis of one or several structural characteristics. At the same time, questions also arise about certain issues within a domain and its relationship types. Therefore, questions relate to domains, too. Likewise, the structural significance of a metric is only given for a domain and a relationship type, which can alternatively be represented as an aggregate view that encompasses two or more domains and relationship types. Lastly, this structural significance is used to provide answers to any initial questions.

In section 6.1, a number of existing frameworks are reviewed. All of these use a generic layer of certain goals or concepts, refining them from an operational level to a measurement level. The most common approach in the management of measurement, the Goal-Question-Metric (GQM) is chosen and adapted, as it is both practical and simultaneously archetypical for other frameworks. Goal orientation is provided by the common goals of process analysis, as reviewed in section 2.2.2.

## 6.1 Existing frameworks to facilitate the analysis of a system

To efficiently employ metrics to assess a process network, the metrics need to be selected according to the user's needs and goals. Measurement frameworks align requirements and their implementation by employing a networked or hierarchical

[74] The strategic level does not necessarily address goals. For example, "Interfaces" does not state a goal but only a concept. Nevertheless, the term "goal" is used to help clarify the GQM scheme and to address the focus of the analysis of a process.

decomposition of the user's requirements down to the entities that make up a solution. In this section, different means of selecting metrics according to the initial specifications are explained.

### 6.1.1 Quality Function Deployment and the House of Quality

Quality Function Deployment is a method to implement requirements in a physical product. The original Japanese name of QFD translates as a property-function-interrelation and refers to its original development at Toyota. QFD is used to identify attributes of a system that are necessary to achieve the characteristics a customer expects and/or requires. Development work can be structured to prioritize those attributes that are of highest relevance to the customer [AKAO 1992]. QFD is commonly used for systems with a low degree of novelty, as knowledge about how requirements can be related to the product attributes is necessary. It can, in general, be applied to any process of structuring how requirements or expectations relate to a system [LINDEMANN 2007, p. 296].

The classic approach proposed by AKAO integrates four aspects by interrelating them via correlation matrices [AKAO 1992]: the customer's requirements, the functions of a product, the quality characteristics of the product, and its components. The House of Quality (HoQ) represents the first of these matrices, relating requirements to quality characteristics (using a DMM); it adds a matrix (i.e., a DSM) to correlate each aspect with itself, pointing to conflicts. Figure 6-1 shows a common House of Quality.

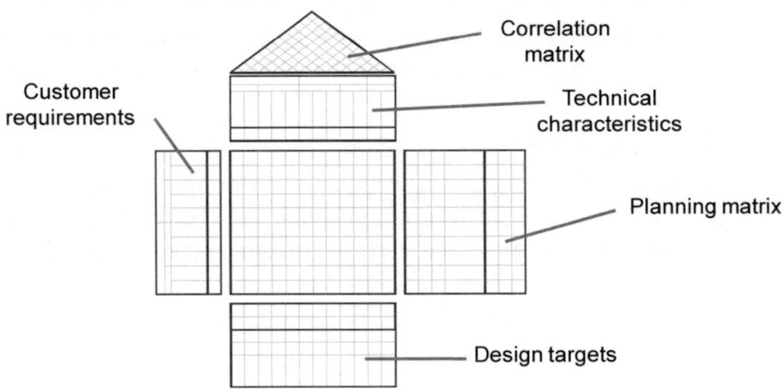


*Figure 6-1: House of Quality*

Quality function deployment can be used at different levels of abstraction, mapping a conceptual or strategic problem to an operational implementation. As such, overall goals, for example, can be related in a structured manner to intermediate functions, and from there to a product's properties, and so on. Therefore, different HoQs can coexist. Generally, QFD provides a means of breaking down a set of conceptual aspects from an operational to a qualitative and even quantitative level. At the same time, it consistently relates these levels to each other. In that, it is similar to the Goal-Question-Metric approach.

### 6.1.2 Goal-Question-Metric

The Goal-Question-Metric approach was developed in the early 1990's. It is a systematic method to set up a quality model in software development, breaking down overall quality goals into intermediate questions and then to metrics to reply to these questions. Returning from the questions to the goals, the measures are interpreted to obtain indications about the software quality. As such, the GQM approach bridges the conceptual level (goals) via an operational level (questions) to a quantitative level (metrics). The metrics serve as concrete and quantifiable entities[75] [BASILI & ROMBACH 1988], as Figure 6-2 shows.

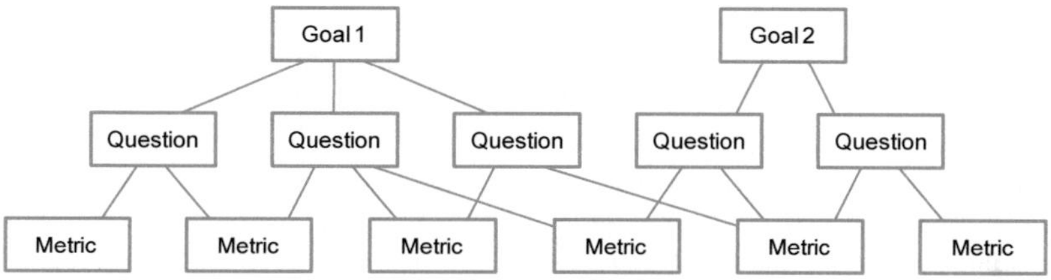


*Figure 6-2: GQM model hierarchical structure [BASILI, et al. 1994]*

A critical part of GQM is the establishment of goals, which are determined using three coordinates: issues, objects, and viewpoints. These are coupled to a purpose that indicates the direction a goal should develop. Table 6-1 shows an example of a GQM application in software development.

*Table 6-1: Example for GQM [BASILI et al. 1994]*

| | | |
|---|---|---|
| **Goal** | Purpose: Improve<br>Issue: the timeliness of<br>Object: change request processing<br>Viewpoint: from the project manager's viewpoint | |
| **Question** | What is the current change request processing speed? | Is the performance of the process improving? |
| **Metric** | • Average cycle time<br>• % of cases outside upper limit | • Subj. rating of manager's satisfaction<br>• Ratio of current and baseline time |

GQM is similar to the Munich Procedural Model [LINDEMANN 2007, p. 45], as it makes overlapping use of basic metrics to answer different questions that relate to different overall goals. As such, it recognizes the fact that individual metrics are not fully independent but rather form a network of metrics, much like the Munich Procedural Model proposes for design methods [LINDEMANN 2003]. At the same

[75] Similar to the earlier Software Quality Metrics approach [BOEHM et al. 1976]

time, the Munich Procedural Model uses seven overall phases (called "elements") that each use an average of five questions to be more concrete about the tasks to be executed during each element. To each question, a number of methods are attributed. However, the model has not been applied to any structure assessment so far.

The GQM approach is sometimes compared to the Balanced Scorecard, which similarly attributes a strategic perspective to individual metrics to measure a company's performance.

### 6.1.3 Balanced Scorecard

The Balanced Scorecard was developed by Robert Kaplan and Peter Norton in 1992 to overcome the one-sided management of companies based only on financial measures [KAPLAN & NORTON 1992]. At the time, the DuPont-System of Financial Control, which focuses on a return on investment calculus, was generally used to strategically steer a company. However, as Kaplan and Norton recognized, human and organizational aspects mattered just as much to develop a sustainable company. MORISAWA illustrates the essence of the Balanced Scorecard as follows [MORISAWA 2002]:

- Achieving a balance among short-term, medium-term, and long-term management objectives through a diverse measurement of performances.
- Creating a sense of understanding by establishing non-financial quantitative indicators (a process index) other than financial indicators.
- Eliminating vagueness by keeping to quantitative indicators.
- Promoting organizational learning through a repeated cycle of hypothesis verification (i.e., hypothesis at the start of a term, correction after the term, and feedback for the next term's plan).
- Providing a common strategic communication platform linking the heads and members of the organization of a company.

Determined to provide a "fast but comprehensive view", the Balanced Scorecard was developed to bring together basic measures for the performance of a company in a comprehensive management report, thus guarding against suboptimization. It links overall strategic perspectives to goals that are to be achieved. For each goal, drivers that enable the company to reach the goal are determined, and suitable "Key Performance Indicators" are attributed that show the level which the driver has achieved. The performance indicators, therefore, rate a goal, differing from the more descriptive viewpoint a metric takes. The performance indicators are obtained by setting up goals for the different perspectives of all stakeholders. Then, for each perspective cause and effect, chains of possible drivers are set up indicating how the goals are to be achieved, linking the customer perception to actions. These cause-and-effect chains are used to determine measures able to represent the level of implementation to satisfy the customer's expectations. Figure 6-3 shows an example of the common four perspectives. However, they can be adapted to suit a company's needs.

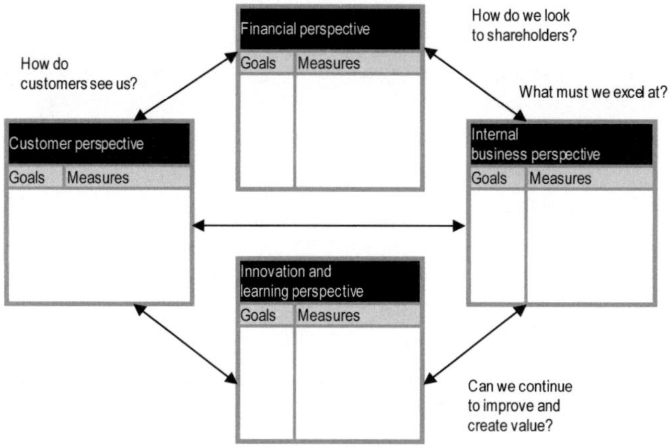


*Figure 6-3: Balanced Scorecard [KAPLAN & NORTON 1992]*

In a so-called "second generation", the Balanced Scorecard was later extended to accommodate the need of changing strategies over time; this so-called strategy map uses a strategic linkage model, making the approach less abstract. Later a vision (called "Destination Statement") was introduced to guide the development of goals [MORISAWA 2002], and the assessment of risks was also incorporated [REICHMANN & FORM 2000].

The Balanced Scorecard is criticized for a number of reasons. It is often seen as too rigid, because it only enables one to see what is previously modeled in the cause-and-effect chains. However, this makes it an efficient tool to navigate changes in a company. Yet, the original intention was to create a comprehensive picture of a company. In fact, the performance indicators are commonly created in a way that they slightly overlap; this overlap helps their efficiency as a controlling tool, as no single indicator can be manipulated without showing up in one or several "neighboring" performance indicators. Furthermore, the process of setting up the performance indicators helps create a common understanding.

The Balanced Scorecard is closely connected to the Goal-Question-Metric approach, as Table 2-10 shows. Different authors use this proximity to develop methodologies that address the advantages of both methods. The Model, Measure, Manage Paradigm [OFFEN & JEFFERY 1997] combines strategic and project issues as in the perspectives of the Balanced Scorecard, breaking down each as done in the GQM approach. In a similar manner, the multi-level Balanced Scorecard uses the GQM across different scorecards for different levels of detail to align perspectives, for example, from a company's level of detail down to a departmental level of detail [BECKER & BOSTELMANN 1999].

*Table 6-2: Balanced Scorecard (BSC) and Goal-Question-Metric (GQM) (based on [BUGLIONE & ABRAN 2000])*

| GQM | BSC | Comment |
|---|---|---|
| Goal | Goal | • GQM goals are project related from an overall perspective<br>• BSC goals relate to a certain level of management within a given perspective |
| Question | Driver | • questions and drivers represent a similar concept<br>• BSC interrelates drivers in a cause-and-effect chain, GQM only lists them |
| Metric | Performance indicator | • comparable<br>• metric is less judgmental about the quality of the measured concept than performance indicator |

### 6.1.4 Directions and requirements

Frameworks allow for a goal-oriented access to methods. They use, generally, matrices that map goals to specific methods; as an intermediary layer, questions can be used to shift the concepts to a more practical level and to specify the information needs. Using common goals from process management, a framework can thus be set up to relate metrics to these goals, similar to the GQM approach, to enable a goal-oriented structural analysis of engineering design processes.

However, no framework provides a comprehensive basis for the interpretation; only the Balanced Scorecard provides access to possible causes behind the measures. There is, thus, the need to extend frameworks to find the way back from the abstract, methodical layer to the deduction of suggestions about how to interpret the results of an analysis.

Therefore, the framework to be designed should fulfill the following requirements, most of which are met best by the GQM scheme, which is most closely focused on guiding the process of using metrics for a goal-oriented analysis (these are derived from the section above and from the requirements shown in Figure 3-1):

- Offer goals as a strategic level of the analysis
- Refine the goals via an intermediate layer that helps to concretize the goals through questions
- Attribute metrics, domains, and relationships to each question that points the user to possible answers
- Provide indications of the possible structural significance of the results obtained from the metrics
- Be modular to incorporate future extensions

## 6.2 Systematic access to the structure of a process

As initially stated, the main problem in the analysis of a large engineering design process is the complexity of this very task, producing a seemingly insurmountable barrier; thus, resorting to a more abstract layer is a necessary intermediary. As Figure 3-3 showed, different aspects that are mutually dependent play a role in this. Given that there are a number of elements to each aspect, there are numerous possibilities that could be recombined; thus, the proposal of a generic approach that integrates all possible aspects is almost impossible and rarely fruitful.

Based on the Structural Process Architecture and the Structural Measurement System, providing for a set of 52 metrics, 6 domains and 26 principal relationship types (based on 71 detailed relationship types), a total 8,42 combinations are possible. The framework suggested here reduces this number to a pragmatic set that serves the basic needs of process management (compare the common goals and concepts of process management on page 66 and the following pages). However, this reduction does not imply that any metric is of lesser relevance, but rather that it only represents a basic subset of the overall solution space that can be extended to suit different goals of process management.

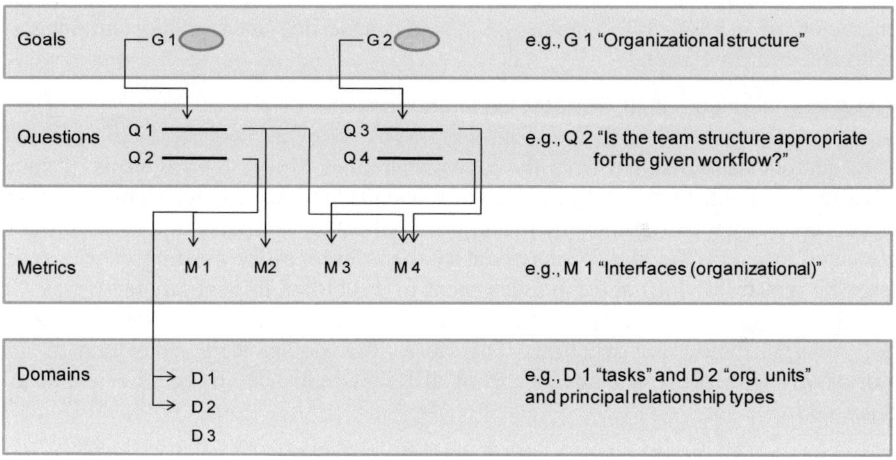


*Figure 6-4: Using goals and questions to identify relevant metrics and domains*

Figure 6-4 illustrates the reduced approach in more detail. For example, if the organizational structure is the focus of the process analysis, the appropriate goal G 1 is selected. This example also indicates why the term "goal" is not fully appropriate (see footnote 74). In the example, goal G 1 proposes two questions Q 1 and Q 2, of which Q 2, "Is the team structure appropriate for the given workflow?", is selected. This question leads to metric M 1 and to domains D 1 and D 2. In many cases, it is possible that several metrics provide answers to one question, and thus several questions can relate to the same metric. In the example, metric M 1 will provide the number of organizational interfaces between the domains tasks and roles, i.e., it will indicate how straightforwardly the human resources are assigned to the individual tasks of the process. If this number is compared to the number of tasks and to the number of roles, the ratio will indicate whether there are few or many assignments per role, indicating the fact that there may be many changes in responsibility for a task. In a second step, if a set of processes is compared to this end, the process that exhibits the highest numbers should then be prioritized for any further analysis.

In the following sections, the main steps are further explained. First, the goals are laid out, then, questions are allocated to each goal, and ultimately, the allocation of metrics, domains, and principal relationship-types is discussed.

### 6.2.1 Goals and questions of structural process analysis

As proposed earlier, there are two ways to approach the analysis of a system, either following a goal-oriented strategy or an opportunistic manner. Maurer defines these two strategies as "define what you need", i.e., the requirements-driven approach, as opposed to "see what you can get", i.e., the opportunistic application of analysis to a system [MAURER 2007, p. 93].

To analyze a process in a goal-oriented way, Table 2-9 (page 71) reviewed common concepts for the structural analysis of a process. These concepts represent abstract classes of analysis; not necessarily all of them have the direct character of a SMART[76] goal, i.e., being **s**pecific, **m**easurable, **a**ttainable, **r**elevant, and **t**ime bound.

The goals were collected from the different concepts of process management, as shown on page 64 and the following. All relevant concepts of process management were detailed from the given references (Table 2-9) in terms of their structural content. For each question, the relevant domains and relationship types, as addressed by the original references, were collected; similarly, the metrics were allocated based on the indications given by the references for existing metrics (see page 80 and following) and the assessment of structures in general (see pages 48 and 52 and following). This attribution is not always ideal, as the literature does not always focus on structure. Therefore, the goals were consolidated in workshops with process managers from different industries to be as relevant as possible.

The goals guide the strategic level of the analysis. Here, the core concepts to be followed during the analysis are set up. As part of the state of the art, seven major goals of process management were identified that are complemented by an eighth goal to integrate the additional aspect of a systematic analysis of Boolean operators to analyze decision points in a process model. The goals are not fully independent, as the attribution of the metrics will show. These goals are:

- Planning
- Resource consumption
- Quality

---

[76] SMART criteria are commonly accepted as the "good" description of goals as part of Management by Objectives [ODIORNE 1980, p. 82].

- Flexibility
- Organizational decomposition
- Interfaces
- Transparency
- Decision making

**Planning** refers to the degree to which the process plan can be pre-determined. As such, it comprises, for example, structural bottlenecks, tasks that can be worked on independently of each other, iterations, or the chances of adherence to a schedule as embedded in the structure of the process. These occur particularly as cycles among points in time, tasks and artifacts, as bridges within the process network, and as densely crosslinked interfaces among several domains. Planning, therefore, particularly addresses the runtime of a process, its critical paths, and its repetitive tasks. The following questions detail this goal:

- To what extent is it possible to incorporate risks into the process planning?

  This question especially addresses the fact that a densely networked process implies a higher risk of delays towards a milestone. Clusters, in particular, and, more generally, iterations are the drivers of such delays. Furthermore, the less linear a process, the more complex its planning to break up cycles among the artifacts or points in time that hinder a linear process flow.

- How can the focus be concentrated on important process steps?

  This question aims at identifying important tasks that have the highest impact on the process flow, being central sinks or distributors of information, thereby coordinating the overall process, and driving and/or controlling iterations.

- What are bottlenecks in the schedule?

  Bottlenecks in the structure are those communication channels or tasks and documents that, if defective or incomplete, hinder the further execution of the process. Therefore, bridge nodes and edges as well as the connectivity of the graph are within the scope of this question.

- What parts of the process are substantially impacted by iterations? What level of uncertainty is handled by the process?

  Iterations are a major driver of costs and delays, although the goal is to improve the quality of an artifact by reworking part of its contents. Therefore, those parts of the process that are impacted by iterations deserve particular attention. Cycles, their start- and end-nodes, their main communication channels, as well as existing and possible clusters, contribute to such iterations. Iterations also point to the level of uncertainty that is inherent to the process, i.e., the degree of novelty of the item that is being developed and the amount of knowledge that needs to be generated as a result.

- What is the stakeholder situation?

  The stakeholder situation is characterized by the number of different domains that are relevant for a process model; therefore, the stakeholder situation typically relates to the size of the network and its different measures.

The **resource consumption** covers aspects like capacities and the utilization of resources that emerge out of the attribution of two domains to each other. Thus, for example, redundant work, the availability of IT systems and other resources, and the homogenous layout of the process are addressed. To this end, structural characteristics such as attainability and sync graphs among tasks, organizational units, and resources are applied. The following questions detail this goal:

- Is the process laid out in a homogeneous manner?

  This question addresses the even distribution of tasks in the process and their allocation to organizational units as well as their inputs and outputs. The interest is to find such tasks and artifacts that collect the knowledge of the process, which will generally cluster in those tasks and organizational units that are the most involved throughout the process. Equally, those organizational units that represent the core competencies of the process can be identified.

- Where is it possible to remove redundancies to reduce waste?

  Multiple allocations to tasks and other entities in the process are an indicator of the unnecessary use of resources; another driver of resource consumption is the frequent change of responsibility, causing additional coordination effort. The metrics grouped under this question, therefore, regard multiple allocations and interruptions in the alignment of different domains.

- Are the resources easily accessible?

  The availability of resources is essential for the efficient execution of any task in a process. Therefore, this question focuses on the reachability of resources within the process network.

The concept of **quality** includes, in particular, the consistency, the integration, and the distribution of information and errors, thereby focusing on the quality of the process, not of the product. By looking at the reachability, the resilience, the hierarchies, and the alignment of the artifacts with the rest of the process network, the propagation of errors, the fragmentation of tasks as well as documents, and the general consistency of information transfer are considered. The following questions detail this goal:

- Does the process allow for the consistent transfer of information?

  Like the accessibility of resources, the continuity of information transfers, i.e., the reachability of one resource from another resource, is the essence of information consistency. This also applies to artifacts that are generated throughout the process.

- Is the documentation in line with the process?

  The alignment of artifacts (representing the intermediate results of the process) and the tasks point to possible issues within the exchange of information throughout the process. Dissimilar structures of these two domains (size, degree distribution, cycles) are an indicator of inefficient documentation.

- What is the risk of error distribution across the process?

  The propagation of information also includes the propagation of errors among the tasks and artifacts. Therefore, short and wide hierarchies point to root nodes within these two domains that have a high impact across the whole process network and that are thus susceptible to collecting incoming errors or to distributing errors rapidly across the process.

The goal of **flexibility** of the process makes use of similar concepts like resource consumption; however, aspects that contribute to the flexibility of a process, for example, redundancy, robustness, and adaptation, commonly consume more resources. As many of these aspects can only be judged from the semantics of the process network, only buffers and the general robustness are regarded closely, evaluating the adjacency and attainability of points in time, tasks, and artifacts in particular. The following questions detail this goal:

- What buffers are available in the process to absorb delays and errors?

  Synchronization points among points in time, tasks, and artifacts that collect information can serve as buffers if used correctly; a node with a high passive degree (i.e., having a high passivity) will collect many inputs before continuing the process. Therefore, these entities need to be identified to be aware of their potential as buffers.

- How robust is the overall process to individual failures?

  The resilience of the overall network facilitates the adaptation of the process in case individual nodes (e.g., key personnel) drop out. Therefore, nodes that could compromise the integrity of the network indicate a lack of flexibility to cope with problems with these entities. Similarly, multiple paths across the overall process point to more flexibility to cope with unforeseen changes in the process structure.

The **organizational decomposition** is intended to establish efficient communication channels by means of a purposeful decomposition of organizational units. Here, coordination of all tasks, the adequate setup of teams, and distinct responsibilities are of interest. Hence, organizational units are focused on, and analyzed for straightforward crosslinking with their tasks, in particular, their internal attainability, clustering, and resilience. The following questions detail this goal:

- Is the organization of workgroups and teams adequate?

  This question addresses the alignment of the process with the organizational setup. The clustering of tasks in the process points to necessary workgroups.

- How well is the organizational structure suited to provide efficient communication?

  The ability of each organizational unit to be able to reach other organizational units is an important driver for communication; therefore, the attainability of organizational units as well as the mean path length is of interest in characterizing the communication within a process; also, bridge nodes and central organizational units are of interest. Similarly, the metrics of this question point to entities that may not be well integrated, being little connected or not reachable at all.

- What is the internal structure of an organizational unit?

  As a socio-technical system, a process is driven to a large extent by opinion leaders and information hubs that coordinate the process, even if they are not the executives that formally manage the process. Therefore, their identification is targeted by the metrics focusing on the centrality and the degree distribution and correlation of the network formed by the organizational units.

**Interfaces** are another important concept in process management. Here, disruptions among resources, artifacts, or organizational units are addressed, as well as errors in the transmission of information, the supply of information in general, and the integration of all organizational units. Hence, hierarchies are of interest, as they point to the propagation and the communication channels that belong across the process. The following questions detail this goal:

- Which entities of the process need to be synchronized?

  This question addresses the need for information exchange among tasks and organizational units; therefore, the degree is of importance, as well as the attainability. The distribution of degrees and their correlation, in particular, point to those entities that are of high importance for the process.

- How fast is communication in the process?

  Like the propagation of errors, the propagation of information is represented by hierarchies across the process, which show what information can reach other entities from its root node. Therefore, the attainability, as well as hierarchies among the tasks and among the organizational units, is examined to characterize the speed of communication.

- What are relevant communication channels?

  While the synchronization within the process takes place at particular tasks or organizational units, there are also characteristic paths within these networks that this question aims to identify. Therefore, path-based metrics are applied.

Furthermore, **transparency** covers the clarity and comprehensiveness of the process, i.e., the degree of complexity of grasping and understanding the process and the involvement of individual entities therein. This transparency affects, of course, all domains of a process network. The following questions detail this goal:

- Are the organizational units aware of their impact on the overall process?

  The higher the degree of crosslinking in the process, the more difficult it is for an individual entity (organizational units, mostly, but also tasks and points in time) to judge the long-range impact of their work. Therefore, the size of the network, the degree of its crosslinking, and its planarity are used to gain insights into this question.

- How transparent is the overall process organization?

  The cognitive ability of humans is limited to comprehending only a few objects at a time; therefore, understanding a highly complex process is very difficult. This question is intended to grant access to cognitive models that allow the evaluation of the degree of complexity of a process network from such a point of view.

Lastly, **decision making** addresses the fact that the structure of a process reveals many decision points, both those that are explicitly modeled as Boolean operators and those that drive iterations, i.e., the start-nodes and end-nodes of cycles that govern a process in particular. As such decisions impact all domains, no particular selection is proposed. The following question details this goal:

- Which decision points have a high impact on the process?

  This question relates to metrics that evaluate the impact of a decision point on the process, mostly through the degree of tasks and business objects. However, the assessment of overall processes is also possible.

Overall, the goals and questions can be used to focus on a more general undertaking towards process improvement. In the author's experience, such projects rarely start with one explicit goal but rather with a vague idea of where the problems that motivate the project might originate from. Therefore, the goals and questions are, above all, intended to point to common problems. However, they should not be understood as a delimiter that indicates the maximum applicability of structural metrics. In fact, during each application, the goals and questions, as well as the set of metrics, domains, and relationship types, should be extended according to the nature of the analysis project.

### 6.2.2 Allocation of metrics, domains and relationship-types

After a principal focus on the analysis is designated using the goals and questions, measurements are used to identify outliers, deduce their significance for the behavior of the actual process, and deduce indications towards possible improvements, thus providing answers to the questions.

To each question, thus, a set of metrics, domains, and relationship types needs to be allocated in a way that it provides answers to the questions. To do so, three research approaches were used that, concurrently, provide a good reference for the allocation of metrics.

First, metrics that have been used to answer one specific question before, as shown in the references, can be allocated for the appropriate questions; for example, McCabe's Cyclomatic number is typically used to compare different control-flows (i.e., different processes or different modules in a process) to their decision structures. Therefore, the allocation of this metric to a question that regards the decision structure of a process is obvious.

Secondly, all metrics were developed for a particular structural significance for each domain of the meta-model (see page 146). This significance is, in each case, based on empirical evidence and on the research that is, in part, shown in the two case studies in the subsequent chapter. The significance given for each metric, therefore, also allows the attribution of individual pairs of metrics and domains (including the principal relationship type) to questions.

Lastly, the metrics embody different, more general, structural characteristics that imply a certain behavior of a domain. Although not detailed in this research, this classification was used to classify the metrics concerning their focus of analysis (robustness, grouping, extent, and propagation; see the tables in appendix 10.7 and the explanation in section 5.2.4 on page 157). Based on this classification, the two methods above of allocating metrics to questions could be cross-verified to see if all relevant metrics had been allocated.

Figure 6-5 shows the simplified S-QGM scheme that results from the attribution of metrics and domains to the questions (the complete framework is given in the appendix; there also are the IDs for goals G01…G08, questions Q01…Q22, and metrics M01…M52). As an example, goal G01 "Interfaces" is detailed with its five questions Q01 through Q05. Question Q01 "To what extent is it possible to incorporate risks into the process planning?" uses metric M40 "Number of cycles", for example, to point to the expected maximum number of iterations for the domain "Tasks". The more iterations there are, the less deterministic the outcome of the process; therefore, a high risk of delays needs to be anticipated and considerable effort needs to be put into the core tasks that are reworked during the iterations. In turn, process planning needs to integrate possible buffers for these iterations, and possibly break up the iterations into distinct phases of rework.

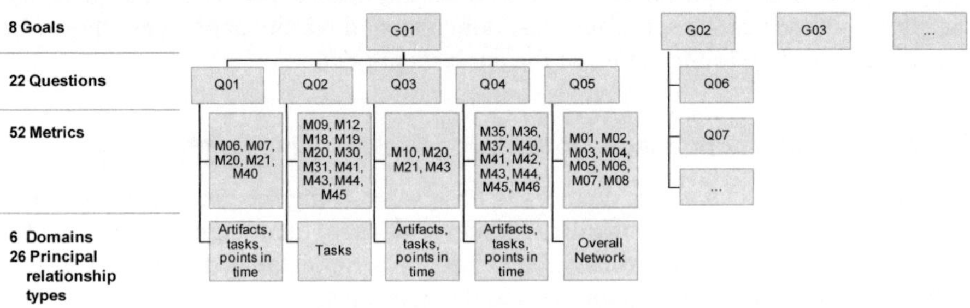


*Figure 6-5: Resulting S-GQM scheme*

Depending on the available native data in the process analysis, either native datasets or aggregate views can be used for the analysis. A combination of both kinds of input data (e.g., a native task-DSM and an aggregate task-DSM that is computed via intermediate artifacts) can be used to further detail the analysis and check for the coherence of the domains among each other. In most analyses, aggregate views will be necessary, as the networks will, in most cases, be multipartite, i.e., the different domains will not be internally networked but only via other domains. In such cases, the domain of reference is provided by the S-GQM, and the aggregation path needs to be chosen carefully to mimic the actual process execution. If, for example, an aggregate view on resources is necessary (e.g., IT systems), these systems will commonly exchange artifacts via the tasks they are allocated to. Therefore, the aggregation path will be from resources to tasks to artifacts to tasks to resources.

### 6.2.3 Identifying structural outliers

As the metrics selected and computed up to here provide a highly condensed picture of the process, they still do not provide detailed information about the process's behavior; however, their main focus is to identify structural outliers that characterize the process's structure. Therefore, this and the subsequent sections are also valid for a process analysis that is not guided by the GQM scheme.

Figure 6-6 shows the basic procedure for working with outliers. To analyze the results from the metrics properly, first, the metrics need to be calculated for the relevant datasets. Then, the results of each metric are regarded individually to identify outliers. Here, the scale of each metric (i.e., the range of values that are possible for the metric) is of interest for running a Pareto analysis to identify outliers properly. Within the scale, outliers are then sought and collected. In the next step, the collected outliers are compared; often, an entity that appears as an outlier in one distribution might also be an outlier in another distribution, making it, therefore, even more relevant to the process. Furthermore, cross-aggregation outliers can only be identified if different metrics are compared in terms of their possible correlation. Lastly, the results need to be described and documented for further use and to serve as input for their interpretation.

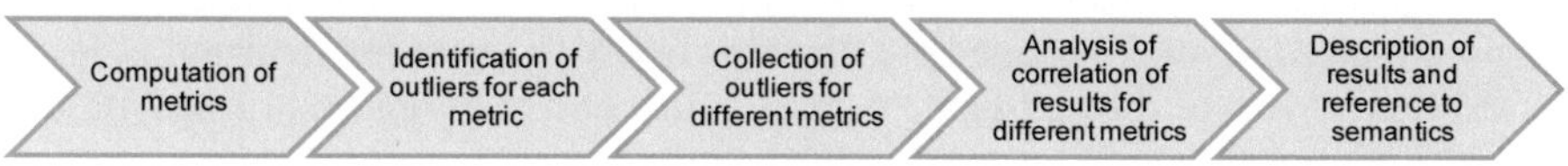


*Figure 6-6: Basic procedure for working with outliers*

### 6.2.4 Structural significance of the outliers

The interpretation of outliers should take place from different perspectives simultaneously, otherwise the actual context of the process is ignored, and the results may not be meaningful.

There are different ways of accessing the meaning of a metric. The basic access to the interpretation of a result is given by the process model itself; in all cases, the outliers should be related to their context and meaning. If a process model contains, for example, the coordination of the process as a distinct task, this task will show up as highly central in the process. While this result is correct, it may be of little use if the principal process flow is being analyzed.

Secondly, the structural context of an entity that appears as a structural outlier should be considered. If, for example, a set of 60 different feedbacks is identified as outliers, these may have a different impact on and relevance for the structure.

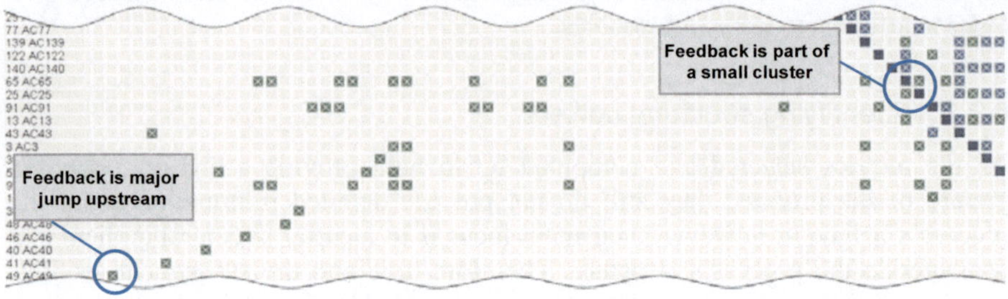


*Figure 6-7: Structural context of feedbacks in task-DSM (part of the case study in section 7.1)*

Figure 6-7 visualizes two different kinds of feedback. While the feedback close to the diagonal of the task-DSM is of little relevance, as it is only part of a short iteration, the other outlier is a major jump back in the process flow.

Thirdly, the structural significance as given by the tables that are part of the description of all metrics can be used. These lists can serve as a comprehensive list of possible interpretations to guide the interpretation. However, they do not represent a complete checklist but only a collection of common interpretations and issues related to each metric, domain, and principal relationship type. In all cases, but especially when using the tables of structural significance, the domain and the relationship type should be reflected in the interpretation to ensure the possible interpretation actually fits the initial data. Particularly when complex aggregations are used (across several intermediate domains or involving many decision points), the expressiveness of the aggregate dataset is low and may not be in line with the principal relationship type assumed in the tables of structural significance.

In the final step, the results and indications of their meaning need to be discussed with relevant staff to deduce conclusions and individual improvement measures. Here, again, the context of the process matters most, as even the results from a well-balanced process analysis (i.e., incorporating many different metrics to ensure a complete picture) do not guarantee that the process improvement is just as comprehensive. To avoid a one-sided suboptimization, the results, therefore, should always be critically reviewed with all possible stakeholders and possibly supported by more detailed methods, for example, simulations for key outliers that undergo improvements in subsequent steps.

## 6.3 Using and adapting the framework

During a process analysis, a framework is employed during the first phase of the procedural model that guides the process analysis (shown in Figure 6-8). Here, the goals are set and appropriate questions are selected according to the individual needs of the analysis project. Each question yields a set of metrics, domains, and relationship types that help produce indications to the questions.

With this input, the system border is basically drawn, and the necessary information can be acquired, either from existing process models or by building a process MDM, as discussed in section 4.5.1. During these two steps, the combination of each metric, domain, and relationship should be reflected in the available data and the information needs that drive the analysis. Certain domains may need refining to better answer the questions or provide information that is not available in sufficient quality (e.g., because a process is so little consolidated that no meaningful process model can be built). Once the process model is available, the results for the metrics can be computed for the relevant domains and relationship types. In the last step, the framework comes into place again, providing guidance to interpret the results obtained. Here, the initial questions should be reconsidered to see if significant answers are provided by the metrics or if the analysis should be re-run with an improved selection of different (and possibly related) metrics, sub-domains, and refined relationship types. To this end, the principles of using domains and relationships, as discussed in Figure 2-3 and the classification of the metrics (Table 5-15), support this process.

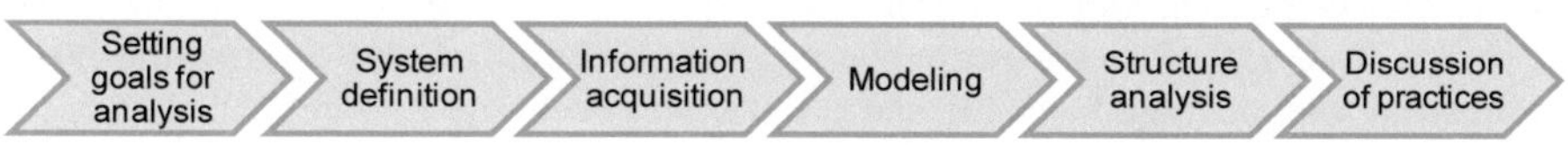


*Figure 6-8: Procedural model for structural process analysis*

The framework is designed in a modular manner and can be extended to different cases as might be necessary. As Figure 3-3 showed, it consists of different aspects that are interrelated, and each can be changed or extended accordingly.

- Goals
- Questions
- Metrics
- Domains
- Relationship types
- Structural significance

Each time, the interdependencies between these aspects need to be reviewed according to the logics discussed in this chapter. Moreover, the interdependencies to the existing entities in the framework should also be reviewed. If, for example, a new structural metric is introduced into a set of metrics to answer a question, it is necessary to examine the informational value that is gained; if, for example, two metrics correlate, the metric should not be introduced, or the two metrics should be marked accordingly.

## 6.4 Conclusion: S-GQM framework for structural analysis

This chapter presented the Structural Goal Question Metric (S-GQM) framework used to integrate an overall approach of structural measurement for process analysis. The framework is designed to address the common goals of process analysis. It makes use of the common principles of pursuing a goal-oriented application of a method. The framework operationalizes the goals by asking questions that embody the relevant facets of each goal.

The development of the goals and questions guided by common goals in process management is defined throughout the state of the art. However, these goals and questions are generally quantitatively evaluated in existing analysis methodologies. Therefore, they had to be mapped for their structural content, which reduces their expressiveness to some extent. The structural focus that is embodied in the framework, therefore, generally relies on case studies that are shown in the next chapter and in the literature review and a recombination of patterns found in the literature on behavioral indications that can be deduced from structure. The framework, therefore, presents a heuristic system (see 76 for the different possible measurement systems that serve as frameworks for systematizing metrics). The goals and questions may not be complete, but they are meant as a guideline for common use cases and to demonstrate the application of the structural metrics. Depending on the context of the individual application, an extension and adaption are probably necessary; therefore, no "out-of-the-box" application was designed. At the same time, however, the hierarchical design of the framework, based on the GQM scheme, allows easy adaptation.

The allocation of the metrics for the goals and questions is based on different sources to ensure a good quality of the framework. However, the allocation of metrics is difficult. In fact, for most questions, many metrics provide partial answers without being fully relevant to the question. Therefore, for each allocation, a compromise between expressiveness and compactness of the framework has to be made. In the framework presented, this choice was used to design a pragmatic framework that embodies the most viable metrics, as discussed in section 5.3.

Similarly, the allocation of domains and relationship types is sufficient for the purpose pursued, as it was guided by the semantics transported in each question. Nevertheless, a detailed review of the domains and relationship types for each analysis project is necessary, as again no "out-of-the-box" application is realistic. This is because the underlying meta-model only serves as a generic frame of reference. Yet, in practice, different domains and relationship types might be available, either because they are relevant to the company being analyzed, or simply because the process model that is used as an initial starting point dictates a different set. To this end, the structural significance can only be deduced from the description of the metrics. This, however, is supported by the framework in a straightforward manner.

# 7. Industrial application of metrics

In this chapter, the analysis of engineering design processes using the goal-oriented application of structural complexity metrics is shown. To do so, two different processes are modeled as Multiple-Domain Matrices, different metrics are selected and computed for the process models, and the findings from the analyses are compared to statements from engineers and managers involved with the processes to validate the findings.

The first case study in 7.1 focuses on the electronic control unit design for a premium class SUV. Here, the development of different highly-integrated mechatronic components takes place in a linearly planned design process that, however, exhibits large iterations. In the process analysis, no particular goal was followed, but rather a general characterization was of interest to identify the core drivers for the process. Therefore, it is used to demonstrate the application of structural metrics in general.

The second case study, shown in section 7.2, regards the process from the introduction to this book, focusing on the development of the body-in-white of a medium-sized premium class sedan. This process can be characterized as a function-oriented mechanic design process, as it produces the sheet metal design of the car that is optimized for different functions, such as vibration, crashworthiness, and passenger safety. Therefore, the interfaces among the sheet metal engineers and the simulation engineers are of particular interest to ensure the efficient transfer of information between the two groups. Thus, the application of the S-GQM framework and the different aspects of aggregating domains are shown.

Both case studies are used to illustrate the use of the Structural Process Architecture, the Structural Measurement System and the Structural Goal-Question-Metric framework that were laid out in chapters 4 to 6 at different levels. Above all, the use of the complexity metrics per se is reviewed to show how they serve their overall purpose of identifying possible weak spots. Then, at a more detailed level, the most common metrics and their individual use are demonstrated. To use the metrics, the application of the MDM-based process modeling and the framework are also reviewed.

## 7.1 Electronic control unit design: General analysis in Automotive Development

This case study demonstrates an example of the analysis of an engineering design process. The process focuses on the design, testing, and integration of the onboard electronic control units of a premium-class Passenger Car[77]. These devices cover all electronically controlled functions of the vehicle, such as engine control, climate control, and the entertainment system. Figure 7-1 shows the typical integration of electronic control units, sensors, and actors in a vehicle.

---

[77] For nondisclosure reasons, the dataset cannot be published.

### 7.1.1 Goals and focus of the project

The case study was carried out in cooperation with one of the biggest producers of premium cars and the world's biggest manufacturer of commercial vehicles. It was driven by the interest to clarify "how complex" the design process for electronic control units was. To this end, the application of structural analysis was intended to show the different facets of complexity within the process to better prioritize future improvements. Their interest in this analysis was to better characterize the fact that development processes in the automotive industry, especially for highly integrated systems, are continuously increasing in complexity and, therefore, need continuous improvement. This complexity shows, in particular, through the complex network structure of the overall structure of the process, the reach and impact of the activities carried out during the process, and the existence of iterations, as these were not explicitly described in the existing process documentation.

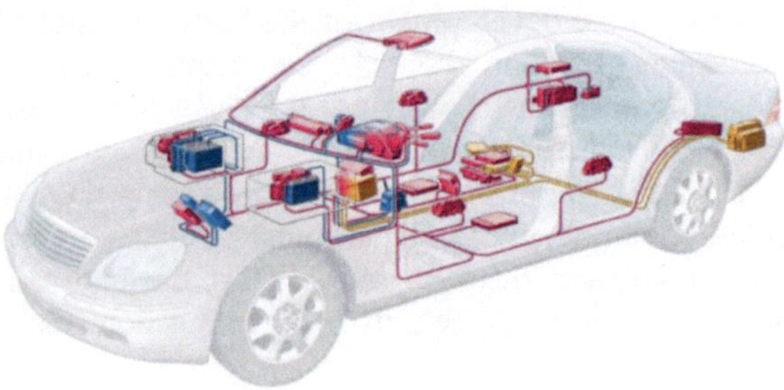

*Figure 7-1: Different onboard controllers in a 2003 Mercedes S-Class [Daimler AG 2010]*

For this purpose, the overall electric/electronic development process, consisting of 15 distinct sub-processes (called "modules" hereafter), was analyzed for possible weak spots taking shape as structural outliers. As such, a Pareto analysis for all relevant computable metrics was carried out to allow a detailed quantified structural comparison of the process entities.

### 7.1.2 The process model used

As many relationships within the process were unknown and not part of the process documentation, the process model was assembled from different sources. Overall, 198 partial models were integrated into one global MDM to represent relevant domains and relationship types. The partial models of the process modules were mostly extracted from individual process models, modeled in UML using Innovator (by MID GmbH), for which an individual export interface was programmed to export the structural content of the process models in a spreadsheet format. This data was available for the domains modules, activities, documents, points in time, and roles. Additionally, data on product attributes was integrated based on information exported from the requirements management system based on Doors (by IBM Telelogic). Figure 7-2 presents the domains and relationship types involved. Because of inconsistencies and a lack of complete information in the given timeframe the project was carried out in, the product attributes had already been excluded at an early phase of the analysis. However, the engineers at the company agreed that the process models themselves were of good quality and should be sufficient to yield a comprehensive picture of the processes. At a later stage, the data concerning the integration of different roles (i.e., human actors in the process) was excluded from the analysis, too, as this data was only partially available, and it was not the core interest in the analysis. Secondly, the density of the aggregate role-role DSM that was computed using the existing data was too dense to deduce any relevant information.

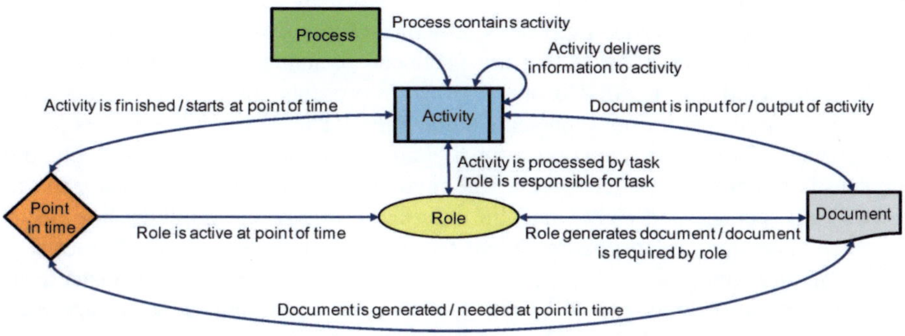


*Figure 7-2: Entity-relationship model of domains and relationship types in the process model*

As can be seen, native data was available on the direct dependencies among the activities, and it was possible to generate different aggregate views on points in time, documents, and activities; each of these models could be attributed to the process modules to which they belonged. Figure 7-3 shows the setup of the process MDM, including the domain "Roles".

| | P | A | R | D | T |
|---|---|---|---|---|---|
| Processes P (P) | | P contains A | | | |
| Activities A (A) | | A delivers information to A | A is processed by R | A has D as output | A finishes at T |
| Roles R (R) | | R is responsible for A | | R generates D | R is active at T |
| Documents D (D) | | D is input for A | D required by R | | D is needed at T |
| Points in time T (T) | | A starts at T | | D is generated at T | |

*Figure 7-3: Meta-MDM of the process model*

After instantiation of the MDM and import of all 198 partial models, the overall MDM[78] consisted of 1860 entities and 7070 native dependencies (all directed). The large number of entities is due to the fact that, according to the company's process modeling guidelines, existing iterations were modeled as repetitive phases, i.e., no upstream flow of information was modeled, but the part of the process that was iterated was modeled a second time. For more iterations, accordingly, the process was repeatedly modeled more times, thus producing a behavioral model rather than a structural model[79]. However, these repeated sections of the process model did bear different names and could not, therefore, be identified. Only the repeated use of documents could, to some extent, indicate iterations.

Figure 7-4 provides an impression of the network of activities of 3 of the 15 process modules. The graph represents how the activities of the design process are interlinked via the documents that are produced or that serve as an input. Each node (numbered for nondisclosure reasons) represents an activity; each edge linking two nodes at a time represents a document. There are 377 activities in the model and 237 different documents. On average, each task has 1.65 edges, i.e., the process is rather densely networked. The modules can be recognized easily in the figure (left, center, and right), and the interfaces among the modules line up in between. The figure also illustrates that the process modules are of different size and structural specificity. The gaps in the process model are not visualized; however, it can be recognized by the fact that the aggregate view on activities linked via documents leaves 51 activities that are not connected to any other activity.

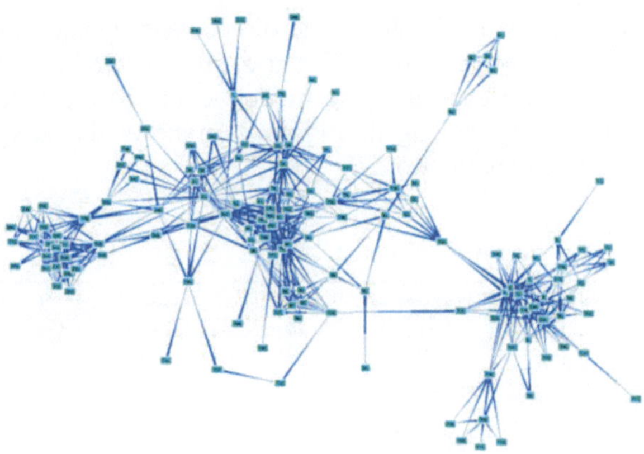

*Figure 7-4: Visualization of network of activities (via documents) as a strength-based graph for three modules*

---

[78] The overall MDM was modeled in both a spreadsheet software and in Loomeo (by Teseon GmbH). All graph depictions are also extracted from Loomeo.

[79] During the analysis, this led to the problem that cycles were hard to identify (see page 164).

### 7.1.3 Analysis and findings

The selection of metrics for a complete representation of the structural complexity of the process was guided by two aspects. On the one hand, a relevant spectrum of metrics and domains was chosen to ensure a broad and balanced picture. On the other hand, domains and aggregate views that were either incomplete (or aggregated from incomplete datasets) or that were too densely networked to conclude any meaningful results (e.g., for roles that represented one complete cluster within the role-role DSM) were excluded. The metrics listed in Table 7-1 were calculated based on this reasoning.

*Table 7-1: Metrics calculated in this case study and domains (native and aggregate)*

| Metric | Activities (native) | Activities (via documents) | Documents (via activities) | Points in time (via activities) | Points in time (via documents) |
|---|---|---|---|---|---|
| **Size and density** | | | | | |
| Number of domains | X | X | X | X | X |
| Number of nodes | X | X | X | X | X |
| Number of classes | X | X | X | X | X |
| Number of interfaces (dom.) | X | X | X | X | X |
| Number of edges per node | X | X | X | X | X |
| Relational density | X | X | X | X | X |
| Number of unconn. nodes | per domain | | | | |
| **Adjacency** | | | | | |
| Activity / Passivity | per module | X | X | X | X |
| Degree distribution | | X | X | X | X |
| Fan criticality | per module | | | | |
| **Attainability** | | | | | |
| Number of reachable nodes | | X | X | X | X |
| Reachability of a node | | X | X | X | X |
| **Closeness** | | | | | |
| Proximity | | X | X | X | X |
| Relative Centrality | | X | X | X | X |
| **Hierarchies** | | | | | |
| Snowball factor | | X | X | X | X |
| Forerun factor | | X | X | X | X |
| **Cycles** | | | | | |
| Number of cycles | per module | per module | per module | per module | |
| Number of cycles per node | per module | per module | per module | per module | |
| Number of cycles per edge | per module | per module | per module | per module | |

As can be seen, the goal was to represent the central categories of structural features (size, density, adjacency, attainability, closeness, hierarchies, cycles, domains) in a balanced way. Within each category, those metrics were chosen that serve as basic counters and represent the active and passive aspects of each node's embedding in the network. Similarly, those domains were used that provide reasonable data quality, which in such a large process is difficult to obtain in an industrial setting, as many different process modelers with a slightly different understanding of the process are involved, generating different models. As it turned out, the data quality of the native DSM of activities was too low to generate purposeful results; this DSM was generated through interviews among engineers that individually were only involved in small parts of the process, which therefore generated a picture too fragmented to provide a coherent process model. Thus, the metrics were generally calculated only for the aggregate views.

During the course of the project, a metric designator was introduced to visualize the datasets for each metric in a simple manner. It represents, on the one hand, whether a metric is calculated for a native or an aggregate dataset and, on the other hand, shows the aggregation used if applicable. For the aggregation, the six different types as proposed by [MAURER 2007, p. 85] were used. See page 47 for more details on this aggregation.

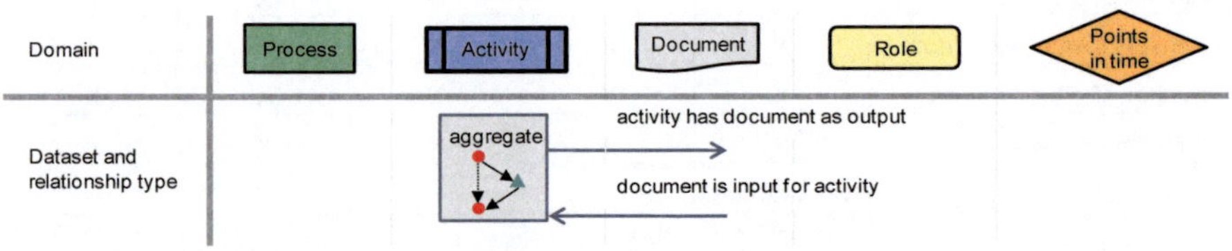


*Figure 7-5: Metric designator used to relate metrics to datasets; the example here shows the attribution of the aggregate view on tasks for the forerun factor*

Table 7-2 characterizes the overall process. Three domains are regarded as the four aggregate views that were agreed upon with the project partners during the kickoff meeting. The process network is quite large, with 710 nodes in total and all distinct, i.e., no nodes are re-used or instantiated several times; however, this information is deceiving, as stated before, because iterations are modeled not as a repeated set of activities but as a second independent process. However, due to the modeling (all data was made anonymous by the company, and the real nature of the entities was only available for a few entities), there are, in fact, activities that are re-used. This can also be seen in the number of edges, which is only slightly larger than the number of nodes for most aggregate views; thus, the process must be quite linear, resulting therefore in a low quota of edges per node. This is also shown by the fact that the relational density is very low. Yet, this picture is partially falsified by the limited quality of the model, in which approximately 15 percent of the nodes are not connected to the model.

*Table 7-2: Size and density of process network*

| | Number of nodes | Number of unconn. nodes | Number of classes | Number of edges | Number of edges per node | Relational density |
|---|---|---|---|---|---|---|
| Modules | 15 | | 15 | | | |
| Documents | 237 | 64 | 237 | | | |
| Documents (via activities) | 237 | | | 435 | 1.8432 | 0.0078 |
| Activities | 377 | 51 | 377 | | | |
| Activities (via documents) | 377 | | | 620 | 1.6445 | 0.00437 |
| Points in time | 96 | 26 | 96 | | | |
| Points in time (via documents) | 96 | | 96 | 135 | 1.40625 | 0.01480 |
| Points in time (via activities) | 96 | | | 65 | 0.6770 | 0.00712 |

From this, a minimal amount of effort for coordination can be deduced, as there is a limited number of interfaces; however, the large number of points in time that are connected to both documents and activities suggests that there is a high risk that these domains are not aligned with each other. Thus, particular regard needs to be paid to a well harmonized organization of these points in time with the needs of the process and the generation of the deliverable documents.

At the same time, the high degree of linearity that the low relational densities point to suggests that there is more potential for concurrent engineering. However, the cascade of consistency in the generation of knowledge about the product (reflected in the linearity of the generated documents) suggests a straightforward and, therefore, easily understandable procedure to generate the documentation.

*Table 7-3: Number of interfaces between domains (native data)*

| Domain | Number of nodes in domain | Modules | Documents | Activities | Points in time |
|---|---|---|---|---|---|
| Modules | 15 | 0 | 0 | 377 | 1 |
| Documents | 237 | 0 | 0 | 279 | 159 |
| Activities | 377 | 0 | 294 | 0 | 296 |
| Points in time | 96 | 1 | 173 | 64 | 0 |

The process model is focused on the activities, as most process models are (compare the description of common modeling methodologies, appendix), as Table 7-3 shows. The activities are strongly linked to documents and points in time; all activities are organized into the different modules, yet not all activities produce documents (or, more generally, any deliverables), and not all are bound to certain points in time. Here, more focus needs to be put on a better model, in particular into integrating the missing edges to generate a meaningful model. The one edge that links one module to one point in time appears to be another modeling error. Again, the low number of interfaces between the different domains points to a possible linear process flow.

The **adjacency** of the structure is reflected in the assessment of the metric **Activity / Passivity,** as shown in the portfolios in Figure 7-6 and following. Here, the majority of all activities has a low activity or passivity in all cases. This takes shape in comparison to the degree distribution (see Figure 7-9), which points to very few activities that are highly influential and that are, at the same time, highly influenced in relation to neighboring activities. There is one activity (node ID 231) that ranks highest in both activity and passivity, while only five other activities (node IDs 14, 78, 81, 227, and 247) make up for the highly critical activities, while all other entities rank considerably lower (left-hand side of Figure 7-6). Thus, these six activities are the core drivers that both coordinate their immediate surroundings and that advance the generation of knowledge of the process. Therefore, these activities merit particular attention during the planning of the process to ensure sufficient resources to lower the risk of generating or distributing errors. However, this picture of the overall process is dominated by one single process module (right-hand side of Figure 7-6) that introduces all of these nodes into the overall process.

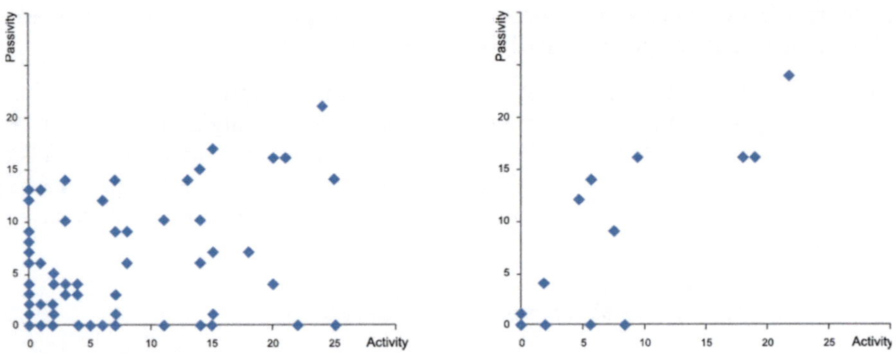


*Figure 7-6: Activity / passivity activities (native) of global process (left) and of a selected module (right)*

The document structure appears similar when compared to the network of activities. Here again most nodes cluster with low values for the activity and passivity metric, and only a few outliers appear. Equally, the aggregate views of documents via activities and of activities via documents (Figure 7-7) are similar, which suggests a good alignment of these two domains. However, each domain has a few outliers that need particular attention, e.g., two highly passive documents (right-hand side of Figure 7-7) that collect more than 20 inputs. As they absorb many possible changes, they are highly error-prone and can possibly be seen as intermediate results of the process.

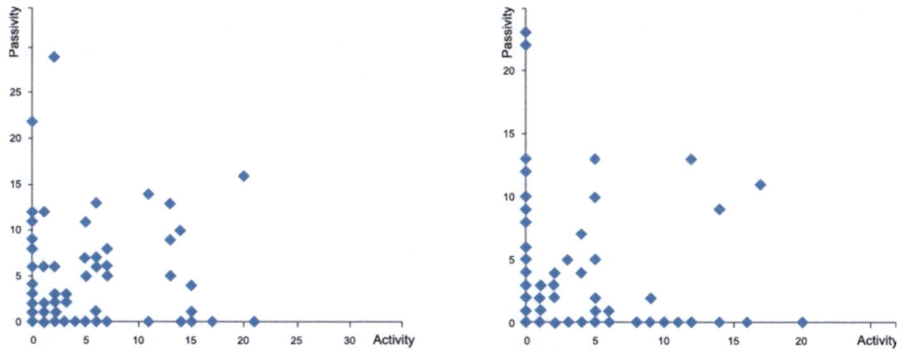


*Figure 7-7: Activities via documents (left) and documents via tasks (right)*

The picture for the points in time (related either via activities or via documents) is spread evenly; there are a number of points in time that serve as starting points (passivity zero) or end points (activity zero) of the process. Those points in time that have a high activity also exhibit a high passivity measure; therefore, these points in time (IDs 32 and 47, noticeably appearing in both parts of Figure 7-8 in the top right corner of the portfolios) are the most important drivers to coordinate the process. Reaching these points in time without delays is, therefore, especially important for the timeliness of the overall process. By recombining and spreading information, however, these points in time are highly susceptible to delays and errors, as they need to serve as buffers (i.e., the process only continues if all input data is available and can be transferred to all subsequent entities).

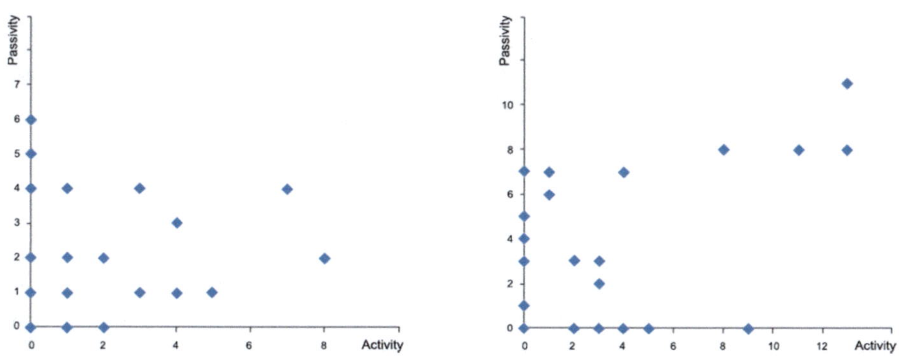


*Figure 7-8: Points in time via activities (left) and via documents (right)*

The **degree distribution** renders the picture more precise. Figure 7-9 shows the distribution of degrees of all activities (via documents) taking shape as a scale-free network. Most nodes have a low degree, and only a few are related to more than one other node actively and passively (i.e., most nodes only have one incident and one outgoing edge). The network therefore has a hub-and-spoke structure and depends highly on a few activities that serve as busses (distributors and sinks at the same time) which ensure the overall transfer of information. These nodes are those that were identified with the activity and passivity metrics.

As these few nodes coordinate the information exchange, the process is unlikely to fail because of random events, e.g., the illness of an engineer, as few activities among all activities are of high importance for the overall process. At the same time, the process is very susceptible to a targeted attack on these nodes, e.g., if the documents relevant to these activities contain errors or if they are deleted on purpose.

Although not shown here, the aggregate view on documents (via activities), for example, exhibits similar properties. This is in line with the results shown for the similar activity and passivity measures, as illustrated in Figure 7-7.

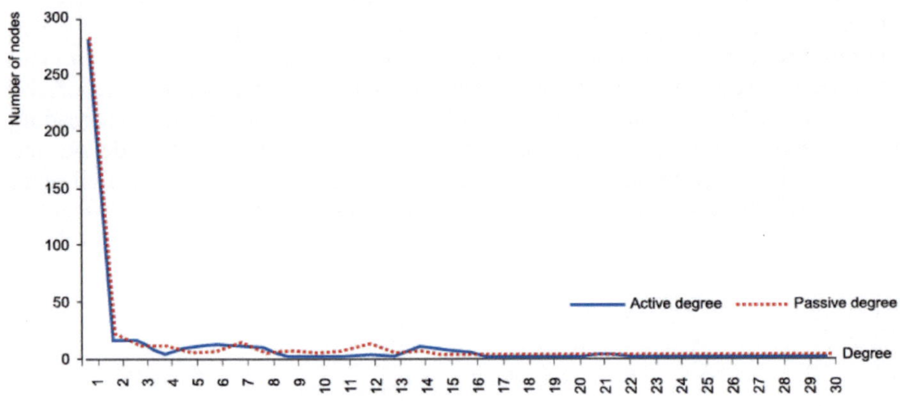


*Figure 7-9: Degree distribution of activities (via documents)*

The **fan criticality** examines the activity and passivity of the modules of the process. Table 7-4 shows the results of the active and passive fan criticality of the activities for the native process network. As can be seen, many modules are minimally connected to the overall process. Some modules (4, 5, 7, 8, 9, and 15) only serve as input or output for the overall process and, therefore, cannot be part of any major iterations. Only three modules have an important input and output relation with the overall process (modules 2, 6, and 10), and, therefore are the most important for the generation of the process outcome, but also the most likely to be delayed or error-prone. At the same time, they necessitate the highest amount of coordination. Again, gaps in the data show, as not all inputs (39 in total, upper row) are balanced by outputs (42 in total).

*Table 7-4: Fan criticality of activities (native) for all modules of the overall process*

| Module | Module 1 | Module 2 | Module 3 | Module 4 | Module 5 | Module 6 | Module 7 | Module 8 | Module 9 | Module 10 | Module 11 | Module 12 | Module 13 | Module 14 | Module 15 |
|---|---|---|---|---|---|---|---|---|---|---|---|---|---|---|---|
| **Fan In** | 0 | 6 | 3 | 0 | 0 | 10 | 4 | 0 | 3 | 5 | 2 | 7 | 0 | 0 | 2 |
| **Fan Out** | 0 | 7 | 4 | 4 | 3 | 10 | 0 | 1 | 0 | 5 | 3 | 2 | 0 | 0 | 0 |

Assessing the **attainability** of the process serves to clarify whether the above analyses, which only regard adjacent nodes, is also true for the mutual impact of nodes across the overall network. Figure 7-10 shows two portfolios for the **number of reachable nodes** and the **reachability of a node**. While the activities (via documents) exhibit a certain degree of clustering in the portfolio, the documents are spread out more evenly. However, the two portfolios use different scales and, therefore, cannot be directly compared.

In the aggregate view on activities (via documents) a few nodes are easily reachable but do not reach any other nodes; these are, therefore, the outcomes of the process. Likewise, some nodes only serve as input for the overall process and cannot be reached and thus cannot be modified during the process. Here, no moving targets are either expected or admitted. There are a few nodes that are both highly reachable and that can reach many other nodes (approximately 25 to 35 each). These activities, again, are central to the information transfer among the activities of the overall process, as they depend highly on other input they share with subsequent activities (ten nodes of 377 nodes: IDs 26, 78, 80, 85, 87, 88, 91, 168, 169, and 170). The high impact of node 78 across the overall network coincides with its local importance, being among the highest activity and passivity measures. All other metrics do not correlate.

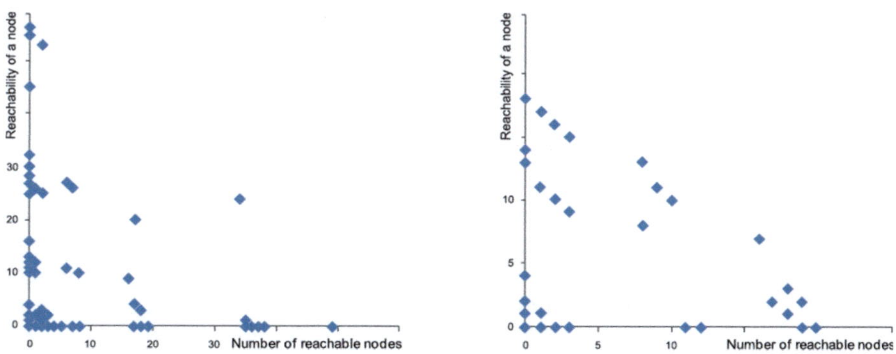


*Figure 7-10: Activities via documents (left) and documents via activities (right)*

Measuring the **closeness** can be done using the **proximity** metric, as shown in Figure 7-11. This proximity puts the attainability of the entities of a process into the perspective of the individual engineer. While the reachability only states if another entity can be reached at all, proximity assesses the distance of this reachability. Here, the measures are normalized to the number of reachable nodes; therefore, a node that only reaches a few nodes in its vicinity, but that can do so using very short paths, will exhibit a high proximity measure.

In the figure, only proximities that are not zero are shown for better visualization. Eight nodes reach a proximity equaling 1 both actively and passively: IDs 14, 59, 60, 61, 83, 98, 114, and 125. Node ID 14 also ranks among those nodes that exhibit a very high activity and passivity and, therefore, is highly influential in its local process module, as it can reach (and be reached by) many other nodes that are situated very close to it. Node ID 78, which appeared as an outlier in the previous metrics, only exhibits a medium proximity (act. 0.69, pass. 0.73), thus is not very centrally located and not as well integrated as suspected. Remarkably, there are many documents (aggregated via activities) that show high active proximity but very low passive proximity. These documents have immediate impact, if changed, on other documents, while depending little on other documents. They can, therefore, be considered very robust in terms of changes.

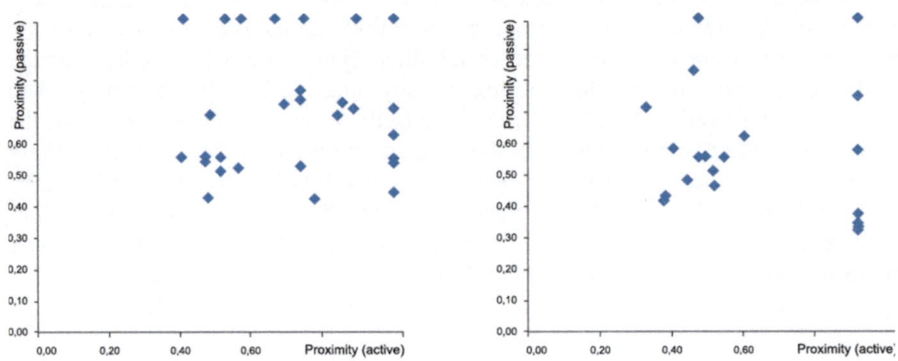


*Figure 7-11: Activities via documents (left) and documents via activities (right), both normalized*

The **relative centrality** assesses a combination of reachability and path length across each node to establish a measure of how central a node is to a network: The more paths that run across a node, the more central it is, if it is simultaneously easily attainable throughout the network. Figure 7-12 shows the histogram of centrality measures for all activities (via documents) of the overall process. Eleven nodes show up as highly central (IDs 20, 92, 114, 125, 156, 171, 191, 216, 221, 252, and 268). These activities are thus most important for the coherence of and communication within the process, serving essentially as a broker for information exchange and the formation of opinions. These activities are, therefore, particularly at risk for errors, as their failure can seriously hinder the function of the process. As a consequence, intensively controlling these tasks is necessary for a positive outcome of the process. These activities also demand attention during process planning. At the same time, there are many activities that are not central to the process. These should be closely examined for their actual contribution to the process, as they are little connected.

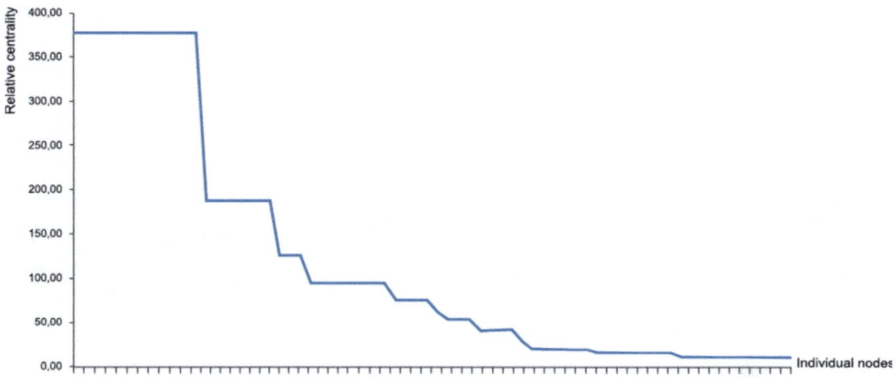


*Figure 7-12: Histogram of relative centrality of individual activities (via documents), only values >10*

**Hierarchies** play an important role in the velocity of propagation of information and errors. Thus, they complement the informational value of the attainability of a node. In contrast to the measure for the number of reachable nodes, the **Snowball factor** weighs the reachable nodes by their distance: The farther away a node, the lower its contribution to the Snowball factor.

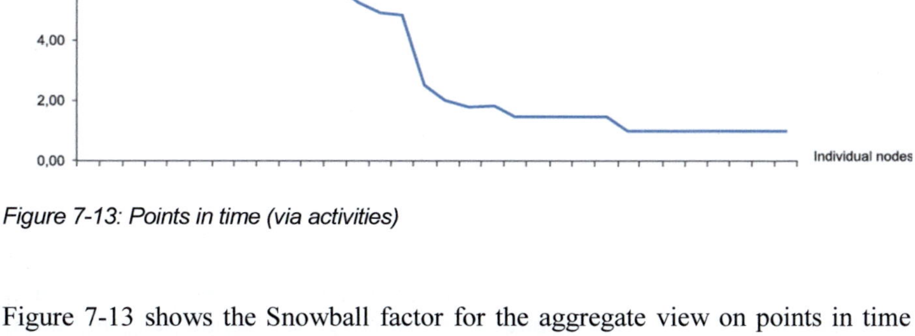


*Figure 7-13: Points in time (via activities)*

Figure 7-13 shows the Snowball factor for the aggregate view on points in time (via activities), and Figure 7-14 regards the aggregate view on points in time (via documents). As can be seen, for the aggregation via activities, very few nodes can be reached; therefore, the process appears relatively robust for a timely execution, as only three nodes (of 96) have a Snowball factor that is greater than six (IDs 47, 56.5, and 32), and twelve further points in time have an average impact on subsequent nodes. Thus, if any of these nodes is delayed, it will rapidly spread this delay over all subsequent points in time.

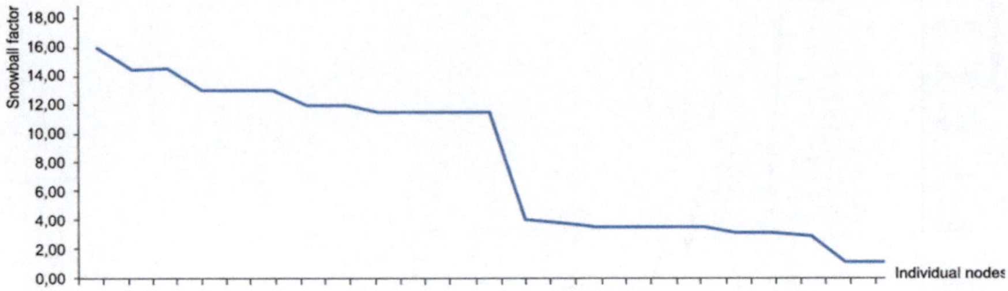


*Figure 7-14: Points in time (via documents)*

Yet, this picture does not hold true if the points in time are aggregated via the documents that are generated as deliverables for the subsequent points in time, as Figure 7-14 shows. Here, twelve points in time have a relatively high Snowball factor (IDs 25.5, 26, 30.5, 32, 32.5, 37, 37.5, 40, 42, 44, 47, and 51), and two of these points in time (IDs 32 and 47) coincide with the aggregate view previously shown. These two nodes are thus most likely to guarantee a timely execution of the process, as they bear the highest risk of delaying a large part of the subsequent points in time.

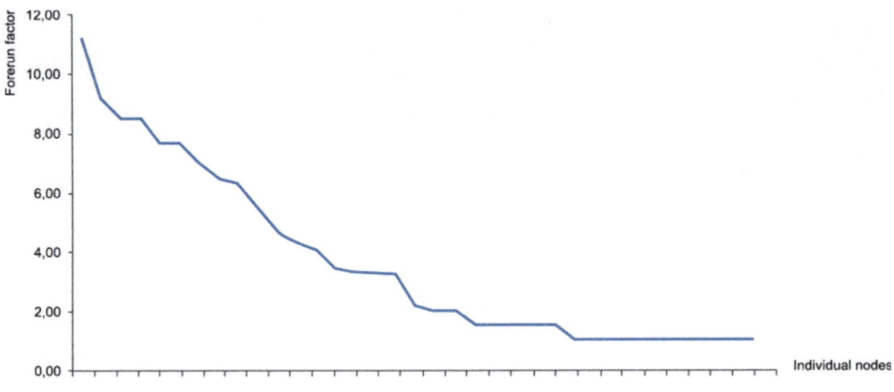


*Figure 7-15: Points in time (via activities)*

The **Forerun factor** similarly regards hierarchies; however, these are incoming hierarchies (i.e., passive ones) while the Snowball factor regards outgoing ones. Figure 7-15 shows a histogram of the Forerun factor for all nodes on the aggregate view for points in time (via activities). Like the results for the Snowball factor, the comparison to the aggregate view on points in time (via documents) yields different results.

While for the aggregation via activities only a few points in time appear to be the most important buffers for delays in the previous process (notably IDs 0, 3, 5, 6, and 32), the aggregation via documents shows again a number of almost equally impacted points in time that are subject to a delay if any previous point in time is delayed (IDs 25.5, 26, 32, 32.5, 33, 37, 42, 47, 51, 57.5, 58, and 60). Again, only point in time ID 32 shows up in both distributions, which indicates that the process planning (i.e., the attribution of points in time to the activities and documents) is not well aligned with the actual process, as these results conflict. On the other hand, point in time ID 32 appears as structural outliers for both incoming and outgoing hierarchies and, therefore, plays a core role in the coordination of the planning. It therefore merits particular attention during planning to ensure the necessary buffers and to install the required measures of process controlling to ensure reaching this point in time according to the schedule.

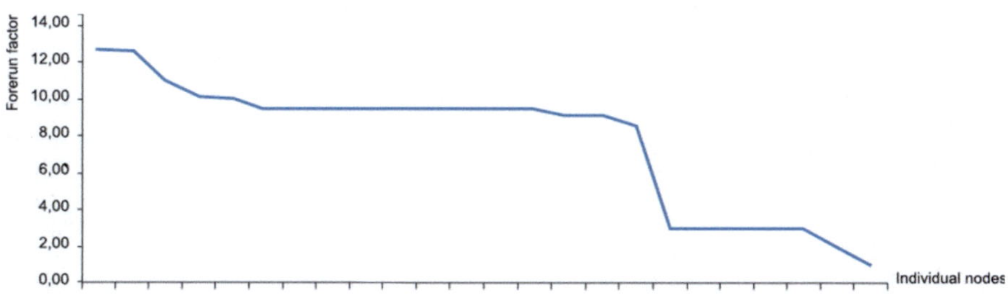


*Figure 7-16: Points in time (via documents)*

An important focus in this case study was to concentrate on iterations that appeared unexpectedly in the process. The measures focusing on **cycles** in the structure (**number of cycles, number of cycles per node**, and **number of cycles per edge**) are both assessed for each process module and for the overall process.

*Table 7-5: Occurrence of cycles in all modules of the process (activities via documents)*

| Module | Module 1 | Module 2 | Module 3 | Module 4 | Module 5 | Module 6 | Module 7 | Module 8 | Module 9 | Module 10 | Module 11 | Module 12 | Module 13 | Module 14 | Module 15 |
|---|---|---|---|---|---|---|---|---|---|---|---|---|---|---|---|
| Number of cycles | 5 | 94 | 0 | 0 | 0 | 84 | 0 | 0 | 0 | 0 | 0 | 0 | 0 | 0 | 0 |

Table 7-5 shows the cycles within each process module; as can be seen, there are a limited number of cycles, and they are of negligible length (two to four, not shown here). This is reasonable, as the partial process models explicitly do not model iterations.

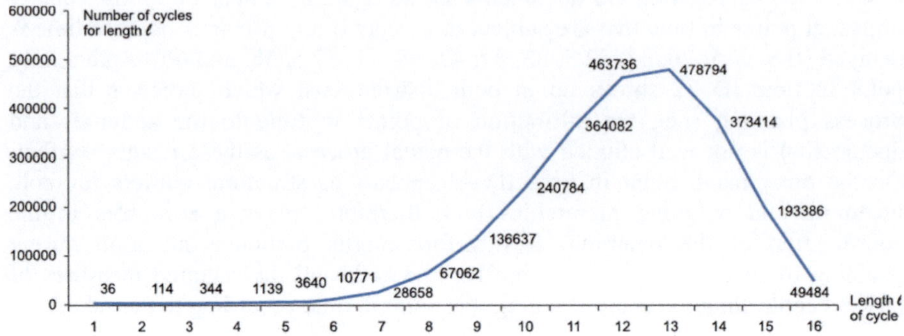


*Figure 7-17: Activities (via documents) for overall process*

However, the process is governed by a large number of cycles. Overall, 2,412,081 cycles can be identified in the aggregate view on activities (via documents) of the overall process. Figure 7-17 shows the number of cycles of different lengths. While there are only a few short cycles (similar to the results in Table 7-5), long cycles up to a length of 16 can be found. A few nodes and edges are the drivers of these cycles; the top five nodes and edges are listed in Table 7-6. As can be seen, the nodes are evenly distributed, each being involved in approximately 200,000 cycles. However, the edge between nodes ID 172 and ID 178 is part of more than two thirds of all cycles and, therefore, of particular importance. The document generating this edge is thus of highest importance to the process.

*Table 7-6: Occurrence of nodes and edges in cycles of activities (via documents)*

| Node ID | Number of cycles per node |
|---|---|
| ID 226 | 2314200 |
| ID 231 | 2276902 |
| ID 243 | 2271232 |
| ID 242 | 2253195 |
| ID 172 | 2122132 |

| Edge (node IDs) | Number of cycles per edge |
|---|---|
| ID 172 - ID 178 | 1453418 |
| ID 23 - ID 255 | 934815 |
| ID 246 - ID 255 | 934815 |
| ID 228 - ID 231 | 845103 |
| ID 244 - ID 239 | 827918 |

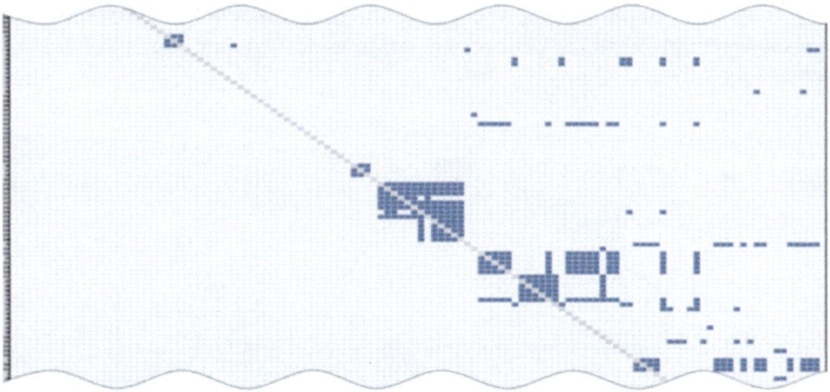

*Figure 7-18: Section of the activity-activity DSM (via documents) exhibiting 74 feedbacks after triangularization*

These cycles are caused by 74 feedbacks (measuring the **number of feedbacks**). Therefore, the process can, in fact, be brought to a highly linear sequence establishing six mostly incomplete clusters that each run the main iterations locally, each being connected to a few other activities.

The graph in Figure 7-19 shows the part of the process graph most impacted by the cycles. In fact, all cycles appear at the interface of two process modules at each time, for example, module 6 on the left-hand side, module 12 on the right-hand side in Figure 7-19. This clarifies why, within the former process organization, iterations were mostly unexpected, as the cycles across the processes were not explicitly known.

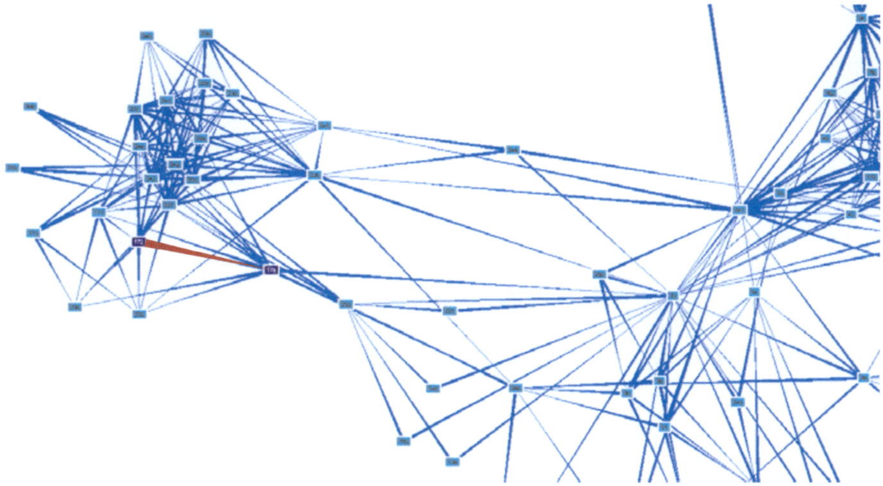

*Figure 7-19: Most occurring edge in all cycles in context of overall process (activities via documents)*

While cycles are most prominent among the activities and documents (not shown here), they also occur among the points in time, as Figure 7-20 shows. This suggests severe problems in process planning, as subsequent points in time cannot precede each other.

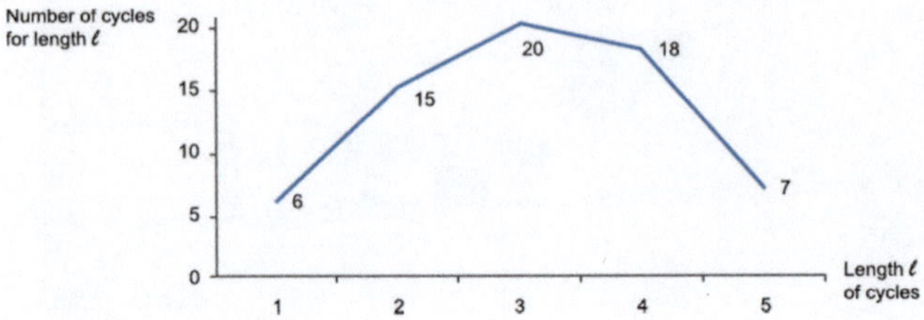


*Figure 7-20: Points in time (via documents): There are 66 cycles among the points in time*

### 7.1.4 Implications and validation

Table 7-7 summarizes the core outliers among the activities of the process being observed. Even considering the low data quality of the process model for which the analysis was conducted, only a few of the 43 activities turned out to be crucial for the function of the overall network consisting of 377 activities. These activities are to be prioritized during process planning and reengineering.

*Table 7-7: Overview of core outliers among activities*

| Node ID | 14 | 20 | 23 | 26 | 32 | 47 | 59 | 60 | 61 | 78 | 80 | 81 | 83 | 85 | 87 | 88 | 91 | 92 | 98 | 114 | 125 | 156 |
|---|---|---|---|---|---|---|---|---|---|---|---|---|---|---|---|---|---|---|---|---|---|---|
| Attainability | | | | X | | | | | | X | X | | | X | X | X | X | | | | | |
| Activity / passivity | X | | | | X | X | | | | X | | X | | | | | | | | | | |
| Proximity | X | | | | | | X | X | X | | | | X | | | | | | X | X | X | |
| Centrality | | X | | | | | | | | | | | | | | | | X | | X | X | X |
| Node in cycles | | | | | | | | | | | | | | | | | | | | | | |
| Part of edge in cycle | | | X | | | | | | | | | | | | | | | | | | | |

| Node ID | 168 | 169 | 170 | 171 | 172 | 178 | 191 | 216 | 221 | 226 | 227 | 228 | 231 | 239 | 243 | 244 | 246 | 247 | 252 | 256 | 268 |
|---|---|---|---|---|---|---|---|---|---|---|---|---|---|---|---|---|---|---|---|---|---|
| Attainability | X | X | X | | | | | | | | | | | | | | | | | | |
| Activity / passivity | | | | | | | | | | | X | | | | | | | X | | | |
| Proximity | | | | | | | | | | | | | | | | | | | | | |
| Centrality | | | | X | | | X | X | X | | | | | | | | | | X | | X |
| Node in cycles | | | | | X | | | | | X | | | | | X | X | | | | | |
| Part of edge in cycle | | | | | X | X | | | | | | X | X | X | | X | X | | | X | |

As the table shows, the listed activities are of a different character, as already seen in the previous section. Only a few appear as structural outliers involving two metrics at the same time. These activities (IDs 14, 78, 114, 125, 172, and 244) are of particular interest to the process. These core outliers, as well as those outliers that showed up during the individual analyses, were reviewed with experts in the company, and the results were found to be meaningful and in line with the expectations of the analysis. Table 7-8 lists the relevant activities and provides a description of what work is conducted at each activity. The activities are ordered in the table as they would take place in the process.

*Table 7-8: Description of activities that showed up during the analysis*

| Entity | Description |
|---|---|
| ID 125 | Definition of functional decomposition as part of the concept for the product architecture |
| ID 47 | Definition of physical interfaces of components as part of overall embodiment concept |
| ID 78 | Initial definition of release-management for software as part of quality assurance |
| ID 178 | Specification of standardized software modules as initial part of software architecture |
| ID 172 | Review of standard software modules for fit and integration as part of software architecture |
| ID 226 | Testing of standard software modules as part of software architecture |
| ID 231 | Testing and validation of overall software in vehicle context as part of software architecture |
| ID 244 | Delivery of standard software to design department and supplier as part of software architecture development |
| ID 32 | Documentation of testing as part of overall quality testing |
| ID 14 | Archiving of verified package as part of configuration management |
| ID 114 | Review for release level as initial part of configuration management |

As the descriptions illustrate, all activities were identified that are core activities of the design of the electronic onboard controllers, including their software. As such, IDs 125 and 47 define the core architecture, ID 78 initializes the release management used to advance the progress of the software code, IDs 178, 172, 226, and 231 are concerned with the standard modules of the software (including testing), and IDs 32 and 114 document the testing process and prepare the release into different configurations of the final vehicle.

In regards to their **degree** (activity, passivity), IDs 125, 47, 78, 14, and 32 are most relevant. Of course, the initial design of the architecture (IDs 125 and 47) both impacts many directly subsequent activities and accesses the knowledge of many other activities; therefore, these activities are at the center of an initial hub-and-spoke-like process structure. During the process, the onboard-bus and its communication interfaces are defined. Therefore, if, for example, a signal is wrongly placed, it is highly likely that the final controller is not functional. Equally, the release-management (initiated during ID 78) is active throughout the process, thus accessing many other activities during the process. Also, the documentation of different tests (ID 32) collects information from various activities, forwarding the results to the relevant engineers. Thus, all of these tasks act as synchronization points during the process. Similarly, the archiving of verified releases (ID with matching documentation, numbering, and versioning) is relevant for all successive tasks and, therefore, shows a high degree, i.e. having many connections to other tasks. During this task, software and hardware are also checked for compatibility.

IDs 125, 78, and 114 also exhibit a tight integration into the overall process, as they not only impact their direct neighbors but are **attainable** throughout the network. The early definition of the functional architecture (ID 125) sets the cornerstone for the subsequent development of the controller. Therefore, it has a high proximity to all other activities, as they detail the functions for serial production. The release management (ID 78) is not only accessed directly by many activities, but its definition is used throughout the process to access, assemble, and control the maturity of different parts of the software and the overall controller design; therefore, its high values for reachability and number of reachable nodes make sense. Lastly, ID 114 indicates a very high centrality. In this activity, the developed controller is reviewed for the fulfillment of the required functionality, and thus the decision as to whether it has sufficiently matured to be embedded in the overall vehicle is finalized. Thus, it brings together all strands of the development process and may start an iteration where necessary. As many activities rely on archived (and therefore released) software revisions, the archiving of software releases also plays a central role.

Lastly, IDs 172, 244 and 226 appear in most cycles, and IDs 172, 178, 244, and 231 are part of the most important information transfers in all **iterations**. The review of software maturity (ID 172) focuses on controlling an externally developed software code that is generated by a supplier and tested at the car manufacturer for the fulfillment of all requirements; therefore, many iteration cycles are run between the actual programming and the testing. The same applies for the overall testing (ID 226). Additionally, changes in the software that are necessary due to findings during testing (ID 231) can necessitate a change in the requirements (ID 178); a review of the requirements may be necessary for the verification of each test. Finally, ID 244 is the task during which externally developed standard software is integrated into those software packages that are developed in-house. The compatibility between these different software packages is highly relevant for a later integration into the vehicle and, therefore, is an essential part of iterations between software suppliers and the software engineers at the company.

The computation of aggregate views of one single domain that is networked via other domains has drawn attention to problems in process planning, for example, such as different cycles among points in time which, by their nature, need to be in a linear sequence and cannot iterate. This implies that the modeling of iterations as repeated sequences within each process module is not purposeful, as the process still exhibits higher-level iterations. This, however, was originally not expected, as the process modeling implies. The reason is that the process, although well-structured into modules, exhibits iterations across the interfaces between the modules, which neither the engineers involved nor the process modelers were apparently suspecting.

### 7.1.5 Reflection

As the case study on electronic control unit development shows, the overall approach works well. As confirmed by the engineers in the company, the core entities of the process and the driving influences were identified precisely. This confirms, on the one hand, the value of structural metrics and, on the other hand, the identification of core drivers of the process by using the concept of structural outliers.

Apart from the metrics on the size and density of the network structure, particularly the metric “degree distribution” (which encompasses, in principle, the activity / passivity and the fan criticality), the centrality, the Snowball and Forerun factors, and the cycle-based metrics (especially the number of cycles per node and per edge) prove to be a good means of spotting entities that are of relevance for the network. Also the structural significance of these metrics could be verified for the domains that were reviewed (activities, documents, points in time) using the aggregate, rather general relationship types, of the model. Obtaining these results confirms the gain that is obtained by using aggregate views for managing the multipartite nature of the network.

The other metrics that were applied (closeness, reachability and number of reachable nodes, number of cycles) showed good results, too; however, these did not deviate much from the results obtained through the other metrics and, therefore, contributed less additional insight. The number of unconnected nodes, finally, helped the quality of the model to be judged, but provided little value in estimating the process's behavior.

Nevertheless, these results indicate that the chosen approach of using structural metrics and structural outliers is able to provide viable results with minimal effort in a systematic manner. The fact that the derived behavior of the process, as identified through the metrics, was confirmed by the company, indicated that the research question could be answered for this case study.

The findings thus confirm the hypotheses initially used in this research: The fact that the process consists of interconnected domains is, in fact, clearly visible in the entity-relationship diagram that depicts the domains and relationship types (Figure 7-2). Equally, the identification of structural patterns within this network could be shown, and the second hypothesis can, therefore, be considered viable, as well. Last, the use of outliers could be demonstrated, although only upper-bound outliers were sought.

## 7.2 Automotive design process at Audi AG: Analysis of interfaces

To come full circle, the initial case study is taken up again. As initially explained, the process shown in Figure 1-2 (page 12) shows an EPC model of the design of the body-in-white of a premium class sedan at Audi AG. This model is now further explored in the following. The model is also part of this book and available through the Springer website[80].

Audi AG is a major German automotive manufacturer, catering especially for the premium segment in nearly all markets. The development process for each car development project follows the overall process specifications, consolidated in a high level process standard. This process is, for each project, broken down into a process and a project plan. The model used here was generated ex-post, i.e. when the project in focus was almost finished. The process regards the development process of a new mid-size sedan, focusing especially on one derivative from the main platform.

[80] See http://extras.springer.com/ for the dataset

The process model was established in the context of a larger project on the improvement of communication between design and simulation departments, as lined out in section 1-2.

### 7.2.1 Goals and focus of the project

To remain consistent with the goals of the overall project, which was focused on the improvement of communication between design and simulation departments, the goal "Interfaces" was chosen for the analysis of the process. Table 7-9 lists the three questions and the 18 assigned metrics belonging to three domains.

*Table 7-9: Questions and assigned metrics for goal "Interfaces"*

| Questions for Goal "Interfaces" | Metrics | Domains |
|---|---|---|
| Which entities of the process need to be synchronized? | • Number of unconnected nodes<br>• Degree correlation (nodes)<br>• Fan criticality<br>• Synch. points / distrib. points<br>• Snowball factor<br>• Forerun factor | • Organizational units<br>• Tasks |
| How fast is communication in the process? | • Number of reachable nodes<br>• Reachability of a node<br>• Proximity<br>• Relative centrality<br>• Snowball factor<br>• Forerun factor | • Organizational units<br>• Tasks |
| What are relevant communication channels? | • Degree distribution<br>• Number of paths<br>• Path length | • Artifacts<br>• Organizational units<br>• Tasks |
| | • Weight of an edge<br>• Centrality of path (centrality)<br>• Centrality of path (degree)<br>• Number of cycles per edge<br>• Number of feedbacks | |

Before the application of the metrics, each metric was reviewed concerning its computability and the relevance of the results. As shown in appendix 10.6, some metrics are too complex to calculate for large models, as the computation time of available algorithms is often not proportional to the number of nodes. In this model, this refers to path-based metrics. As there are many paths possible across the overall process, path-based metrics could not be computed for the example; therefore, such results are excluded. Similarly, only an estimate of cycles was possible due to the high degree of crosslinking in the model, which was calculated for a simplified model, from which all entities were removed that clearly do not contribute to the core process (i.e., the strategic process planning during the very early phases and the final calculus of fatigue and endurance). Lastly, instead of the metrics "Synchronization points / distribution points", the metric "Activity /

passivity" was chosen, as it similarly processes the degree of an entity in order to judge its integration into the network. However, as the process model also contains Boolean operators (54 decision points modeled as OR-connectors), the "Activity / passivity" is more suitable for analyzing the model.

Lastly, the domain "IT systems" was also integrated into the analysis, as the communication within the process strongly relied on the exchange of models between engineering software tools (Computer-Aided Design and Computer-Aided Engineering tools).

Before the metrics are detailed, however, the available model is explained. As the following section will show, no native datasets were available for the domains in question; therefore, different aggregate views were deduced.

### 7.2.2 The process model used

The process chart as initially shown (Figure 1-2, page 12) was build from interviews with 68 engineers in all involved departments (see organizational chart, Figure 1-4) during approximately four months. Each individual interview was modeled, and the model was fed back to the interviewee to verify the partial model before its integration into the overall context. The model was discussed again in a series of workshops with relevant management. The model thus collected and consolidated a considerable amount of knowledge on how to run such an engineering design process.

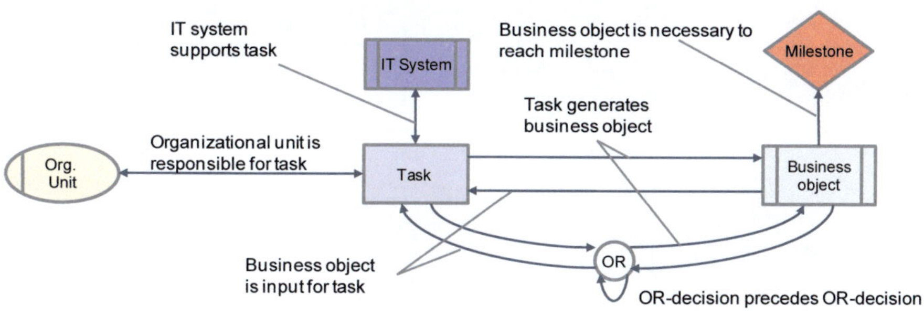


*Figure 7-21: Entity-relationship model of domains and relationship types in the process model*

Figure 7-21 shows its meta-model of available native data for the initial analysis. The actual organizational setup (i.e., the reflexive relationship type between organizational entities as shown in Figure 1-4) was not integrated into the model, as it changed several times during the course of the actual project. As the figure shows, the principal process flow is an alternating sequence of tasks and business objects. IT systems and organizational units are allocated to the tasks. In the native model, these allocations are not directed; for the structural analysis, the undirected relationship types were, however, converted into bidirectional relationship types to enable the method of "path searching" (see page 124) to create aggregate views.

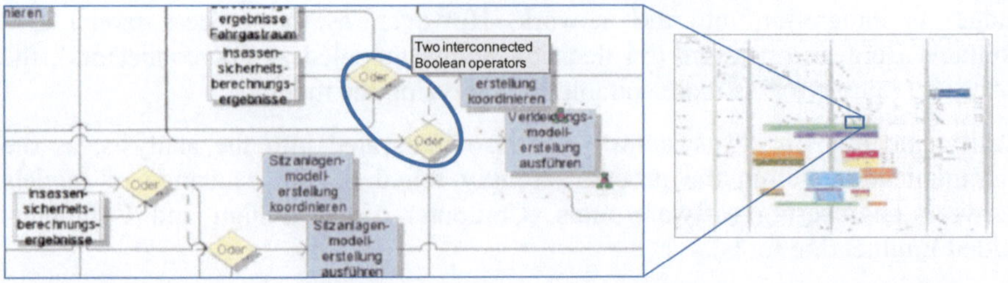


*Figure 7-22: Section of original process model (in German) with two successive Boolean operators*

Furthermore, the model uses Boolean operators in the principal process flow. However, AND operators are not used, but if a task generates two business objects at a time, two relationships are simply instantiated. If, however, one or the other business objects is created (exclusive decisions are not differentiated, therefore ORs also represent possible XORs), an OR-connector is inserted. Decision points are exclusively splits or joins; if a split-decision follows a join-decision, two successive OR-connectors are used. This occurs only four times in the model. Figure 7-22 provides an example from the original process model.

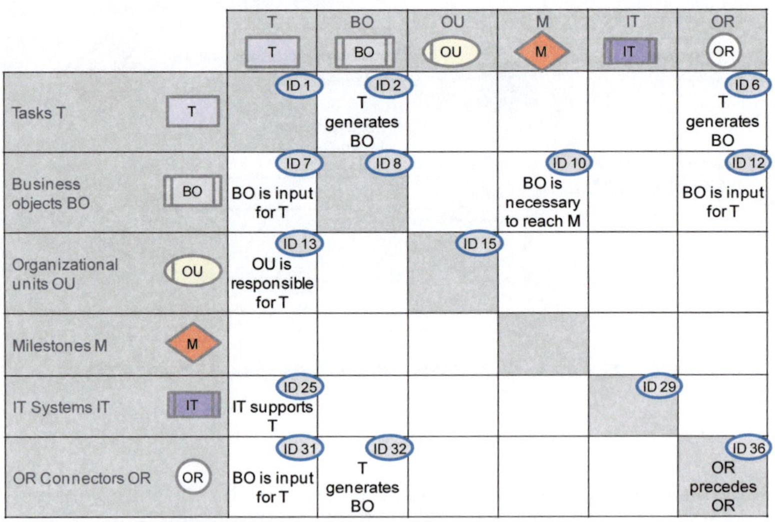


*Figure 7-23: Meta-MDM of the process model including IDs for all partial matrices*

From this data, a MDM was generated, whose meta-MDM is shown in Figure 7-23. IDs 1, 8, 15, and 29 are empty, as these actually are the aggregate views that are needed for the metrics as chosen for the goal "Interfaces". All other DMMs that have an ID are native datasets exported from the original process modeling tool (ARIS Toolset by IDS Scheer AG).

Four aggregate views are needed; each view could, theoretically, be computed in several ways. However, only those aggregations that mimic the actual process execution were chosen; therefore, tasks were aggregated via the intermediate business objects. Similarly, business objects were aggregated via the intermediate tasks. Organizational units and IT systems each were aggregated via the tasks and intermediate business objects.

To explain the impact of the OR-connectors, the aggregation for tasks is further detailed. In the process model, a maximum of two OR connectors occur between any set of tasks and business objects. Thus, three different aggregations are possible, as shown in Figure 7-24: no OR, one OR, or two ORs between a task and a business object, both for the mapping from tasks to business objects and vice versa. For the creation of a complete aggregate view for tasks, therefore, $3^2$ cases need to be combined to create a network that spans all possible combinations, as IDs 1.1a through 1.3a can be combined with IDs 1.1b through 1.3b. Therefore, the aggregate view of tasks calculates as shownin Table 7-10. Here, all three mappings of tasks to business objects are combined with all three possible mappings of business objects to tasks, generating nine intermediate results that each are task-task DSMs. The superposition of these nine DSMs then generates the complete aggregation, which takes shape as a DSM with ID 1.

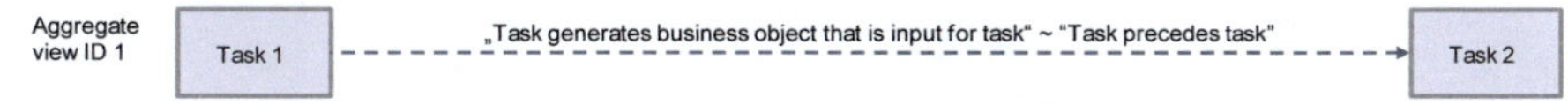


*Figure 7-24: Possible aggregations for different constellations of OR connectors*

The relationship type is the same for all three cases; although connectors are treated as an additional domain, this does not change the relationship type; therefore all three cases can be superposed to generate one overall aggregate view, as shown in Table 7-10. Ultimately, the superposition of all nine intermediate matrices that originate from the computation generate the aggregate task-DSM (ID 1), which includes all paths across all OR-connectors.

*Table 7-10: All possible aggregations for the process model as shown in Figure 7-24*

| | **ID 1.1b (ID 7)** | **ID 1.2b (ID 12 · ID 31)** | **ID 1.3b (ID 12 · ID 36 · ID 31)** |
|---|---|---|---|
| **ID 1.1a = (ID 2)** | ID 1.1a · ID 1.1b = ID 2 · ID 7 | ID 1.1a · ID 1.2b = ID 2 · ID 12 · ID 31 | ID 1.1a · ID 1.3b = ID 2 · ID 12 · ID 36 · ID 31 |
| **ID 1.2a = (ID 6 · ID 32)** | ID 1.2a · ID 1.1b = ID 6 · ID 32 · ID 7 | ID 1.2a · ID 1.2b = ID 6 · ID 32 · ID 12 · ID 31 | ID 1.2a · ID 1.3b = ID 6 · ID 32 · ID 12 · ID 36 · ID 31 |
| **ID 1.3a = (ID 6 · ID 36 · ID 32)** | ID 1.3a · ID 1.1b = ID 6 · ID 36 · ID 32 · ID 7 | ID 1.3a · ID 1.2b = ID 6 · ID 36 · ID 32 · ID 12 · ID 31 | ID 1.3a · ID 1.3b = ID 6 · ID 36 · ID 32 · ID 12 · ID 36 · ID 31 |

This procedure similarly applies to the other possible aggregate views. Business objects can be aggregated via intermediate tasks and OR connectors, creating a DSM with the ID 8, and organizational units can be aggregated via tasks as ID 29. Therefore, for organizational units, the aggregation shown above can be applied and extended to create an organizational unit DSM. The same can be done for IT systems for the matrix with ID 36. All of these networks are fully coherent for the case study, as the model quality of the initial process map is of very good quality. Figure 7-25 visualizes the task-network as a strength-based graph. As can be seen, most tasks are well integrated into a general body that iterates and does not exhibit any clear structure at all, while a few tasks stick out as start- or end-nodes. At the very center, the process revolves around task AC 65, which will later be identified as one of the core tasks (the coordination of simulating the crash of the vehicle).

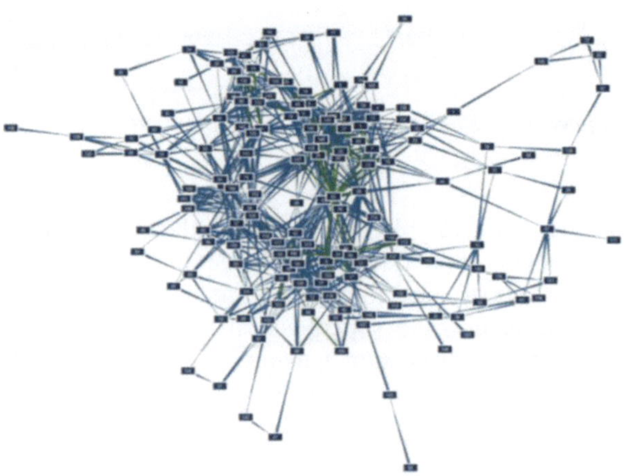

*Figure 7-25: Visualization of activity network (via business objects)*

Overall, four aggregate views are thus calculated, which are summed up in Table 7-11. Here, the basic metrics which show the size of each resulting network are also given. The domain "organizational units" in particular sticks out. Here, all entities are densely networked.

*Table 7-11: Overview of computed aggregate views*

| **Aggregate view** | Tasks | Business objects | Organizational units | IT Systems |
|---|---|---|---|---|
| **Aggregation via** | Business objects and OR-connectors | Tasks and OR-connectors | Aggregate view on tasks | Aggregate view on tasks |
| **Number of nodes** | 160 | 134 | 14 | 27 |
| **Number of edges (multigraph)** | 799 | 392 | 92 (931) | 284 (1506) |
| **Edges per node (multigraph)** | 4.99 | 2.98 | 6.57 (66.5) | 10.52 (55.78) |
| **Relational density (multigraph)** | 0.031 | 0.022 | 0.51 (5.11) | 0.40 (2.14) |
| **Unconn. nodes** | 0 | 0 | 0 | 0 |

This dense network poses a large problem for the metrics, as these are designed to assess the existence of a relation between two entities. However, in the aggregate view "organizational units", multiple paths between any two entities exist, i.e., the resulting graph is a multigraph with up to 80 edges between two nodes. If this multigraph is converted into a binary DSM (i.e., redundant edges and reflexive relations along the diagonal are removed), 85 edges remain. The problem occurs with the other aggregate views, too, but to a much lesser extent. Tasks only exhibit three multigraphs of magnitude 3 and thirty-two of magnitude 2. Business objects have 119 extra paths if the multigraph is not converted into a binary DSM. Their conversion into a binary DSM, therefore, does not change the overall quality of the network to a large extent (as a comparison, the largest degree of a node is 32 for a binary DSM for tasks). Therefore, binary matrices are used that represent only the existence of a relationship and neglect multigraphs. Only organizational units and IT Systems are significantly impacted by multiple paths; therefore, Table 7-11 provides both values for the binary DSM and for the multigraphs for these two aggregate views.

### 7.2.3 Analysis and findings

As Table 7-12 shows, those metrics that detail the goal "Interfaces" were selected for this case study. The metric "Proximity" was omitted in the case study, as its message is transmitted, in part, by the combination of the reachability and the centrality metrics. Furthermore, the cycles could only be estimated, as the large and densely connected network was not computable[81] by any reasonable effort (simulations always crashed after 120 GB temporary files were written). Therefore, only the cycles for the aggregate view on tasks were estimated by computing the maximum computable cycle length of 11. Equally, the development of cycles and the involved nodes and edges for this length confirm with their trend the estimate that is used here. Similarly, cycles for artifacts were estimated up to length 15.

*Table 7-12: Overview of metrics calculated in this case study and assigned domains (all aggregate views)*

| | Tasks | Bus. Objects | Org. units | IT systems |
|---|---|---|---|---|
| Activity / passivity | X | X | | |
| Degree distribution | X | X | | |
| Fan criticality | X | | | |
| Number of reachable nodes | X | X | X | X |
| Reachability of a node | X | X | X | X |
| Relative centrality | X | X | | |
| Snowball factor | X | X | | |
| Forerun factor | X | X | | |
| Number of cycles | X | X | | |
| Number of cycles per node | X | X | | |
| Number of cycles per edge | X | X | | |

Appendix 10.9 (page 398) lists the complete results from the case study regrouped into four tables for each aggregate view. Here, only selected results are shown. These are regrouped by the three questions suggested by goal G01 "Interfaces."

**Which entities of the process need to be synchronized?**

Overall, the process exhibits a strong hub-and-spoke-like structure, as the degree distribution for tasks shows (Figure 7-26, left-hand side). There are many tasks that are minimally connected, while only a few nodes have a high degree of up to 32. These tasks especially are the main hubs that drive the synchronization. The degree correlation chart provides further insight. Here, mostly lower degrees correlate, and most are connected to a few nodes of medium degree (10 to 14).

[81] See also appendix 10.6

Interestingly, there is only one major synchronization point that shows both a high incident and an outgoing degree. Task AC 65 ("Coordinate simulation of crash") has 16 incident and 30 outgoing edges. All other tasks that serve as collection or distribution points do not have both functions at the same time. Therefore, communication at these points can be coordinated with comparably low effort (e.g., using checklists for the tasks in question, such as task AC 32 "Release cockpit" or AC 66 "Simulate crash"), as they mostly channel communication. However, the improvement of the process should be centered on AC 65 in a more detailed way, as it impacts many other tasks simultaneously.

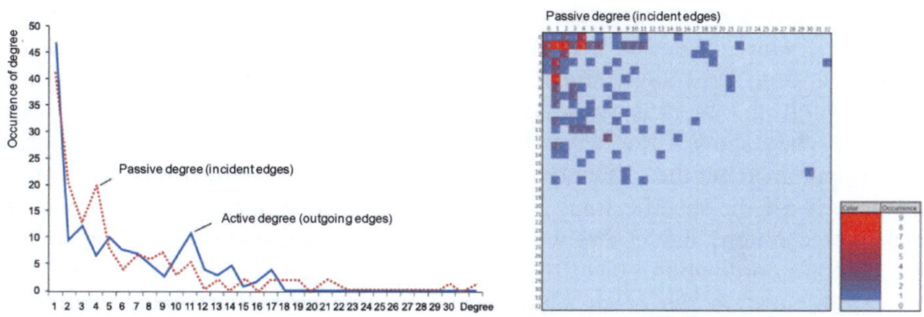


*Figure 7-26: Degree distribution (left) and correlation (right) for aggregate view on tasks (via artifacts)*

The other metrics confirm the central role of task AC 65. It also shows the highest results for the metrics concerning its relative centrality (top outlier), the snowball factor and the forerun factor (second-ranking outlier in both), and it is involved in most iterations (top outlier). Therefore, it can be deduced that this task transports an important part of the core knowledge about the development of the body-in-white, which is often about safety aspects in case of an accident.

This is also confirmed by the fact that the simulation of a crash (AC 66) has an equally high impact not only in the analysis above as to its degree (it shows up as highly passive, collecting 17 different input artifacts into one overall simulation model), but also in its snowball factor (top outlier), which indicates that throughout the process, the results from this task significantly impact the subsequent development. Therefore, when improving this task, subsequent tasks and their stakeholders not only need to be integrated (only one task is actually directly connected in its wake), but all tasks that rely on the actual results from this simulation. In fact, task AC 66 is able to reach 109 different other tasks of all 160 tasks in the process, thus impacting the process to an important degree. Nevertheless, this ranking of reachability is only an average value, as the initial tasks of the process are able to reach all tasks in the process. As a standalone-value, the reachability is, therefore, only of limited expressiveness.

Furthermore, task AC 43 "Setup simulation model for crash" also shows up with the top outlier for the forerun and is significantly involved in iterations (third outlier in cycles, most important outlier among cycles per edge, i.e., communication channels across which iterations are run). During this task, the simulation model for task AC 66 is prepared, collecting many artifacts that are generated in the antecedent process. This task, therefore, has the potential to detect errors that occurred at an earlier stage and that will show up during the integration into an overall model. This integration takes shape in a comparatively low snowball factor, i.e., the information is channeled in a single stream and only spreads out throughout the process at a later stage.

Like the number of unconnected nodes, which is zero, the fan criticality is similar for all sub-processes present in the process. Therefore, no additional information can be deduced from this metric.

Although not shown here, the aggregate view on artifacts exhibits similar properties, which center on the simulation models, especially the crash task, and the results from these simulations. The simulation results for the crash task do, in fact, exhibit the highest snowball factor and the highest relative centrality; however, they show a low value for the number of reachable nodes, indicating their importance but the difficulty of accessing the results. Furthermore, important documents are the specifications that impact many subsequent tasks, as expected, and the technology model that was mentioned in the introduction. This model is the central coordination object that collects all changes throughout the process. Therefore, next to the crash simulation data, this artifact is among the most important for better communication among the departments involved.

**How fast is communication in the process?**

The speed of communication among the tasks was already explained, in part, with the previous question that relies equally on the snowball factor and forerun factor. In addition, the communication among the departments involved is considered. To do so, the aggregate view on organizational units is computed via tasks and artifacts, thus assessing the direct information exchange among the departments involved. Here, it is not the development department but the simulation department, which is responsible for safety applications, that is revealed as the most important outlier for its centrality. This indicates that this department is the driving force to settle conflicts among components and create the central opinions and final concepts of a large part. This is confirmed by the metric snowball factor, for which this organizational unit equally scores highest. Therefore, the department is well embedded in the process. However, it cannot easily generate these results, as the low forerun factor points to a high effort for the collection of relevant information. This analysis also holds true for the other two simulation departments that are involved, although these two organizational units show lower values in all metrics.

By contrast, the body-in-white department is the classic engineering department, producing the sheet metal design of the vehicle's body. Although not as central to the overall process (i.e., a low value for its centrality), this department is much better integrated into the process, showing a well-balanced snowball and forerun factor; therefore, the efficient transfer of both input and output information is much better assured.

This picture is somewhat archetypical for such development scenarios, as an empirical study found in [HERFELD 2007] shows. While the development of the product's functions is actually run in the simulation departments, the "classic" embodiment design engineers are still seen as the driving force in a process, thus the process is centered on their work.

The graph of the aggregate view on organizational units (Figure 7-27) confirms this picture. Clearly, the safety application department takes a central role, while the departments focused on mechanical component design exist as outsiders.

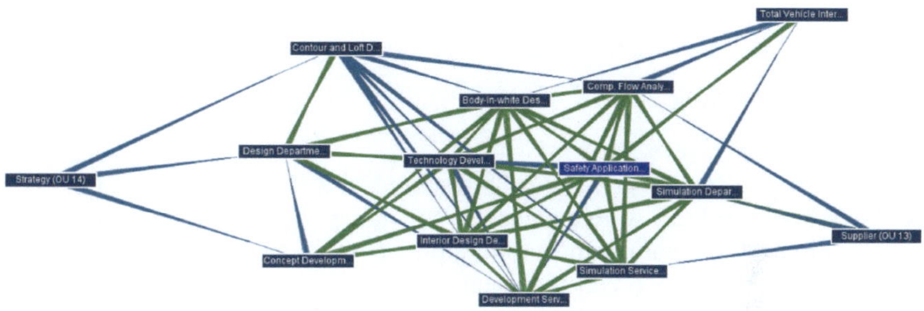


*Figure 7-27: Relations among organizational units via the principal process flow (i.e., via tasks and artifacts)*

**What are relevant communication channels?**

The computation of path-based metrics could not be calculated due to computational reasons; therefore, the relative centrality was used to deduce indications. The original aggregate view on departments, although no metric per se, also provides further insight into the communication structure.

Figure 7-28 shows the computed aggregate view on organizational units, as it results from the necessary matrix multiplications. As can be seen, the matrix presents a multigraph, i.e., the values in each cell indicate the number of paths between each pair of organizational units. Here, for example, the simulation department (OU 10) has 38 different communications channels to the body-in-white design department (OU6); however, when examined the other way around, only 16 communication paths exist. This confirms the picture from the previous question, stating that collecting information to build simulation models demands a high effort, while the dissemination of the results is much easier.

| | OU 1 | OU 2 | OU 3 | OU 4 | OU 5 | OU 6 | OU 7 | OU 8 | OU 9 | OU 10 | OU 11 | OU 12 | OU 13 | OU 14 |
|---|---|---|---|---|---|---|---|---|---|---|---|---|---|---|
| Concept Development Department (OU 1) | | 2 | | 13 | | 1 | 1 | | 1 | | | | | |
| Design Department (OU 2) | | | | 1 | 4 | 2 | 2 | | | | | 4 | | |
| Total Vehicle Intergration Department (OU 3) | | | | | | 1 | | | 1 | | | | | |
| Technology Development Department (OU 4) | 8 | 6 | | | | 24 | 20 | 12 | | 19 | 27 | 5 | | |
| Contour and Loft Department (OU 5) | | 3 | | 3 | | 2 | 3 | 2 | | | 1 | 2 | | |
| Body-in-white Design Department (OU 6) | 5 | 4 | | 16 | | | 4 | 11 | 1 | 16 | 27 | 3 | | |
| Interior Design Department (OU 7) | 7 | 6 | | 15 | | 3 | | 6 | 1 | 23 | 18 | 10 | | |
| Comp. Flow Analysis Department (OU 8) | | | 2 | 2 | | 18 | 8 | | 6 | 37 | 27 | 7 | | |
| Safety Applications Department (OU 9) | 2 | 1 | 2 | 10 | | 11 | 19 | 12 | | 32 | 19 | 3 | 3 | |
| Simulation Department (OU 10) | | | 3 | 9 | | 38 | 25 | 34 | 26 | | 46 | 15 | | |
| Simulation Service Provider (OU 11) | | | | 3 | | 12 | 5 | 17 | 14 | 42 | | 4 | | |
| Development Service Provider (OU 12) | | | | | | 4 | 11 | 4 | | 13 | 12 | | | |
| Supplier (OU 13) | | | | | | | | 3 | 6 | 8 | 3 | | | |
| Strategy (OU 14) | 2 | 2 | | | 2 | | | | | | | | | |

*Figure 7-28: Multiple paths between organizational units (aggregate view as multigraph, via tasks and artifacts)*

Additionally, the interfaces among IT resources[82] were examined. The necessary interfaces between them were deduced as an aggregate view via tasks and business objects. Figure 7-29 shows the multigraph of interfaces that can be computed; again, the number in each cell represents the number of paths, thus indicating the need for an interface. As can be seen, IT systems RE 4, RE 6, RE 12, RE 16, RE 22, and RE 27 are of high importance; all except RE 27 (the text editor that is used to customize simulation models before they are submitted to being solved) are directly related to the transfer of geometry data into the setup of simulation models. Here, the need for interfaces can, therefore, be directly deduced.

| | RE 1 | RE 2 | RE 3 | RE 4 | RE 5 | RE 6 | RE 7 | RE 8 | RE 9 | RE 10 | RE 11 | RE 12 | RE 13 | RE 14 | RE 15 | RE 16 | RE 17 | RE 18 | RE 19 | RE 20 | RE 21 | RE 22 | RE 23 | RE 24 | RE 25 | RE 26 | RE 27 |
|---|---|---|---|---|---|---|---|---|---|---|---|---|---|---|---|---|---|---|---|---|---|---|---|---|---|---|---|
| Abaqus (RE 1) | | 4 | 4 | 3 | 3 | 3 | 1 | 2 | 4 | | | 2 | | | | 2 | 4 | | | | 3 | 2 | | | | | 2 |
| Abaqus Viewer (RE 2) | | | | 17 | | 4 | | | | | | 12 | 2 | 2 | | 15 | | 3 | | | | 15 | 2 | | | | 11 |
| Animator (RE 3) | | | | 25 | | 9 | | | | 1 | 1 | 18 | 5 | 5 | | 22 | 1 | 3 | | | | 22 | 2 | | | | 17 |
| ANSA (RE 4) | 2 | | 1 | | | 2 | | | | | | 8 | 3 | 3 | | 10 | | 1 | 5 | 2 | 1 | 9 | | 4 | | 2 | 9 |
| Ansa Meta (RE 5) | | | | 13 | | 3 | | | | | | 9 | 2 | 2 | | 11 | | 2 | | | | 11 | 2 | | | | 9 |
| CATIA (RE 6) | | 1 | 2 | 31 | 1 | | 1 | 3 | 1 | | | 28 | 5 | 5 | | 28 | 3 | | | 2 | 3 | 28 | 1 | | | | 28 |
| Covise (RE 7) | | | | 7 | | 3 | | | | | | 5 | 2 | 2 | | 6 | | 1 | | | | 6 | 1 | | | | 5 |
| Ensight (RE 8) | | | | 16 | | 5 | | | | | | 11 | 4 | 4 | | 13 | | 2 | | | | 13 | 2 | | | | 11 |
| EVA (RE 9) | | | | 22 | | 5 | | | | | | 17 | 2 | 2 | | 20 | | 3 | | | | 20 | 2 | | | | 16 |
| Phalanx (RE 10) | | | | | | 1 | | | | | 1 | | | | | 1 | 1 | | | | | 1 | | | | | |
| FEMfat (RE 11) | | | | | | 1 | | | | 2 | | | | | | 1 | 1 | | 1 | | | 1 | | | | | |
| Mesher 2 (RE 12) | 2 | | 1 | 3 | | | | | | | | | | | | 3 | | | 5 | 2 | 1 | 3 | | 4 | | | 3 |
| Surface Modeler 1 (RE 13) | | | | 3 | | | | | | | | | | 5 | | | | | | | | | | | | 2 | 2 |
| Surface Modeler 2 (RE 14) | | | | 3 | | | | | | | | | 5 | | | | | | | | | | | | | 2 | 2 |
| Surface Modeler 3 (RE 15) | | | | 1 | | 6 | | | | | | | 1 | 1 | | | | | | | | | | | | | |
| Medina (RE 16) | 2 | | 1 | 5 | | 1 | | | | 1 | 2 | 5 | | | | | | | 7 | 2 | 1 | 6 | | 4 | | | 6 |
| Office (RE 17) | | | | 26 | | 7 | | | | | | 20 | 4 | 4 | | 23 | | 3 | | | | 23 | 2 | | | | 19 |
| Mesher 1 (RE 18) | | | | | | | | | | | | | | | | 1 | | | | | | | | | | | 1 |
| Nastran (RE 19) | | 4 | 7 | 7 | 3 | 8 | 3 | 7 | 5 | 2 | 2 | 6 | | | | 7 | 10 | | | 4 | 9 | 7 | | | | | 6 |
| Pam Crash (RE 20) | | 3 | 4 | 10 | 2 | 6 | 2 | 4 | 3 | 1 | 1 | 8 | 2 | 2 | | 9 | 6 | | | | 5 | 9 | | | | | 7 |
| Pam View (RE 21) | | | | 24 | | 10 | | | | 1 | 1 | 17 | 6 | 6 | | 21 | 1 | 3 | | | | 21 | 2 | | | | 16 |
| Patran (RE 22) | 2 | | 1 | 3 | | | | | | 1 | 1 | 3 | | | | 4 | | | 6 | 2 | 1 | | | 4 | | | 4 |
| Powerflow (RE 23) | | 2 | 2 | 1 | 2 | 1 | 1 | 2 | 2 | | | | 1 | 1 | | | 2 | | | | 2 | | | | | | 1 |
| Permas (RE 24) | | 3 | 4 | 6 | 2 | 5 | 2 | 4 | 4 | | | 5 | | | | 5 | 6 | | | 2 | 5 | 5 | | | | | 5 |
| Concept Modeler (RE 25) | | | | 1 | | | | | 1 | | | | | | | 1 | | | 1 | | | | | 1 | | | |
| Star (RE 26) | | | 1 | 1 | | 1 | 1 | 2 | | | | | 2 | 2 | | | 2 | | | 2 | 2 | | | | | | |
| Text Editor (RE 27) | 2 | | | 3 | | | | | | | 1 | 3 | 1 | 1 | | 4 | | | 6 | 1 | | 4 | | 4 | | 2 | |

*Figure 7-29: Necessary interfaces among IT systems based on an aggregate view on tasks (via artifacts)*

[82] All IT systems mentioned are registered trademarks by the respective companies.

Converting the matrix from Figure 7-29 into a DSM allows the different metrics that elicit the mutual attainability of different IT systems to be computed. For the snowball factor, Pam Crash (RE 20), the central tool for crash simulation, scores highest, thus supporting the previous analysis that the artifacts related to crash simulation are a core element of the efficient collaboration in the process; here, an interface to subsequent systems will raise the usability of the crash simulation results. The necessary interfaces are, above all, to link to RE 4, RE 16, RE 22, and RE 12 (in descending order of importance according to interface matrix, Figure 7-29). In terms of its forerun factor, Ansa (RE 4) is the most important IT system; this tool is used to support the setup of so-called input decks, i.e., the setup of simulation models that are then solved using Nastran, for example. The interface to this tool, therefore, supports the collection of information that was identified as an important key to better collaboration in the process. As the interface matrix illustrates, the tool collects information from many different systems; therefore, the import of geometry data is not only needed, but the management of different partial models are needed to set up a simulation model (e.g., boundary conditions, materials, reference date, etc.). Ultimately, Catia (RE 6) appears as key outlier (upper bound) for its relative centrality. This is not surprising, as it is the core tool to design the sheet metal body of the body-in-white.

### 7.2.4 Implications and validation

The core findings of this case study point to a limited set of tasks, artifacts, organizational units, and IT systems that appear as the most important structural outliers. These findings were reviewed with engineers along with a series of discussions and workshops, and the results largely coincide with the engineers' intuitive understanding of their work and involvement in the process that was reviewed. On the whole, all results were judged meaningful, and the three questions that guided the operationalization of the goal "interfaces" were deemed correct by all engineers. Initially, it was suggested by the engineers in the company that the risk in planning and the consistent transfer of information should be considered as further questions; however, the engineers later dismissed these as too vague to be answered from the structure, as they rely much more on the actual content of the process.

All entities that were identified in the case study were also designated by the engineers independently from the results of the metrics. This confirms the concept of the structural metrics and the structural outliers, as well as the scope of the goal "interfaces and the related aspects of structural process analysis." However, the order of importance obtained through this case study was judged differently from the results of the outliers, which prescribe a certain priorization of the entities. In contrast to the structural outliers, the three most important tasks were identified as:

1. Support development of body structure
2. Coordinate simulation of crash
3. Coordinate simulation of passenger safety

All of these tasks also became top upper-bound outliers in this case study; however, the order suggested by the engineers deviated slightly.

Likewise, as important artifacts, the following three entities were designated in the company independently from metrics:

1. Interior lining concept
2. Body structure model
3. Simulation results crash

Again, these artifacts were identified as top outliers, however, in a different order. In discussion, the reasons for this different weighing in industry were that all engineers had more background information available on the entities that did not show up in the structure, such as the actual informational content of each artifact, the cost determination of a particular task, or the criticality of timing of information availability. Furthermore, certain political aspects and their own involvement in certain tasks caused the engineers to weigh the entities according to other standards, not just the structure of the process.

### 7.2.5 Reflection

As already seen with the first case study (see section 7.1.5), the results can be judged viable and meaningful, which again confirms the concept of the structural metrics, the identification of important entities using structural outliers, and the underlying MDM-based modeling.

The process modeling proved to be very useful in the analysis of the process model in EPC notation that already existed. The process model was assembled from a dataset that was exported out of the original modeling tool (ARIS Toolset by IDS Scheer AG) and completed with further data that was acquired externally. In fact, the organizational structure among the different departments was originally not part of the model, and neither were the interfaces among the IT systems, which were later added as two additional DSMs (not shown in the entity-relationship diagram in Figure 7-21). These were only used later in the project to determine missing interfaces.

The goal-oriented selection of the metrics, domains, and relationship types proved viable, too. Nevertheless, discussions showed that such a S-GQM framework can only be of limited prescriptive use, as each project requires different answers and therefore different metrics to answer these questions. To this end, the framework provided a good starting point for the analysis.

As useful metrics that provided good insights into the process and good structural significance, the combination or reachability / number of reachable nodes and the Snowball / Forerun factor proved to be especially useful in judging how an entity is embedded in the overall process. Furthermore, the degree distribution, encompassing the synchronization / distribution points and the fan criticality, was very helpful to assess the individual impact of an entity in its immediate environment. These measures seem to be the most useful and offer a good first overview if combined with the relative centrality. Also, the number of paths, as found through the computation of aggregate views in the hypergraphs, showed the tight coupling of certain parts of the process and sufficiently answered the initial interest in this process analysis.

Unfortunately, the cycle-based metrics could only be estimated due to limitations in computing power and appropriate algorithmic support. However, the adapted approximation for a lower number of cycles appears reasonable, as computations with a simplified network showed. However, this issue needs further exploration.

## 7.3 Conclusions from the case studies

In general, the results of the case studies verify many speculations that could not be clarified before within the two companies. To this end, the findings from the metrics point to the core drivers of the process, and their individual review points to a set of activities that all appear meaningful. This applies to all domains that were reviewed during the case studies, even if only examples of the results could be shown. Within both case studies, the results were confirmed by engineers in the companies as being correct. Therefore, the metrics have proven their **meaningfulness**[83]. This demonstrates that both the overall concept and the individual metrics are viable means of analysis which generate meaningful results.

Secondly, all metrics that were used have been able to differentiate those entities of higher relevance from those of lower relevance. As was discussed in the findings of each case study, it can be deduced that the **representation** is also purposeful, and the relevant scales provide a sufficient basis to compare the entities within a process. This contributes to the applicability of the hypothesis of identifying structural outliers via the scale of the metrics, which prove viable, as well.

Thirdly, the different results of the metrics indicate that the metrics do not correlate, as they point to different nodes that, for each metric, appear as most important. This makes it possible to conclude that the **uniqueness** of each metric is sufficient to identify different entities with the different structural characteristics that each individual metric represents.

At the same time, it could be seen that the different layers of the overall networks exhibit similar properties (e.g., having hub-and-spoke like structures, being linear, revolving around a few core entities, having a high potential for more concurrent engineering through more intense networking, etc.) but are not well aligned. This was demonstrated in both case studies and across the different domains within each of them. For example, in the first case study, the fact that only 237 documents appear as results of the overall 377 activities underscores this fact. However, the exclusion of iterations in the process modules relativizes this observation, as many activities will, in fact, appear twice or more, although they cannot be identified as such. This verifies that processes can be understood as network-like structures exhibiting different structural patterns.
Finally, the use of the overall S-GQM framework could be demonstrated. As the case study shows, it provides a good starting point for a first analysis; however, its application necessitates detailed reflection to choose metrics, domains, and relationship types for a specific problem. It should, therefore, generally be understood as a guideline which aids in the planning of a process analysis.

[83] The metrics are reviewed based on the core requirements of a good measurement (meaningfulness, representation, and uniqueness), as discussed in section 2.3.1 (page 76).

# 8. Conclusions and outlook

In this section, the results presented in this book are reviewed, followed by a discussion of the strengths and weaknesses of the approach and the implications that can be drawn from them for both academic and industrial applications.

## 8.1 Summary of results

This research was run to design a method that supports the purposeful and systematic description and analysis of an engineering design process in terms of the structure of relations of its entities. The intent of this analysis is to alert a user to possible weak spots that merit further attention for potential improvement. To achieve this, the research is based on the hypothesis that a process is a network of entities and relationships of different types, within whose constellation certain meaningful patterns can be identified that can be related to the behavior of the process. While these structural characteristics occur throughout the process, possible weak spots in a process can be identified by looking at structural outliers, representing the most peculiar patterns in a process.

As a basis of this goal, the review of contributions from system theory, graph-theory, matrix-based methods for structural complexity management, network theory, process management, and software engineering showed that a systematic method for a goal-oriented analysis and improvement of engineering design process is still missing. Yet, existing methods provide a good basis to construct a comprehensive solution based on the needs of process management (i.e., the modeling and the goal-orientation), from the management of structures (i.e., dependency modeling and complexity metrics), and the structured analysis of complex systems (i.e., procedural models and frameworks).

The solution is based on a goal-oriented analysis procedure that guides the complete process of analyzing a structure. It uses three constituents: a modeling method for processes, a comprehensive set of structural metrics to assess the model, and a framework that provides a goal-oriented selection of the structural metrics and the necessary parts of the process that need to be modeled.

The process is modeled using multiple-domain matrices that were extended to incorporate the necessary constructs to fully represent any common process model (i.e., the different domains and relationship types, attributes, and logic operators). The modeling method is supported by a meta-model, the Structural Process Architecture, to facilitate the recombination of different existing process models as well as to guide the information collection when modeling a process.

Based on this meta-model, which also provides the semantics of the structure of common process models, the existing body of complexity metrics was adapted and extended to a comprehensive set of 52 structural metrics, the Structural Measurement System. Each metric is based on a detailed theoretical reasoning in line with measurement theory; furthermore, most metrics are based on empirical evidence. All metrics are completed by a description of their structural significance that supports drawing inferences about the behavior of a process subjected to analysis.

The metrics and modeling of the process are systematized in the general framework, the Structural Goal Question Metric framework. This framework is guided – on the strategic level – by goals common to process management. Each goal is detailed by a set of questions that can be answered using the structural metrics and their structural significance. To do so, each question has been allocated the relevant structural metrics as well as the necessary semantics, represented as domains and relationship types provided by the meta-model.

The overall approach was demonstrated using three case studies from automotive development. All three confirm the applicability of the hypotheses and the viability of the approach as well the individual constituents as shown.

## 8.2 Reflection

In this section, a reflection on the strengths and weaknesses is presented first, followed by the individual implications that can be drawn for the industrial application of the solution presented here and the implications for research.

### 8.2.1 Strengths and weaknesses

The general **analysis procedure** is designed to complement existing approaches for process analysis. Therefore, it is not meant to replace existing methods, which it is not able to do. While existing approaches for process analysis focus mostly on quantitative models, the approach presented using structural metrics investigates qualitative models, i.e., a representation of the structure of a process. Thus, the approach presented is not a method to support process planning as such, as it does not provide any direct means of planning the interaction of any entities of a process. Nevertheless, it is able to provide indications about how the structure of a process will impact its behavior and, therefore, helps clarify the structure of the process plan.

The analysis procedure examines all aspects of analyzing a process. Thus, its completeness is relevant for the approach to be applicable in an industrial context. The goal-orientation, the modeling scheme, the analysis method, and the systematic access to the significance of the results obtained during the analysis are, therefore, integrated as a whole.

The actual implementation of the approach in a company was not reviewed in detail in this research. However, the case studies show that the approach is viable and provides an effective and efficient means of obtaining good results. As it is based on commonly accepted approaches for the methodical analysis of a system, the approach can, therefore, be considered suitable to fill the gap detailed by the initial research question.

The goal of establishing a **Structural Process Architecture** as a structural process modeling method as the first of three constituents to the overall approach is based on the hypothesis that a process consists of multiple layers of a network. This hypothesis was verified through the literature review, which showed how almost all common process models couple the entities of the model not only through the primary flow of information but via other different relationship types, e.g., the organizational structure or IT interfaces. The hypothesis was also shown to be valid during both case studies, showing that indirect relationships via intermediate domains can be condensed meaningfully to provide better insight into such complex network structures.

This Structural Process Architecture was designed using Multiple-domain matrices (MDM), and it serves as a semantic basis for the metrics and as orientation when assembling different existing process models or modeling a process from scratch. The matrix-based modeling is in line with common approaches to the management of structural complexity, using existing modeling schemes, and adapting and extending them to the needs of process management. It is, furthermore, able to represent large systems in a manageable fashion, even though the visualization is not intuitive. Nevertheless, large matrices are hard to handle, and obtaining results always requires extensive tool support, especially spreadsheet software.

The design of the **Structural Measurement System** was undertaken to create a means of systematically obtaining a comprehensive picture of an engineering design process. A metrics-based approach was chosen, as it offers a means of condensing the information into a reduced form which can be easily and systematically applied to all entities and relationships of a complex system. The structural metrics are based on existing metrics taken from comparable environments (especially software programs or workflow design), or on structural characteristics that so far have not been evaluated numerically but are based on previous application in similar systems. In doing so, it was possible to base all metrics on existing empirical evidence as to the validity of their application and on experience about the extent to which interpretations are possible. The viability of the results that were obtained in the two case studies confirms their use and the second hypothesis, which states that it is possible to draw inferences about the behavior of a system by analyzing its structure. This was also confirmed in the literature that was reviewed.

However, the completeness of the approach was not verified from all possible angles. The development of the approach was guided by the identified goals of process management as well as by the spectrum of available structural characteristics that were collected from different disciplines and matched to the domains and properties of processes (such as iterations, workgroups, etc.). However, an inverse approach was not taken, i.e., no structural characteristics and structural metrics were sought starting from a set of process properties that were to be analyzed.

The structural metrics grant access to the behavior of a process without the detailed modeling of the actual behavior by providing an estimate based on structural patterns and their significance. This inference, however, has less depth than other approaches, such as the simulation of a process, a value stream analysis, or path costing. Such methods, although more detailed, provide deeper insight but necessitate a significantly higher amount of modeling effort and focus more on single issues. The question remains unsolved as to whether it is reasonable to expect that enough information can be captured to describe either the structure or the behavior sufficiently "accurately" for analysis. Practical experience indicates that it often takes many months even for a small process to be modeled as correctly as possible. Still, the degree to which such models are "good" remains unknown. Nevertheless, the approach shown here can be used to guide more detailed efforts, as the structural metrics make it possible for such undertakings to focus on more sophisticated methods only as a second step.

To do so, the structural metrics are also able to work with models of average quality. Of course, any analysis should be based on complete modeling of a process. Only if sound data is used, can the results be fully trusted. Yet, process models in industry are rarely complete, mostly because it is either too time-consuming to obtain a complete model or because the procedures that are modeled often cannot be turned into one coherent model as there is no single way that things are done. For such incomplete data, the approach has proven possible, as even for sub-sets of a process relevant results can be found.

Nevertheless, the use of structural metrics is only suitable for large processes, as a certain minimum population of nodes and edges is necessary to obtain results, whose outliers can be evaluated with good quality. The design of the structural metrics has shown that they are not viable for small models of less than approximately 40 entities. However, with larger models, the third hypothesis works well, as the case studies have shown, even though only upper-bound outliers have been sought and evidence for the other three kinds of structural outliers have not been shown.

Unfortunately, no absolute judgment whether a process is "good" or "bad" can be derived from the application of the metrics. However, tendencies are possible, although even this is subject to how each company wants to develop or how things are done in that company. Therefore, generalizing about each metric's structural significance is limited. Overall, the approach still requires a deep understanding of the principles of the metrics to interpret the results correctly, as the structural significance of the metrics is significantly impacted by the chosen domains and relationship types, and no standardized interpretation is possible.

Thus, a high risk exists that the metrics can be misleading if applied by an unskilled user. As a consequence, the application of the metrics necessitates a critical reflection of the application, implementation, and interpretation each time the metrics are used. In fact, a small change in the structure can cause major changes in the results of the metrics [BIEDERMANN et al. 2009]. This risk of misunderstanding the metrics can be accompanied by the risk of manipulation. As the outcome can change significantly for even minor changes within the structure, a user could adapt the results to his personal advantage. Therefore, several "overlapping" metrics should be used at all times to cross-verify the metrics among each other.

Lastly, mathematical inadequacies still exist in the formulation of the structural metrics. There is still the inability to work with very densely populated networks and multigraphs, as the case studies have demonstrated. Furthermore, the computation time for some metrics, especially cycle-based and path-based metrics, is still high, and algorithms are still insufficient; yet, there are only limited means of estimating the metrics, even though in many cases the computation of a complete solution is not necessary (compare the cycles in case study 7.2).

As the last constituent of the solution, the setup of a **Structural Goal Question Metric framework** to guide a process analysis project is designed to aid the selection of appropriate metrics in relation to a chosen goal of process analysis. It was implemented using the GQM scheme. However, certain simplifications were used. In particular, the formulation of goals, which is an important aspect of the original scheme, was not used, as the framework that was designed here is designed to have a wider focus of application. However, the framework was extended to also guide the interpretation, using the structural metrics allocated to the structural metrics, an aspect that is not part of the original GQM scheme.

The benefit gained from the framework was demonstrated in case study 7.2. Although the selection of metrics to be allocated to a goal and its questions is difficult and at times fuzzy, the framework serves as a good starting point for any analysis. In comparison to the current state of the art, the framework is the most complete method in this field of research.

However, the framework is lacking a generic analysis mode. Often, process analysis in industry is done "because there is something wrong", and a general analysis for possible problems is needed. To do so, a basic set of metrics is required that is not part of the framework.

Also, the framework does not provide clear guidance about the use of native and aggregate datasets, but only about the necessary domains, whose targeted selection remains unsolved. Another level could be introduced to better differentiate the use of aggregate views from the use of native data.

With all these constituents, the research question as laid out in chapter 1 can be sufficiently answered. All three hypotheses have proven viable and correct, and the requirements have been met. Still, the inadequacies that were discussed in the above paragraph prevail, hindering, however, only certain areas of the applicability of the method that was developed.

### 8.2.2 Implications for industry

The approach shown changes the paradigm of process improvement to some extent, as engineers in any company need to understand that today's processes are not as linear as they used to be and that they are part of a dense network of activities. While graph and network theory still are too abstract for many individuals to fully understand, the detailed description of different characteristics can help individual engineers to better place themselves in these networks. In design methodology, this trend has already begun to establish itself [GAUSEMEIER et al. 2006].

Thus, management and engineers in industry are offered a new approach that extends current methodologies for systematically analyzing existing processes for possible improvements by looking at the network of relationships across the whole process organization.

The overall approach is oriented to the needs of industrial practice. Despite its high degree of abstraction, it allows processes at a pragmatic level to be analyzed and indications that can directly support process improvement to be deduced. The high degree of abstraction is necessary to analyze a process top down as well as to compare the results of different analyses across several processes to strategically guide process improvement activities. This enables management to base decisions on more than a gut feeling, which is often the primary source of decision in industry [GIGERENZER 2007]. To better support such decisions, using structural metrics can provide a tool similar to a Balanced Scorecard for process improvement activities, as it provides access to the cause-effect relationships in a structure (i.e., spotting an outlier using metrics, then looking into the actual structure of the process).

Unlike many existing approaches in process management, the structural metrics presented in this research require minimal effort in data acquisition and computation for results that point to improvement potentials that – in a second step – can be analyzed further. Thus, the approach presented is relevant, as it is able to rely on existing process models that are already available in many companies. From these models, patterns that govern a process can be extracted, and knowledge about the typical behavior of these processes can be uncovered and submitted to further analysis as to its implications about how a process is commonly run. To this end, it matters little if the process model is not the most recent, as engineering design processes vary little from one development project to the next if the product architectures remain similar. Typically, the patterns (for example, the relevance of certain tasks or the centrality of core product models) will prevail for many generations of a product. Therefore, knowledge about the importance and impact of core entities of a process are directly transferable.

In general, a process analysis using the structural measurement system for engineering design processes can be applied for various purposes:

- Comparing different processes at a given point in time to prioritize the investment of resources into process improvement and rework. For example, a process manager might wish to compare a number of processes he is responsible for. To determine which of them is the most complex and thus bears the highest potential to cause errors, a complexity metric is useful to identify the most complex process to start improvement with.

- Tracing changes over time to schedule possible improvements. For example, an organizational setup may grow more and more complex over time, as new teams are introduced. To trace the degree of complexity, a process engineer can employ structural metrics to better estimate the degree of stability and suitability of the architecture. The metrics can aid quality assurance and the maintenance of such systems.

- Assessing complex process structures at an abstract level to estimate the amount of effort. For example, in project planning, a linear timeline is desirable to guarantee a smooth process execution. If, however, the tasks are interlinked in a way that no ideal sequence can be reached (e.g., as triangularization for a DSM would provide), an analysis of the structure can support the process planner to judge how much effort might be needed for communication during process runtime.

- Identification of improvement potential and of error-prone entities of a process organization. For example, if all information is routed through a person who is highly central to the process, there is a risk that if this person falls ill or changes his or her employer, the process disintegrates.
- Assessment of the human cognitive ability to understand a process. The more complex a process becomes, the more complex the interaction with such a process, and it is not perceived as transparent. Assessing how easily a system can be comprehended (e.g., the flowchart of a process or the various states of a product and their mutual dependencies) can serve to design it better and to judge how users will interact with it.

However, structural metrics do need to be handled with care. Only if they are well-accepted and if their impact is understood, they can be usefully employed. Otherwise, there is a high risk that they will be misleading, as they represent a much reduced picture of the process.

### 8.2.3 Implications for Research

While the implications above are true for research as well, the formal modeling and evaluation of different characteristics of design processes makes it possible to describe in detail what concurrent engineering actually is like. Otherwise descriptions often remain vague. In other terms, the presented research helps making "patterns" in engineering design processes [WYNN et al. 2007] [EPPINGER 2001] [BADKE-SCHAUB & GEHRLICHER 2003] become clearer and accessible.

While there is no "perfect" process, the numerical analysis of processes makes comparisons easier. In the long run, using structural metrics makes it possible to compare a number of processes for characteristics of "good" and "bad" processes. As an intermediate step, the creation of a "footprint" of different kinds of engineering processes is definitely within reach, using a pre-defined set of structural metrics for a standardized comparison, for example, a process that is centered on a few people who are highly knowledgeable, as opposed to a process during which a new product development is undertaken and no detailed know-how is available. This might occur in different task distributions, their different tendencies to rely on iterations, and the changing centrality of staff involved. However, such an analysis will also rely on datasets that model processes at a comparable level of detail with a comparable modeling scope (e.g., how iterations are resolved in the model).

## 8.3 Outlook

Despite the effort invested in this research, some items remain unsolved and represent opportunities for future research.

The formalization of structural metrics was omitted here, in part, to provide a comprehensive overview of the existing basis. However, the metrics were only described textually, and a formalized description remains to be shown. This also includes the completion of the set of necessary algorithms. Using a mathematical description, the mutual interdependencies and correlation of the metrics could also be explored, which would contribute to the classification of the metrics and their allocation to relevant solution principles as well as the framework. A possible long-term vision for this undertaking could be a formal algebra for structural analysis that provides a complete set of properties, rules, and operations that can be done using structural characteristics.

To further classify the metrics, an interesting step would be to approach the problem tackled in this research by "coming from the other side". As the approach was developed by starting from the available means of structural analysis and mapping them to relevant aspects of a process's behavior, a complete solution can only be obtained by classifying all relevant properties and allocating the necessary means of analysis for them.

Next, the use of metrics with logic operators has not been shown, although it has been part of this research [GÜRTLER et al. 2009]. However, even the work that has been done so far is limited to assessing the degree of an entity. Nevertheless, its working principles can be transferred to the attainability within a graph in the same way, opening up its transfer to all other structural characteristics as well. At the same time, such effort needs to be accompanied by the adaptation of rules for interpretation and, more generally, the different structural significance that the metrics bear if they do not consider an existing structure but one that *could* become a structure (in the case of OR or XOR operators).

Furthermore, the methodical management of aggregate views can be extended. In this research, only those aspects of creating aggregate views were explored that were needed in the given context (path-searching for DSMs). The details of creating the other possible aggregate views were not explored. This concerns, in particular, the management of interacting with the emerging aggregate relationship types that are difficult to handle. To this end, action-based research to formulate ontologies which help define suitable, more compact relationship types than the one described on page 127 are desirable. Furthermore, a general framework to guide the goal-oriented aggregation of different domains is still needed, which helps choose relevant inputs for the aggregation under a certain goal and which prescribes a domain of reference, the domains to be integrated into the domain of reference, its relationship types, and the more compact aggregate relationship type. This aggregational framework could be integrated into the GQM scheme to close the existing gap in selecting the right dataset for answering a specific question.

The interaction of the approach developed with the product architecture has only been touched upon, as it is not the direct focus of this research. The alignment of the product architecture and the process architecture has not been regarded in more detail, even though an adaptor to the product was created through the domain "product attribute" in the meta-MDM.

# 9. References

AHN et al. 2007

Ahn, Y.-Y.; Han, S.; Kwak, H.; Moon, S.; Jeong, H.: Analysis of topological characteristics of huge online social networking services. In: Proceedings of the 16th International World Wide Web Conference, Banff, Alberta, Canada, 08.-12-05.2007. New York, NY: ACM 2007, pp. 835-844. ISBN: 978-1-59593-654-7.

AICHELE 1997

Aichele, C.: Kennzahlenbasierte Geschäftsprozessanalyse. Dissertation, Universität Saarbrücken, 1996. Wiesbaden: Gabler 1997. ISBN: 3-409-12173-0.

AKAO 1992

Akao, Y.: QFD - Quality Function Deployment. Landsberg a. L.: Moderne Industrie 1992. ISBN: 3-478-91020-6.

ALBERS et al. 2005

Albers, A.; Burkardt, N.; Meboldt, M.: Spalten Problem Solving Methodology In The Productdevelopment. In: Samuel, A. E. et al. (Eds.): Proceedings of the 15th International Conference on Engineering Design, ICED'05, Melbourne, Australia, 15.-18.08.2005. Barton ACT: Engineers Australia 2005. ISBN: 0-85825-788-2.

ALBERT & BARABASI 2002

Albert, R.; Barabasi, A.-L.: Statistical mechanics of complex networks. Reviews of Modern Physics 74 (2002) 1, pp. 47-97.

ALBERT et al. 2000

Albert, R.; Jeong, H.; Barabasi, A.-L.: Error and Attack Tolerance of Complex Networks. Nature 406 (2000) 6794, pp. 378-382.

ALLWEYER 2008

Allweyer, T.: BPMN - Business Process Modeling Notation. Norderstedt: Books on Demand 2008. ISBN: 978-3-8370-7004-0.

AMERI et al. 2008

Ameri, F.; Summers, J.; Mocko, G. M.; Porter, M.: Engineering design complexity: an investigation of methods and measures. Research in Engineering Design 19 (2008) 2-3, pp. 161-179.

ANDERL & TRIPPNER 2000

Anderl, R.; Trippner, D.: STEP - Standard for the Exchange of Product Model Data. Stuttgart: Teubner 2000. ISBN: 3-519-06377-8.

ANDERSON 2007

Anderson, C.: The Long Tail - How endless choice is creating unlimited demand. London: Random House Publ. 2007. ISBN: 978-1-8441-3851-7.

ANDERSON 1972

Anderson, P. W.: More is different. Science 177 (1972) 4047, pp. 393-396.

ARGYRIS & SCHÖN 1978

Argyris, C.; Schön, D. A.: Organizational Learning: A Theory of Action Perspective. Reading, MA: Addison-Wesley, 1978. ISBN: 0-201-00174-8.

ARGYRIS et al. 1985

Argyris, C.; Putnam, R.; McLain Smith, D.: Action Science: Concepts, Methods, and Skills for Research and Intervention. San Francisco: Jossey-Bass 1985. ISBN: 0-87589-665-0.

ART AND SCIENCE FACTORY 2009

Art And Science Factory: Map of Complexity Science. Cleveland, OH: Art & Science Factory 2009. URL: http://www.art-sciencefactory.com/complexity-map_feb09.html - Access Date: 04.03.2009.

AUE & DUSCHL 1982

Aue, A.; Duschl, R.: SSADM & GRAPES: Two Complementary Major European Methodologies for Information Systems Engineering. Berlin: Springer 1982. ISBN: 3-540-55380-0.

AZUMA & MOLE 1994

Azuma, M.; Mole, D.: Software Management Practice and Metrics in the European Community and Japan: Some Results of a Survey. Journal of Systems and Software 26 (1994) 1, pp. 5-18.

BADICA & FOX 2005

Badica, C.; Fox, C.: On the Application of WF-Nets for Checking Hybrid IDEF0-IDEF3 Business Process Models. In: Yakhno, T. (Ed.): Proceedings of the 3rd International Conference on Advances in Information Systems, ADVIS 2004, Izmir, Turkey, 20.-22.10.2004. Lecture Notes in Computer Science, 3261. Berlin: Springer 2005, pp. 543-553. ISBN: 3-540-23478-0.

BADKE-SCHAUB & GEHRLICHER 2003

Badke-Schaub, P.; Gehrlicher, A.: Patterns of decisions in design: leaps, loops, cycles, sequences and meta-processes. In: Folkeson, A. et al. (Eds.): Proceedings of the 14th International Conference on Engineering Design, ICED'03, Stockholm, Sweden, 19.-21.08.2003. Glasgow: The Design Society 2003. ISBN: 1-904670-00-8.

BALASUBRAMANIAN et al. 2005

Balasubramanian, S.; Gupta, M.: Structural metrics for goal based business process design and evaluation. Business Process Management Journal 11 (2005) 6, pp. 680-694.

BALDWIN & CLARK 2000

Baldwin, C. Y.; Clark, K. B.: Design Rules - The Power of Modularity. Cambridge, MA: MIT Press 2000. ISBN: 0-262-02466-7.

BANATHY 1997

Banathy, B. H.: A Taste of Systemics. Pocklington: International Society for the Systems Sciences 1997. URL: http://www.isss.org/taste.html - Access Date: 28.02.2009.

BAR-YAM 1997

Bar-Yam, Y.: Dynamics of complex systems. Reading, MA: Addison-Wesley 1997. ISBN: 0-8133-4121-3.

BARABÁSI 2003

Barabási, A.-L.: Linked - How Everything is Connected to Everything Else and What it Means for Business, Science, and Everyday Life. New York, NY: Plume 2003. ISBN: 0-452-28439-2.

BARABÁSI & ALBERT 1999

Barabási, A.-L.; Albert, R.: Emergence of Scaling in Random Networks. Science 286 (1999) 5439, pp. 509-512.

BARNETT & LEWIS 1998

Barnett, V.; Lewis, T.: Outliers in Statistical Data. Chichester: Wiley 1998. ISBN: 0-471-93094-6.

BASHIR 1999

Bashir, A. H.: Metrics for Design Projects: A Review. Design Studies 20 (1999) 3, pp. 263-277.

BASHIR & THOMSON 1999

Bashir, A. H.; Thomson, V.: Estimating Design Complexity. Journal of Engineering Design 10 (1999) 3, pp. 247-257.

BASILI et al. 1994

Basili, V. R.; Caldiera, G.; Rombach, H. D.: Goal Question Metric Paradigm. In: Marciniak, J. J. (Ed.): Encyclopedia of Software Engineering. New York, NY: Wiley 1994, pp. 528-532. ISBN: 0-471-54004-8.

BASILI & ROMBACH 1988

Basili, V. R.; Rombach, H. D.: The TAME Project: Towards Improvement-Oriented Software Environments. IEEE Transactions on Software Engineering 14 (1988) 6, pp. 758-773.

BAUMBERGER 2007

Baumberger, C.: Methoden zur kundenspezifischen Produktdefinition bei individualisierten Produkten. Dissertation, Technische Universität München, 2007. München: Dr. Hut 2007. ISBN: 978-3-89963-660-4.

BECK & STUHR 2008

Beck, C.; Stuhr, O.: STAN - Strukturanalyse für Java. JavaSPEKTRUM 5 (2008) pp. 44-49.

BECKER & PFEIFFER 2008

Becker, J.; Pfeiffer, D.: Solving the Conflicts of Distributed Process Modelling - Towards an integrated Approach. In: Golden, W. et al. (Eds.): Proceedings of the 16th European Conference on Information Systems, ECIS 2008, Galway, Ireland, 09.-11.06.2008. Galway: ECIS 2008. ISBN: 978-0-9553159-2-3.

BECKER et al. 2003

Becker, J.; Kugeler, M.; Rosemann, M.: Process Management - A Guide for the Design of Business Processes. Berlin: Springer 2003. ISBN: 3-540-43499-2.

BECKER & BOSTELMANN 1999

Becker, S. A.; Bostelmann, M. L.: Aligning Strategic and Project Measurement Systems. IEEE Software 16 (1999) 3, pp. 46-51.

BECKER et al. 1995

Becker, J.; Rosemann, M.; Schütte, R.: Grundsätze ordnungsgemäßer Modellierung. Wirtschaftsinformatik 37 (1995) 5, pp. 435-445.

BEER 1972

Beer, S.: Brain of the Firm - The Managerial Cybernetics of Organization. London: Allen Lane 1972. ISBN: 0-7139-0219-1.

BELHE & KUSIAK 1996

Belhe, U.; Kusiak, A.: Modeling Relationships Among Design Activities. ASME Journal of Mechanical Design 118 (1996) 4, pp. 454-460.

BELHE & KUSIAK 1995

Belhe, U.; Kusiak, A.: Resource Constrained Scheduling of Hierarchically Structured Design Activity Networks. IEEE Transactions on Engineering Management 42 (1995) 2, pp. 150-158.

BENSON 2007

Benson, A.: Qualitätssteigerung in komplexen Entwicklungsprojekten durch prozessbegleitende Kennzahlensysteme. Dissertation, Technische Universität Hamburg-Harburg, 2007. Göttingen: Cuvillier 2007. ISBN: 978-3-86727-295-7.

BEST & WETH 2009

Best, E.; Weth, M.: Geschäftsprozesse optimieren - Der Praxisleitfaden für erfolgreiche Reorganisation. Wiesbaden: Gabler 2009. ISBN: 978-3-8349-1384-5.

BICHLMAIER 2000

Bichlmaier, C.: Methoden zur flexiblen Gestaltung von integrierten Entwicklungsprozessen. Dissertation, Technische Universität München, 2000. München: Utz 2000. ISBN: 3-89675-710-5.

BICHLMAIER & GRUNWALD 1999

Bichlmaier, C.; Grunwald, S.: PMM - Process Module Methodology for Integrated Design and Assembly Planning. In: Proceedings of the ASME Design Engineering Technical Conferences, DETC, 4th Design for Manufacturing Conference, Las Vegas, Nevada, 12.-15.09.1999. New York, NY: ASME 1999. ISBN: 0-7918-1974-4.

BIEDERMANN et al. 2009

Biedermann, W.; Kreimeyer, M.; Lindemann, U.: Measurement System to Improve Data Acquistion Workshops. In: Kreimeyer, M.; Fadel, G.; Lindemann, U. (Eds.): Proceedings of 10th International Design Structure Matrix Conference – DSM'09. Munich: Hanser 2009.

BIEDERMANN & LINDEMANN 2008

Biedermann, W.; Lindemann, U.: Cycles in the Multiple-Domain Matrix – Interpretation and Applications. In: Kreimeyer, M. et al. (Eds.): Proceedings of the 10th International DSM Conference, Stockholm, 11.-12.11.2008. Munich: Hanser 2008, pp. 25-34. ISBN: 978-3-446-41825-7.

BIENA 2008

Biena, W.: Graphs and Networks. In: Finke, G. et al. (Eds.): Operations Research and Networks. London: Wiley 2008, pp. 29-69. ISBN: 978-1-84821-092-9.

BLESSING & CHAKRABARTI 2009

Blessing, L. T. M.; Chakrabarti, A.: DRM, a Design Research Methodology. Berlin: Springer 2009. ISBN: 978-1-84882-586-4.

BLESSING 2002

Blessing, L. T. M.: What is this thing called Design Research? In: Proceedings of the 2002 International CIRP Design Seminar, Hong Kong, 16.-18.05.2002. Hong Kong: Hong Kong University of Science and Technology 2002.

BOARDMAN & SAUSER 2006

Boardman, J.; Sauser, B.: System of Systems - the meaning of of. In: Proceedings of the 2006 IEEE/SMC International Conference on System of Systems Engineering, Los Angeles, CA, 24.-26.04.2006. Piscataway, NJ: IEEE Operations Center 2006, pp. 6-12. ISBN: 1-4244-0188-7.

BOEHM 1988

Boehm, B.: A Spiral Model of Software Development and Enhancement. IEEE Computer 21 (1988) 5, pp. 61-72. ISSN: 0018-9162.

BOEHM 1981

Boehm, B. W.: Software Engineering Economics. Englewood Cliffs, NJ: Prentice-Hall 1981. ISBN: 0-13-822122-7.

BOEHM et al. 1976

Boehm, W.; Brown, J. R.; Lipow, M.: Quantitative Evaluation of Software Quality. In: Proceedings of the 2nd International Conference on Software Engineering, ICSE'76, San Francisco, California, 1976. Los Alamitos, CA: IEEE Computer Society Press 1976, pp. 592-605.

BOLLINGER 1995

Bollinger, T.: What can happen when metrics make the call. IEEE software 12 (1995) 1, p. 15.

BOLLOBÁS 1981

Bollobás, B.: Degree Sequences of Random Graphs. Discrete Mathematics 33 (1981) pp. 1-19.

BONJOUR 2008

Bonjour, E.: Contributions à l'instrumentation du métier d'architecte système: de l'architecture modulaire du produit à l'organisation du système de conception. Besançon: Habilitation de l'Université de Franche-Comté 2008.

BRAHA & BAR-YAM 2004

Braha, D.; Bar-Yam, Y.: Topology of Large-scale Engineering Problem-solving Networks. Physical Review E 69, 016113-1-7.

BRANDES & ERLEBACH 2005

Brandes, U.; Erlebach, T.: Introduction. In: Brandes et al. (Eds.): Network Analysis - Methodological Foundations. Lecture Notes in Computer Science, 3418. Berlin: Springer 2005, pp. 1-6. ISBN: 3-540-24979-6.

BRAUN & LINDEMANN 2007a

Braun, S. C.; Lindemann, U.: A Multilayer Approach for Early Cost Estimation of Mechatronical Products. In: Bocquet, J.-C. (Ed.): Proceedings of the 16th International Conference on Engineering Design, ICED'07, Paris, 28.-31.08.2007. Glasgow: The Design Society 2007. ISBN: 1-904670-02-4.

BRAUN & LINDEMANN 2007b

Braun, T.; Deubzer, F.: New Variant Management Using Multiple-Domain Mapping. In: Lindemann, U. et al. (Eds.): Proceedings of the 9th International DSM Conference, Munich, 16.-18.10.2008. Aachen: Shaker 2007, pp. 363-372. ISBN: 978-3-8322-6641-7.

BROOKS 1987

Brooks, F. P.: No Silver Bullet - Essence and Accidents of Software Engineering. IEEE Computer 20 (1987) 4, pp. 10-19.

BROWNING 2009

Browning, T. R.: The Many Views of a Process: Towards a Process Architecture Framework for Product Development Processes. Systems Engineering 12 (2009) 1, pp. 69-90.

BROWNING 2002

Browning, T. R.: Process Integration Using the Design Structure Matrix. Journal of Systems Engineering 5 (2002) 3, pp. 180-193.

BROWNING & EPPINGER 2002

Browning, T. R.; Eppinger, S. D.: Modeling Impacts of Process Architecture on Cost and Schedule Risk in Product Development. IEEE Transactions on Engineering Management 49 (2002) 4, pp. 428-442.

BROWNING & RAMASESH 2007

Browning, T. R.; Ramasesh, R. V.: A Survey of Activity Network-Based Process Models for Managing Product Development Projects. Production and Operations Management 16 (2007) 2, pp. 217-240.

BROWNING 2001

Browning, T. R.: Applying the Design Structure Matrix to System Decomposition and Integration Problems: A Review and New Directions. IEEE Transactions on Engineering Management 48 (2001) 3, pp. 292-306.

BUGLIONE & ABRAN 2000

Buglione, L.; Abran, A.: Balanced Scorecard and GQM: What are the Differences? In: Proceedings of the 3rd European Software Measurement Conference, FESMA-AEMES 2000, Madrid, 18.-20.10.2000. Madrid: AEMES 2000. ISBN: 84-688-7161-3.

BULLINGER & SCHREINER 2001

Bullinger, H.-J.; Schreiner, P.: Business Process Management Tools - Eine evaluierende Marktstudie über aktuelle Werkzeuge. Stuttgart: Fraunhofer IRB 2001. ISBN: 3-8167-5629-8.

BULLINGER & WARSCHAT 1996

Bullinger, H.-J.; Warschat, J. (Eds.): Concurrent Simultaneous Engineering Systems - The Way to Successful Product Development. London: Springer 1996. ISBN: 3-540-76003-2.

BURGIN 1982

Burgin, M.: Generalized Kolmogorov Complexity and Duality in Theory of Computations. Notices of the Russian Academy of Sciences 25 (1982) 3, pp. 19-23.

BURR et al. 2003

Burr, H.; Deubel, T.; Vielhaber, M.; Haasis, S.; Weber, W.: Challenges for CAx and EDM in an international automotive company. In: Folkeson, A. et al. (Eds.): Proceedings of the 14th International Conference on Engineering Design, ICED'03, Stockholm, Sweden, 19.-21.08.2003. Glasgow: The Design Society 2003. ISBN: 1-904670-00-8.

CAMI & DEO 2008

Cami, A.; Deo, N.: Techniques for analyzing dynamic random graph models of web-like networks: An overview. Networks 51 (2008) 4, pp. 211-255.

CANTAMESSA et al. 2006

Cantamessa, M.; Milanesio, M.; Operti, E.: Value Chain Structure and Correlation Between Design Structure Matrices In: ElMaraghy, H. A. et al. (Eds.): Advances in Design. London: Springer 2006, pp. 303-313. ISBN: 978-1-84628-004-7.

CARDOSO 2007

Cardoso, J.: Business Process Quality Metrics: Log-Based Complexity of Workflow Patterns. Lecture Notes in Computer Science 4803/2007 (2007) pp. 427-434.

CARDOSO 2006

Cardoso, J.: Approaches to Compute Workflow Complexity. In: Leymann, F. et al. (Eds.): Dagstuhl Seminar Proceedings. The Role of Business Processes in Service Oriented Architectures, Schloss Dagstuhl, 16.-21.07.2006. Wadern: Internationales Begegnungs- und Forschungszentrum für Informatik, IBFI 2006.

CARDOSO et al. 2006

Cardoso, J.; Mendling, J.; Neumann, G.; Reijers, H. A.: A Discourse on Complexity of Process Models. In: Eder, J. et al. (Eds.): Proceedings of the Business Process Management Workshops, BPM 2006, 2nd International Worshop on Business Process Intelligence, BPI 2006, Vienna, Austria, 04.-07.09.2006. Lecture Notes in Computer Science, 4103. Berlin: Springer 2006, pp. 117-128. ISBN: 3-540-38444-8.

CARDOSO 2005a

Cardoso, J.: Control-flow Complexity Measurement of Processes and Weyuker's Properties. Proceedings of World Academy of Science, Engineering and Technology, PWASET 8 (2005), pp. 213-218.

CARDOSO 2005b

Cardoso, J.: How to Measure the Control-flow Complexity of Web Processes and Workflows. In: Fischer, L. (Ed.): The Workflow Handbook. Lighthouse Point: Future Strategies 2005, pp. 199-212. ISBN: 0-9703509-8-8.

CHALUPNIK et al. 2008

Chalupnik, M. J.; Wynn, D. C.; Eckert, C. M.; Clarkson, P. J.: Investigating Design Process Performance Under Uncertainty. In: Horváth, I. et al. (Eds.): Proceedings of the 7th International Symposium on Tools and Methods of Competitive Engineering, TMCE 2008, Izmir, Turkey, 21.-25.04.2008. Delft: TU Delft 2008, pp. 1061-1074. ISBN: 978-90-5155-044-3.

CHERNIAVSKY & SMITH 1991

Cherniavsky, J. C.; Smith, C. H.: On Weyuker's Axioms for Software Complexity Measures. IEEE Transactions on Software Engineering 17 (1991) 6, pp. 636-638.

CHO & EPPINGER 2001

Cho, S.-H.; Eppinger, S. D.: Product Development Process Modeling Using Advanced Simulation. In: Allen, J. K. et al. (Eds.): Proceedings of the ASME 2001 Design Engineering Technical Conferences and Computers and Information in Engineering Conference, DETC'01, 13th International Conference on Design Theory and Methodology, DTM, Pittsburgh, PA, 09.-12.09.2001. New York, NY: ASME 2001. ISBN: 0-7918-3544-8.

CHU et al. 2003

Chu, D.; Strand, R.; Fjelland, R.: Theories of Complexity. Complexity 8 (2003) 3, pp. 19-30.

CLARK 1989

Clark, K. B.: Project Scope and Project Performance: The Effect of Parts Strategy and Supplier Involvement on Product Development. Management Science 35 (1989) 10, pp. 1247-1263.

CLARKSON & ECKERT 2005

Clarkson, P. J.; Eckert, C.: Design Process Improvement - A review of current practice. Berlin: Springer 2005. ISBN: 1-85233-701-X.

CLARKSON & HAMILTON 2000

Clarkson, P. J.; Hamilton, J. R.: 'Signposting', A Parameter-driven Task-based Model of the Design Process. Research in Engineering Design 12 (2000) 1, pp. 18-38.

COATES et al. 2000

Coates, G.; Whitfield, R. I.; Duffy, A. H.; Hills, B.: Coordination Approaches and Systems - Part II: An Operational Perspective. Research in Engineering Design 12 (2000) 2, pp. 73-89.

COOPER & EDGETT 2005

Cooper, R. G.; Edgett, S. J.: Lean, Rapid, and Profitable - New Product Development. New York, NY: Basic Books 2005. ISBN: 0-9732827-1-1.

CRAWLEY & COLSON 2007

Crawley, E.; Colson, J.: The Projection Relationship between Object Process Models (OPM) and Design System Matrices (DSM). In: Lindemann, U. et al. (Eds.): Proceedings of the 9th International DSM Conference, Munich, 16.-18.10.2008. Aachen: Shaker 2007, pp. 137-150. ISBN: 978-3-8322-6641-7.

DAENZER & Huber 2002

Daenzer, W. F.; Huber, F. (Eds.): Systems Engineering - Methodik und Praxis. Zürich: Industrielle Organisation 2002. ISBN: 3-85743-998-X.

DAIMLER AG 2010

Daimler Global Media Site, http://media.daimler.com/dcmedia/home/d, viewed on 10 August 2010.

DALAL et al. 2004

Dalal, N. P.; Kamath, M.; Kolarik, W. J.; Sivaraman, E.: Toward an Integrated Framework for Modeling Enterprise Processes. Communications of the ACM 47 (2004) 3, pp. 83-87.

DANEVA et al. 1996

Daneva, M.; Heib, R.; Scheer, A.-W.: Benchmarking Business Process Models. IWi Heft Nr. 136. Saarbrücken: Universität des Saarlandes, Institut für Wirtschaftsinformatik, IWi 1996.

DANILOVIC & BROWNING 2007

Danilovic, M.; Browning, T. R.: Managing complex product development projects with design structure matrices and domain mapping matrices. International Journal of Project Management 25 (2007) 3, pp. 300-314.

DANILOVIC & BROWNING 2004

Danilovic, M.; Browning, T.: A Formal Approach for Domain Mapping Matrices (DMM) to Complement Design Structuring Matrices (DSM). In: Proceedings of the 6th International Design Structure Matrix Workshop, Cambridge, UK, 12.-14.09.2004. Cambridge: University of Cambridge 2004.

DANILOVIC & SANDKULL 2002

Danilovic, M.; Sandkull, B.: Managing Complexity and Uncertainty in a Multiproject Environment. In: Proceedings of the 5th Conference of the International Research Network on Organizing by Projects, Rotterdam. Rotterdam: Erasmus University 2002.

DANILOVIC & BÖRJESSON 2001

Danilovic, M.; Börjesson, H.: Participatory Dependence Structure Matrix Approach. In: Proceedings of the 3rd Dependence Structure Martix (DSM) International Workshop, Cambridge, MA, 29.-30.10.2001. Cambridge, MA: Massachusetts Institute of Technology 2001.

DAVENPORT 1993

Davenport, T. H.: Process Innovation - Reengineering Work Through Information Technology. Boston: Harvard Business School Press 1993. ISBN: 0-87584-366-2.

DE BRUIN et al. 2000

de Bruin, B.; Verschut, A.; Wierstra, E.: Systematic Analysis of Business Processes. Knowledge and Process Management 7 (2000) 2, pp. 87-96.

DE WECK 2007

de Weck, O. L.: On the Role of DSM in Designing Systems and Products for Changeability. In: Lindemann, U. et al. (Eds.): Proceedings of the 9th International DSM Conference, Munich, 16.-18.10.2008. Aachen: Shaker 2007, pp. 311-323. ISBN: 978-3-8322-6641-7.

DEMARCO 1978

DeMarco, T.: Structured Analysis. New York: Yourdon 1978.

DEMING 1994

Deming, W. E.: Out of the Crisis. Cambridge: Cambridge University Press 1994. ISBN: 0-911379-01-0.

DEUBZER et al. 2007

Deubzer, F.; Kreimeyer, M.; Herfeld, U.; Lindemann, U.: A Strategy for efficient collaboration in virtual product development environments. In: Bocquet, J.-C. (Ed.): Proceedings of the 16th International Conference on Engineering Design, ICED'07, Paris, 28.-31.08.2007. Glasgow: The Design Society 2007. ISBN: 1-904670-02-4.

DIEHL 2009

Diehl, H.: Systemorientierte Visualisierung disziplinübergreifender Entwicklungsabhängigkeiten mechatronischer Automobilsysteme. Dissertation, Technische Universität München, 2009. München: Dr. Hut 2009.

DIESTEL 2006

Diestel, R.: Graph Theory. Graduate Texts in Mathematics, 173. Berlin: Springer 2006. ISBN: 978-3-540-26183-4.

DIETRICH 2001

Dietrich, A.: Selbstorganisation - Management aus ganzheitlicher Perspektive. Dissertation, Universität Graz, 2000. Wiesbaden: Gabler 2001. ISBN: 3-8244-7406-9.

DINSMORE & CABANIS-BREWIN 2006

Dinsmore, P. C.; Cabanis-Brewin, J.: The AMA Handbook of Project Management. Toronto: AMACom Books 2006. ISBN: 0-8144-7271-0.

DIXON et al. 1988

Dixon, J. R.; Duffey, M. R.; Irani, R. K.; Meunier, K. L.; Orelup, M. F.: A Proposed Taxonomy of Mechanical Design Problems. In: Tipnis, V. A. et al. (Eds.): Proceedings of the 1988 ASME International Computers in Engineering Conference and Exhibition, 31.07.-04.08.1988, San Francisco, CA. New York: ASME 1988, pp. 41-46.

DONG 2002

Dong, Q.: Predicting and Managing System Interactions at Early Phase of the Product Development Process. Ph.D. Thesis, Massachusetts Institute of Technology, 2002. Cambridge, MA: Massachusetts Institute of Technology 2002.

DOROGOVTSEV & MENDES 2002

Dorogovtsev, S. N.; Mendes, J. F.: Evolution of Networks. Advances in Physics 51 (2002), pp. 1079-1187.

DRUCKER 2007

Drucker, P. F.: The Practice of Management. Amsterdam: Elsevier 2007. ISBN: 978-0-7506-8504-7.

DUMKE & LEHNER 2000

Dumke, R.; Lehner, F.: Software-Metriken - Entwicklungen, Werkzeuge und Anwendungsverfahren. Wiesbaden: Gabler 2000. ISBN: 3-8244-7120-5.

EBEN et al. 2008

Eben, K.; Biedermann, W.; Lindemann, U.: Modeling Structural Change over Time – Requirements and First Methods. In: Kreimeyer, M. et. al. (Eds.): Proceedings of the 10th International DSM Conference, Stockholm, 11.-12.11.2008. Munich: Hanser 2008, pp. 15-24. ISBN: 978-3-446-41825-7.

EICHINGER et al. 2006

Eichinger, M.; Maurer, M.; Lindemann, U.: Using Multiple Design Structure Matrices. In: Marianovic, D. (Ed.): Proceedings of the 9th International Design Conference, DESIGN 2006, Dubrovnik, Croatia, 15.-18.05.2006. Zagreb: University of Zagreb, Faculty of Mechanical Engineering and Naval Architecture 2006, pp. 229-236. ISBN: 953-6313-79-0.

EPPINGER 2001

Eppinger, S. D.: Patterns of product development Interactions. In: Culley S. et al. (Eds.): 13th International Conference on Engineering Design, ICED'01, Glasgow, UK, 21.-23.08.2001. Bury St. Edmunds: Professional Engineering 2001, pp. 283-290. ISBN: 1-86058-354-7.

ERDOES & RÉNYI 1959

Erdoes, P.; Rényi, A.: On Random Graphs I. Publicationes Mathematicae 6 (1959), pp. 290-297.

EUROPEAN FOUNDATION FOR QUALITY MANAGEMENT 1995

European Foundation for Quality Management: Selbstbewertung, Richtlinien für Unternehmen. Brüssel: EFQM 1995.

FAHRWINKEL 1995

Fahrwinkel, U.: Methode zur Modellierung und Analyse von Geschäftsprozessen zur Unterstützung des Business Process Reengineering. Dissertation, Universität Paderborn, 1995. Paderborn: HNI 1995. ISBN: 3-931466-00-0.

FINK & HAMPP 2005

Fink, M.; Hampp, T.: Eine Untersuchung zum Metrikeinsatz In der Industrie. In: Büren, G. et al. (Eds.): Proceedings of the DASMA Software Metrik Kongress, MetriKon 2005, Kaiserslautern, 15.-16.11.2005. Aachen: Shaker 2005, pp. 33-42. ISBN: 3-8322-4615-0.

FINKE 2008

Finke, G.: Operations Research and Networks. London: Wiley 2008. ISBN: 978-1-84821-092-9.

FLURSHEIM 1977

Flursheim, C.: Engineering Design Interfaces: A management philosophy. London: The Design Council 1977. ISBN: 0-85072-051-6.

FORRESTER 1977

Forrester, J. W.: Industrial Dynamics. Cambridge, MA: MIT Press 1977. ISBN: 0-262-06003-5.

FREEMAN 1978

Freeman, L. C.: Centrality in Social Networks - Conceptual Clarification. Social Networks 1 (1978) 3, pp. 215-239.

FREISLEBEN 2001

Freisleben, D.: Gestaltung und Optimierung von Produktentwicklungsprozesen mit einem wissensbasierten Vorgehensmodell. Magdeburg: PhD Thesis Universität Magdeburg 2001.

FRUCHTERMAN & REINGOLD 1991

Fruchterman, T.; Reingold, E.: Graph drawing by force-directed placement. Software: Practice and Experience 21 (1991) 11, pp. 1129-1164.

GAITANIDES et al. 1994

Gaitanides, M.; Scholz, R.; Vrohlings, A.; Raster, M.: Prozeßmanagement - Konzepte, Umsetzungen und Erfahrungen des Reengineering. München: Hanser 1994. ISBN: 3-446-17715-9.

GÄRTNER et al. 2008

Gärtner, T.; Rohleder, N.; Schlick, C. M.: Simulation of Product Change Effects on the Duration of Development Processes Based on the DSM. In: Kreimeyer, M. et al. (Eds.): Proceedings of the 10th International DSM Conference, Stockholm, 11.-12.11.2008. Munich: Hanser 2008, pp. 199-208. ISBN: 978-3-446-41825-7.

GAUSEMEIER et al. 2006

Gausemeier, J.; Hahn, A.; Kespohl, H. D.; Seifert, L.: Vernetzte Produktentwicklung - Der erfolgreiche Weg zum Global Engineering Network. München: Carl Hanser 2006. ISBN: 3-446-22725-3.

GAUSEMEIER & FINK 1999

Gausemeier, J.; Fink, A.: Führung im Wandel - Ein ganzheitliches Modell zur zukunftsorientierten Unternehmensgestaltung. München: Carl Hanser 1999. ISBN: 3-446-21079-2.

GEIGER 2000

Geiger, O.: Kennzahlenorientiertes Entwicklungscontrolling - Ein ganzheitliches, kennzahlenbasiertes Planungs-, Steuerungs- und Kontrollinstrument zur Analyse des Entwicklungsbereichs industrieller Unternehmen. Dissertation, Technische Universität Braunschweig, 2000. Aachen: Shaker 2000. ISBN: 3-8265-5942-8.

GHANI et al. 2008

Ghani, A. A. A.; Wei, K. T.; Muketha, G. M.; Wen, W. P.: Complexity Metrics for Measuring the Understandability and Maintainability of Business Process Models using Goal-Question-Metric (GQM). International Journal of Computer Science and Network Security 8 (2008) 5, pp. 219-225.

GIGERENZER 2007

Gigerenzer, G.: Gut Feelings. New York: Viking 2007. ISBN: 978-0-670-03863-3.

GIRVAN & NEWMAN 2002

Girvan, M.; Newman, M. E.: Community Structures in Social and Biological Networks. Proceedings of the National Academy of Sciences 99 (2002) 12, pp. 8271-8276.

GOLDENFELD & KADANOFF 1999

Goldenfeld, N.; Kadanoff, L. P.: Simple Lessons from Complexity. Science 284 (1999) 5411, pp. 87-89.

GRAEBSCH et al. 2007

Graebsch, M.; Lindemann, U.; Weiss, S.: Lean Development in Deutschland. München: Dr. Hut 2007. ISBN: 978-3-89963-496-9.

GRIFFIN 1993

Griffin, A.: Metrics for Measuring product Development Cycle Time. Journal of Product Innovation Management 10 (1993) pp. 112-125.

GRONBACK 2006

Gronback, R.: Model Validation: Applying Audits and Metrics to UML Models. In: Proceedings of the 2004 Borland Conference, BORCON 2004, San Jose, CA, 11.-15.09.2004. URL: http://conferences.embarcadero.com/article/32089 - Access Date: 29.02.2009.

GROSS & YELLEN 2005

Gross, J. L.; Yellen, J.: Graph Theory and its Applications. Boca Raton: Chapman & Hall/CRC 2005. ISBN: 1-584-88505-X.

GRUHN & LAUE 2007a

Gruhn, V.; Laue, R.: On experiments for measuring cognitive weights for software control structures. In: Zhang, D. et al. (Eds.): 6th IEEE International Conference on Cognitive Informatics, ICCI 2007, Lake Tahoe, CA, 06.-08.08.2007. Piscataway, NJ: IEEE 2007, pp. 116-119. ISBN: 978-1-424-41327-0.

GRUHN & LAUE 2007b

Gruhn, V.; Laue, R.: What business process modelers can learn from programmers. Science of Computer Programming 65 (2007) 1, pp. 4-13.

GRUHN & LAUE 2006a

Gruhn, V.; Laue, R.: Complexity Metrics for Business Process Models. In: Abramowicz, W. et al. (Eds.): Proceedings of the 9th International Conference on Business Information Systems, BIS 2006, Klagenfurt, Austria, 31.05.-02.06.2006. Bonn: Gesellschaft für Informatik, GI 2006, pp. 1-12. ISBN: 3-88579-179-X.

GRUHN & LAUE 2006b

Gruhn, V.; Laue, R.: Komplexitätsmetriken für Geschäftsprozessmodelle. In: Mayr, H. C. et al. (Eds.): Proceedings of the Modellierung 2006, Innsbruck, Austria, 22.-24.03.2006. Bonn: Gesellschaft für Informatik, GI 2006, pp. 289-292. ISBN: 3-88579-176-5.

GRUHN et al. 2006c

Gruhn, V.; Laue, R.; Meyer, F.: Berechnung von Komplexitätsmetriken für ereignisgesteuerte Prozessketten. In: Nüttgens, M. et al. (Eds.): Proceedings of the 5th GI Workshop und Arbeitskreistreffen Geschäftsprozessmanagement mit Ereignisgesteuerten Prozessketten, EPK 2006, Vienna, Austria, 30.11.-01.12.2006. Bonn: Gesellschaft für Informatik, GI 2006, pp. 189–202.

GÜRTLER et al. 2009

Gürtler, M.; Kreimeyer, M.; Lindemann, U.: Extending the Active Sum / Passive Sum Measure to Include Boolean Operators: A Case Study. In: Kreimeyer, M.; Fadel, G.; Lindemann, U. (Eds.): Proceedings of 10th International Design Structure Matrix Conference, DSM'09. Munich: Hanser 2009.

GUILLAUME & LATAPY 2004

Guillaume, J.-L.; Latapy, M.: Bipartite Structure of all Complex Networks. Information Processing Letters 90 (2004) 5, pp. 215-221.

GUIZZARDI et al. 2002

Guizzardi, G.; Pires, L. F.; Sinderen, M. J. v.: On the role of Domain Ontologies in the design of Domain-Specific Visual Modeling Languages. In: Proceedings of the 17th ACM Conference on Object-Oriented Programming, Systems, Languages, and Applications, OOPSLA 2002, 2nd Workshop on Domain-Specific Visual Languages, Seattle, WA, 04.-08.11.2002. New York, NY: ACM Press 2002. ISBN: 1-581-13471-1.

HALES & GOOCH 2004

Hales, C.; Gooch, S.: Managing Engineering Design. London: Springer 2004. ISBN: 1-85233-803-2.

HALL 1963

Hall, A. D.: A Methodology for Systems Engineering. Princeton: Van Nostrand 1963.

HALSTEAD 1977

Halstead, M. H.: Elements of software science. New York, NY: North Holland 1977. ISBN: 0-444-00215-4.

HAMMER & CHAMPY 2003

Hammer, M.; Champy, J.: Reengineering the Corporation - A Manifesto for Business Revolution. New York: HarperCollins 2003. ISBN: 0-06-055953-5.

HARRINGTON 1991

Harrington, H. J.: Business Process Improvement – The Breakthrough Strategy for Total Quality, Productivity and Competitiveness. New York: McGraw-Hill 1991. ISBN: 0-07-026768-5.

HARRISON & MAGEL 1981

Harrison, W. A.; Magel, K. I.: A Complexity Measure based on Nesting Level. ACM SIGPLAN Notices 16 (1981) 3, pp. 63-74.

HATCHUEL & WEIL 2003

Hatchuel, A.; Weil, B.: A new Approach of Innovative Design: an Introduction to C-K Theory. In: Folkeson, A. et al. (Eds.): Proceedings of the 14th International Conference on Engineering Design, ICED'03, Stockholm, Sweden, 19.-21.08.2003. Glasgow: The Design Society 2003. ISBN: 1-904670-00-8.

HAYES 2000a

Hayes, B.: Graph Theory in Practice: Part I. American Scientist 88 (2000) 1, pp. 9-13.

HAYES 2000b

Hayes, B.: Graph Theory in Practice: Part II. American Scientist 88 (2000) 2, pp. 104-109.

HAWKINS 1980

Hawkins, D. M.: Identification of Outliers. Chapman & Hall, London 1980. ISBN: 0-412-21900-X

HEISENBERG 2007

Heisenberg, W.: Physics and Philosophy - The Revolution in Modern Science. New York: HarperPerennial 2007. ISBN: 978-0-06-120919-2.

HEISIG et al. 2008

Heisig, P.; Grebici, K.; Ariyo, L.; Caldwell, N. H. M.; Clarkson, P. J.: Towards an Integrated Product – Process – Rationale Framework for the Product Life Cycle. In: Darlington, M. et al. (Eds.): Proceedings of the KIM Project Conference 2008, Knowledge & Information Management Through Life, University of Reading, 02.-03.04.2008. Bath: University of Bath 2008.

HENDERSON & CLARK 1990

Henderson, R.; Clark, K.: Architectural Innovation: The Reconfiguration of Existing Product Technologies and the Failure of Established Firms. Administrative Science Quarterly 35 (1990) 1, pp. 9-30.

HENRY et al. 1981

Henry, S.; Kafura, D.; Harris, K.: On the Relationships Among Three Software Metrics. In: Proceedings of the 1981 ACM Workshop/Symposium on Measurement and Evaluation of Software Quality, College Park, Md., 25.-27.03.1981. New York: ACM 1981, pp. 81-88. ISBN: 0-89791-038-9.

HERFELD 2007

Herfeld, U.: Matrix-basierte Verknüpfung von Komponenten und Funktionen zur Integration von Konstruktion und Simulation. Dissertation, Technische Universität München, 2007. München: Dr. Hut 2007. ISBN: 978-3-89963-545-4.

HERFELD et al. 2006

Herfeld, U.; Kreimeyer, M.; Deubzer, F.; Frank, T.; Lindemann, U.; Knaust, U.: Verknüpfung von Komponenten und Funktionen zur Integration von Konstruktion und Simulation in der Karosserieentwicklung. In: Proceedings of the 13th Tagung Berechnung und Simulation im Fahrzeugbau, Würzburg, 27.-28.09.2006. VDI-Berichte, 1967. Düsseldorf: VDI-Verl. 2006, pp. 259-276. ISBN: 3-18-091967-1.

HEYMANN 2005

Heymann, M.: "Kunst" und Wissenschaft in der Technik des 20. Jahrhunderts - Zur Geschichte der Konstruktionswissenschaft. Zürich: Chronos 2005. ISBN: 3-0340-0723-X.

HÖFFERER 2007

Höfferer, P.: Achieving Business Process Model Interoperability using Metamodels and Ontologies. In: Proceedings of the 15th European Conference on Information Systems, ECIS 2007, St. Gallen, 07.-09.06.2007. St. Gallen: University of St. Gallen 2007, pp. 1620-1631.

HOLLAND 1996

Holland, J. H.: Hidden Order - How Adaptation Builds Complexity. Reading, MA: Addison-Wesley, 1996. ISBN: 0-201-40793-0.

HORNBY 2007

Hornby, G. S.: Modularity, Reuse, and Hierarchy: Measuring Complexity by Measuring Structure and Organization. Complexity 13 (2007) 2, pp. 50-61.

HORVÁTH 2003

Horváth, P.: Controlling. München: Vahlen 2003. ISBN: 3-8006-2992-5.

IDS SCHEER 2007

IDS Scheer: Business Process Report 2007. Saarbrücken: IDS Scheer AG 2007.

IEEE 1991

IEEE: Standard Glossary of Software Engineering Terrninology. IEEE Std 610.12-1990. New York: IEEE 1991. ISBN: 1-559-37067-X.

ILIE et al. 2008

Ilie, D.; Fischer, F.; Lindemann, U.: Analysis of the Information Environment in the Context of Target and Requirements Management in the Automotive Industry. In: Proceedings of the 2008 ASME International Design Engineering Technical Conferences and Computers and Information in Engineering Conference, IDETC/CIE 2008, 28th Computers and Information in Engineering Conference, New York City, NY, 03.-06.08.2008. New York, NY: ASME 2008, pp. 659-670. ISBN: 978-0-7918-4327-7.

INCOSE 2007

INCOSE: Systems Engineering Handbook v3.1. San Diego: INCOSE 2007.

JACKSON 1991

Jackson, M. C.: Systems Methodology for the Management Sciences. New York: Plenum 1991. ISBN: 0-306-43877-1.

JARRATT 2004

Jarratt, T. A. W.: A Model-based Approach to Support the Management of Engineering Change. Ph.D. Thesis, University of Cambridge, 2004. Cambridge: University of Cambridge 2004.

KAIN et al. 2008

Kain, A.; Kirschner, R.; Lindemann, U.; Wastian, M.; Schneider, M.; Klendauer, R.; Gunkel, J.: An Approach to Model Time Dependent Process-Stakeholder Networks. In: Kreimeyer, M. et al. (Eds.): Proceedings of the 10th International DSM Conference, Stockholm, 11.-12.11.2008. Munich: Hanser 2008, pp. 71-82. ISBN: 978-3-446-41825-7.

KAPLAN & NORTON 1992

Kaplan, R. S.; Norton, D. P.: The Balanced Scorecard - Measures that drive Performance. Harvard Business Review 70 (1992) 1, pp. 71-79.

KARNIEL & REICH 2009

Karniel, A.; Reich, Y.: From DSM-based planning to Design Process Simulation: A review of process-scheme logic verification issues. IEEE Transactions on Engineering Management 56 (2009) 4, pp. 636-649.

KARNIEL & REICH 2007

Karniel, A.; Reich, Y.: Coherent Interpretation of DSM Plan to PDP Simulation. In: Bocquet, J.-C. (Ed.): Proceedings of the 16th International Conference on Engineering Design, ICED'07, Paris, 28.-31.08.2007. Glasgow: The Design Society 2007. ISBN: 1-904670-02-4.

KAUFFMAN 1993

Kauffman, S. A.: The Origins of Order - Self-Organization and Selection in Evolution. New York: Oxford University Press 1993. ISBN: 0-19-507951-5.

KEIJZER et al. 2007

Keijzer, W.; Kreimeyer, M.; Schack, R.; Lindemann, U.; Zäh, M.: Vernetzungsstrukturen in der Digitalen Fabrik. München: Dr. Hut 2007. ISBN: 3-89963-378-4.

KEIJZER 2007

Keijzer, W. C.: Wandlungsfähigkeit von Entwicklungsnetzwerken. Dissertation, Technische Universität München, 2007. München: Technische Universität München 2007.

KELLER et al. 2005

Keller, R.; Eger, T.; Eckert, C. M.; Clarkson, P. J.: Visualising change propagation. In: Samuel, A. E. et al. (Eds.): Proceedings of the 15th International Conference on Engineering Design, ICED'05, Melbourne, Australia, 15.-18.08.2005. Barton ACT: Engineers Australia 2005. ISBN: 0-85825-788-2.

KERNLER 1996

Kernler, H.: PPS-Controlling. Wiesbaden: Gabler 1996. ISBN: 3-409-92294-6.

KIM et al. 2002

Kim, B. J.; Yoon, C. N.; Han, S. K.; Jeong, H.: Path Finding Strategies in Scale-free Networks. Physical Review E 65 (2002) 2, id. 027103.

KLEINFELD 2002

Kleinfeld, J. S.: The Small World Problem. Society 39 (2002) 2, pp. 61-66.

KLUTH et al. 2008

Kluth, M.; Ahlemann, F.; Teuteberg, F.: SEMAT – Ein Werkzeug zur ontologiebasierten Analyse und zum Vergleich von Prozessmodellen. In: Loos, P. et al. (Eds.): Proceedings of the Modellierung betrieblicher Informationssysteme, MobIS 2008, Saarbrücken, Germany, 27.-28.11.2008. Bonn: Gesellschaft für Informatik, GI 2008, pp. 128-147. ISBN: 978-3-88579-235-2.

KNEUPER 2007

Kneuper, R.: CMMI - Verbesserung von Softwareprozessen mit Capability Maturity Model Integration. Heidelberg: dpunkt 2007. ISBN: 3-89864-373-5.

KNOWLEDGE BASED SYSTEMS 1992

Knowledge Based Systems: IDEF3 method report. College Station, TX: Knowledge Based Systems, Inc. 1992.

KÖNIG et al. 2008

König, C.; Kreimeyer, M.; Braun, T.: Multiple-Domain Matrices as a Framework for Systematic Process Analysis. In: Kreimeyer, M. et al. (Eds.): Proceedings of the 10th International DSM Conference, Stockholm, 11.-12.11.2008. Munich: Hanser 2008, pp. 131-244. ISBN: 978-3-446-41825-7.

KOPPEL 1987

Koppel, M.: Complexity, Depth and Sophistication. Complex Systems 1 (1987) 6, pp. 1087-1091.

KORNMEIER 2007

Kornmeier, M.: Wissenschaftstheorie und wissenschaftliches Arbeiten - Eine Einführung für Wirtschaftswissenschaftler. Heidelberg: Physica-Verlag 2007. ISBN: 978-3-7908-1918-2.

KORTLER et al. 2009

Kortler, S.; Kreimeyer, M.; Lindemann, U.: A planarity-based complexity metric. In: Proceedings of the 17th International Conference on Engineering Design, ICED'09, Stanford, CA, 24.-27.08.2009. Glasgow: The Design Society 2009.

KREIMEYER ET AL. 2010

Kreimeyer, M.; Wynn, D. C.; Clarkson, P. J.; Lindemann, U.: Profiling PD Processes by Combining Structural Analysis and Simulation. In: Proceedings of 11th International Design Conference DESIGN 2010. Glasgow: The Design Society 2010, pp. 1131 – 1142.

KREIMEYER 2008

Kreimeyer, M.: Elements of Research on Engineering Design Processes (Presentation in Process SIG). In: Marjanovic, D. et al. (Eds.): Proceedings of the 10th International Design Conference, DESIGN 2008, Dubrovnik, Croatia, 19.-22-05.2008. Glasgow: The Design Society 2008. ISBN: 953-6313-90-1.

KREIMEYER et al. 2008a

Kreimeyer, M.; Daniilidis, C.; Lindemann, U.: A Framework to Classify Process Improvement Projects. In: Marjanovic, D. et al. (Eds.): Proceedings of the 10th International Design Conference, DESIGN 2008, Dubrovnik, Croatia, 19.-22-05.2008. Glasgow: The Design Society 2008, pp. 951-958. ISBN: 978-953-6313-89-1.

KREIMEYER et al. 2008b

Kreimeyer, M.; Deubzer, F.; Herfeld, U.; Lindemann, U.: Strategies Towards Maturity Of Product Information Objects To Manage Concurrent Engineering Processes. In: Horváth, I. et al. (Eds.): Proceedings of the 7th International Symposium on Tools and Methods of Competitive Engineering, TMCE 2008, Izmir, Turkey, 21.-25.04.2008. Delft: TU Delft 2008, pp. 925-940. ISBN: 978-90-5155-044-3.

KREIMEYER et al. 2008c

Kreimeyer, M.; König, C.; Braun, T.: Structural Metrics to Assess Processes. In: Kreimeyer, M. et al. (Eds.): Proceedings of the 10th International DSM Conference, Stockholm, 11.-12.11.2008. Munich: Hanser 2008, pp. 245-258. ISBN: 978-3-446-41825-7.

KREIMEYER et al. 2008d

Kreimeyer, M.; Braun, S.; Gürtler, M.; Lindemann, U.: Relating two Domains via a Third - an Approach to Overcome Ambiguous Attributions using Multiple-Domain Matrices. In: Proceedings of the 2008 ASME International Design Engineering Technical Conferences and Computers and Information in Engineering Conference, IDETC/CIE 2008, 28th Computers and Information in Engineering Conference, New York City, NY, 03.-06.08.2008. New York, NY: ASME 2008. ISBN: 978-0-7918-4327-7.

KREIMEYER et al. 2007a

Kreimeyer, M.; Deubzer, F.; Danilovic, M.; Fuchs, S. D.; Herfeld, U.; Lindemann, U.: Team Composition to Enhance Collaboration Between Embodiment Design and Simulation Departments. In: Bocquet, J.-C. (Ed.): Proceedings of the 16th International Conference on Engineering Design, ICED'07, Paris, 28.-31.08.2007. Glasgow: The Design Society 2007. ISBN: 1-904670-02-4.

KREIMEYER et al. 2007b

Kreimeyer, M.; Eichinger, M.; Lindemann, U.: Aligning Multiple Domains of Design Processes. In: Bocquet, J.-C. (Ed.): Proceedings of the 16th International Conference on Engineering Design, ICED'07, Paris, 28.-31.08.2007. Glasgow: The Design Society 2007. ISBN: 1-904670-02-4.

KREIMEYER et al. 2007c

Kreimeyer, M.; Eichinger, M.; Lindemann, U.: Assessment of Process Networks Using Graph and Network Theory Based Key Figures. In: Van Velden, D. (Ed.): Proceedings of the 4th Future Business Technology Conference, FUBUTEC 2007, Delft, the Netherlands, 25.-27.04.2007. Ghent: EUROSIS-ETI 2007, pp. 13-19. ISBN: 978-90-77381-33-5.

KREIMEYER et al. 2006a

Kreimeyer, M.; Herfeld, U.; Deubzer, F.; Dequidt, C.; Lindemann, U.: Function-driven product design in virtual teams through methodical structuring of requirements and components. In: Proceedings of the 8th Biennial ASME Conference on Engineering Systems Design and Analysis, ESDA 2006, Torino, Italy, 04.-07.07.2006. New York: ASME 2006. ISBN: 0-7918-3779-3.

KREIMEYER et al. 2006b

Kreimeyer, M.; Herfeld, U.; Deubzer, F.; Lindemann, U.: Effiziente Zusammenarbeit von Konstruktions- und Simulationsabteilungen in der Automobilindustrie. CiDaD Working Paper Series 1 (2006) 2.

KREIMEYER et al. 2006c

Kreimeyer, M.; Heymann, M.; Lauer, W.; Lindemann, U.: Die Konstruktionsmethodik im Wandel der Zeit - Ein Überblick zum 100sten Geburtstag von Prof. Wolf Rodenacker. Konstruktion (2006) 10, pp. 72-74.

KREIMEYER et al. 2006d

Kreimeyer, M.; Neumüller, K.; Lindemann, U.: A Management Model for the Design Process in Vehicle Safety Development. In: Marianovic, D. (Ed.): Proceedings of the 9th International Design Conference, DESIGN 2006, Dubrovnik, Croatia, 15.-18.05.2006. Glasgow: The Design Society 2006, pp. 1525-1532. ISBN: 953-6313-79-0.

KREIMEYER et al. 2005

Kreimeyer, M.; Deubzer, F.; Herfeld, U.; Lindemann, U.: A Survey on Efficient Collaboration of Design and Simulation in Product Development. In: Proceedings of the 23rd CADFEM Users' Meeting 2005 International Congress on FEM Technology, Bonn, 09.-11.11.2005. Gräfing: CADFEM GmbH 2005.

KRISHNAN et al. 1997

Krishnan, V.; Eppinger, S. D.; Whitney, D. E.: A Model-Based Framework to Overlap Product Development Activities. Management Science 43 (1997) 4, pp. 437-451.

KÜHN 2004

Kühn, H.: Methodenintegration im Business Engineering. Dissertation, Universität Wien, 2004. Wien: Universität Wien 2004.

KUSIAK & WANG 1993

Kusiak, A.; Wang, J.: Efficient Organizing of Design Activities. International Journal of Production Research 31 (1993) 4, pp. 753-769.

KUSIAK et al. 1995

Kusiak, A.; Wang, J.; He, W.; Feng, C.-X.: A Structured Approach for Analysis of Design Processes. IEEE Transactions on Components, Packaging, and Manufactoring Technology - Part A 18 (1995) 3, pp. 664-673.

LACHNIT 1976

Lachnit, L.: Zur Weiterentwicklung betriebswirtschaftlicher Kennzahlensysteme. Zeitschrift für betriebswissenschaftliche Forschung 28 (1976) 4, pp. 216-230.

LANGER et al. 2009

Langer, S.; Kreimeyer, M.; Müller, P.; Lindemann, U.; Blessing, L.: Entwicklungsprozesse hybrider Leistungsbündel - Evaluierung von Modellierungsmethoden unter Berücksichtigung zyklischer Einflussfaktoren. In: Thomas, O. et al. (Eds.): Dienstleistungsmodellierung - Methoden, Werkzeuge und Branchenlösungen. Heidelberg: Physica-Verlag 2009, pp. 71-87. ISBN: 978-3-7908-2098-0.

LATVA-KOIVISTO 2001

Latva-Koivisto, A. M.: Finding a complexity measure for business process models. Research report. Helsinki: Helsinki University of Technology, Systems Analysis Laboratory 2001.

LÄUCHLI 1991

Läuchli, P.: Algorithmische Graphentheorie. Basel: Birkhäuser 1991. ISBN: 3-7643-2663-8.

LEE & YOON 1992

Lee, G. S.; Yoon, J.-M.: An Empirical Study on the Complexity Metrics of Petri Nets. Microelectronics and Reliability 32 (1992) 3, pp. 323-329.

LEE 2003

Lee, T.: Complexity Theory in Axiomatic Design. Ph.D. Thesis, Massachusetts Institute of Technology, 2003. Cambridge, MA: Massachusetts Institute of Technology 2003.

LI et al. 2005

Li, L.; Alderson, D.; Tanaka, R.; Doyle, J. C.; Willinger, W.: Towards a Theory of Scale-Free Graphs: Definition, Properties, and Implications. Internet Mathematics 2 (2005) 4, pp. 431-523.

LIBERATI et al. 2007

Liberati, M.; Munari, F.; Racchetti, P.; Splendiani, T.: Social Network Techniques Applied to Design Structure Matrix Analysis. The Case of a New Engine Development at Ferrari SpA. In: Lindemann, U. et al. (Eds.): Proceedings of the 9th International DSM Conference, Munich, 16.-18.10.2008. Aachen: Shaker 2007, pp. 35-47. ISBN: 978-3-8322-6641-7.

LIMA 2007

Lima, M.: VisualComplexity.com. URL: http://www.visualcomplexity.com - Access Date: 04.03.2009.

LINDEMANN et al. 2009

Lindemann, U.; Maurer, M.; Braun, T.: Structural Complexity Management. Berlin: Springer 2009. ISBN: 978-3-540-87888-9.

LINDEMANN 2007

Lindemann, U.: Methodische Entwicklung technischer Produkte. Berlin: Springer 2007. ISBN: 3-540-37435-3.

LINDEMANN 2003

Lindemann, U.: Methods are Networks of Methods. In: Folkeson, A. et al. (Eds.): Proceedings of the 14th International Conference on Engineering Design, ICED'03, Stockholm, Sweden, 19.-21.08.2003. Glasgow: The Design Society 2003, pp. 625-626. ISBN: 1-904670-00-8.

LIU et al. 2003

Liu, X.; Wang, Y.; Jiang, S.: A Metrics based Task Analysis Model for Design Review Planning. Design Studies 24 (2003) pp. 375-390.

LOCH et al. 2003

Loch, C.; Mihm, J.; Huchzermeier, A.: Concurrent Engineering and Design Oscillations in Complex Engineering Projects. Concurrent Engineering: Research and Applications 11 (2003) 3, pp. 733-750.

LÓPEZ-MESA et al. 2004

López-Mesa, B.; Eriksson, S.; Thompson, G.: The Decomposition and Linkage of Design Methods and Problems. In: Marjanovic, D. (Ed.): Proceedings of the 8th International Design Conference, DESIGN 2004, Dubrovnik, Croatia, 17.-20.05.2004. Glasgow: The Design Society 2004, pp. 367-376. ISBN: 953-6313-59-6.

LORENZ 2009

Lorenz, M.: Handling of Strategic Uncertainties in Integrated Product Development. Dissertation, Technische Universität München, 2008. München: Dr. Hut 2009. ISBN: 978-3-89963-923-0.

LUCE & KRUMHANSL 1988

Luce, R. D; Krumhansl, C. L.: Measurement, scaling, and psychophysics. In: Atkinson, R. C.; Herrnstein, R. J.; Lindzey, G.; Luce, R. D. (Eds.): Stevens' Handbook of Experimental Psychology. New York: John Wiley and Sons 1988, pp. 3-74. ISBN: 0-471-04207-2.

LUKAS et al. 2007

Lukas, M.; Gärtner, T.; Rohleder, N.; Schlick, C. M.: A Simulation Model to Predict Impacts of Alterations in Development Processes. In: Lindemann, U. et al. (Eds.): Proceedings of the 9th International DSM Conference, Munich, 16.-18.10.2008. Aachen: Shaker 2007, pp. 127-136. ISBN: 978-3-8322-6641-7.

MAIER et al. 2008

Maier, A.; Kreimeyer, M.; Lindemann, U.; Clarkson, P. J.: Reflecting communication: a key factor for successful collaboration between embodiment design and simulation. Journal of Engineering Design 20 (2009) 3, pp. 265-287.

MAIER 2007

Maier, A.: A grid-based Asessment Method of Communcation in Engineering Design. Ph.D. Thesis, University of Cambridge, 2007. Cambridge: University of Cambridge 2007.

MALIK 2003

Malik, F.: Strategie des Managements komplexer Systeme. Habilitation, Universität St. Gallen, 1978. Bern: Haupt 2003. ISBN: 3-258-06684-1.

MARCA & MCGOWEN 1988

Marca, D. A.; McGowen, C. L.: SADT - Structural Analysis and Design Technique. New York: McGraw-Hill 1988. ISBN: 0-07-040235-3.

MASLOV & SNEPPEN 2002

Maslov, S.; Sneppen, K.: Specificity and Stability in Topology of Protein Networks. Science 296 (2002) 5569, pp. 910-913.

MAURER 2007

Maurer, M.: Structural Awareness in Complex Product Design. Dissertation, Technische Universität München, 2007. München: Dr. Hut 2007. ISBN: 978-3-89963-632-1.

MAURER et al. 2006

Maurer, M.; Pulm, U.; Eichinger, M.; Lindemann, U.: Extending Design Structure Matrices and Domain Mapping Matrices by Multiple Design Structure Matrices. In: Proceedings of the 8th Biennial ASME Conference on Engineering Systems Design and Analysis, ESDA 2006, Torino, Italy, 04.-07.07.2006. New York: ASME 2006. ISBN: 0-7918-3779-3.

MAURER et al. 2004

Maurer, M.; Pulm, U.; Lindemann, U.: Utilization of graph constellations for the development of customizable product spectra. In: Proceedings of the 4th International ICSC Symposium on Engineering of Intelligent Systems, EIS 2004, Funchal, Madeira, Portugal, 29.02.-02.03.2004. Millet, Alberta: ICSC Interdisciplinary Research Canada 2004. ISBN: 3-906454-35-5.

MCCABE 1976

McCabe, T. J.: A Complexity Measure. IEEE Transactions on Software Engineering 2 (1976) 4, pp. 308-320.

MCCORMICK et al. 1972

McCormick, J. W. T.; Schweizer, P. J.; White, T. W.: Problem Decomposition and Data Reorganization by a Clustering Technique. Operations Research 20 (1972) 5, pp. 993-1009.

MCQUAID 1997

McQuaid, P. A.: The profile metric and software quality. In: Proceedings of the International Conference on Software Quality. Montgomery, AL, 06.-08.10.1997. Milwaukee, WI: American Society for Quality 1997, pp. 245-252.

MENDLING 2008

Mendling, J.: Metrics for Process Models. Lecture Notes in Business Information Processing, 6. Berlin: Springer 2008. ISBN: 978-3-540-89223-6.

MENDLING et al. 2007

Mendling, J.; Reijers, H. A.; Cardoso, J.: What Makes Process Models Understandable? In: Alonso, G. et al. (Eds.): Proceedings of the 5th International Conference on Business Process Management, BPM 2007, Brisbane, Australia, 24.-28.09.2007. Berlin: Springer 2007, pp. 48-63. ISBN: 978-3-540-75182-3.

MERRIAM-WEBSTER ONLINE DICTIONARY 2009

Merriam-Webster Online Dictionary: Structure. URL: http://www.merriam-webster.com/dictionary/structure - Access Date: 05.02.2009.

MERTINS & JOCHEM 1998

Mertins, K.; Jochem, R.: MO²GO. In: Bernus, P.; Mertins, K.; Schmidt, G. (Eds.): Handbook on Architectures of Information Systems. Berlin: Springer 1998. ISBN: 3-540-64453-9.

MILLER et al. 2006

Miller, W. D.; McCarter, G.; Hayenga, C. O.: Modeling Organizational Dynamics. In: Proceedings of the 2006 IEEE/SMC International Conference on System of Systems Engineering, Los Angeles, CA, 24.-26.04.2006. Piscataway, NJ: IEEE Operations Center 2006, pp. 94-99. ISBN: 1-4244-0188-7.

MILO et al. 2002

Milo, R.; Shen-Orr, S.; Itzkovitz, S.; Kashtan, N.; Chklovskii, D.; Alon, U.: Network Motifs: Simple Building Blocks of Complex Networks. Science 298 (2002) 5594, pp. 824-827.

MINGERS 1994

Mingers, J.: Self-Producing Systems - Implications and Applications of Autopoiesis. Berlin: Springer 1994. ISBN: 0-306-44797-5.

MINNEMANN 1991

Minnemann, S.: The Social Construction of a Technical Reality - Empirical Studies of Group Engineering Practice. Ph.D. Thesis, Stanford University, 1991. Palo Alto, CA: Stanford University 1991.

MOODY 2003

Moody, J.: Race, School Integration, and Friendship Segregation in America. American Journal of Sociology 197 (2003) 3, pp. 679-716.

MORASCA 1999

Morasca, S.: Measuring attributes of concurrent software specifications in petri nets. In: Proceedings of the 6th International Software Metrics Symposium, METRICS'99, Boca Raton, FL, 04.-06.11.1999. Los Alamitos, CA: IEEE Computer Society 1999, pp. 100-110. ISBN: 0-7695-0403-5.

MORISAWA 2002

Morisawa, T.: Building Performance Measurement Systems with the Balanced Scorecard Approach. Tokyo: Nomura Research Institute 2002. NRI Papers No. 45 (2002) April 1.

MUTSCHELLER 1996

Mutscheller, A. M.: Vorgehensmodell zur Entwicklung von Kennzahlen und Indikatoren für das Qualitätsmanagement. Dissertation, Universität St. Gallen, 1996. St. Gallen: Universität St. Gallen 1996.

NASA 1995

NASA: NASA Systems Engineering Handbook, SP-6105. Washington, D. C.: National Aeronautics and Space Administration 1995.

NAVLAKHA 1987

Navlakha J. K.: A Survey of System Complexity Metrics. The Computer Journal 30 (1987) 3, pp. 233-238.

NEWELL 1990

Newell, A.: Unified Theories of Cognition. Cambridge, MA: Harvard University Press 1990. ISBN: 0-674-92099-6.

NEWMAN 2003a

Newman, M. E.: The Structure and Function of Complex Networks. SIAM Review 45 (2003) 2, pp. 167-256.

NEWMAN 2003b

Newman, M. E.: Mixing patterns in networks. Physical Review E 67 (2003) 2, id. 026126.

NEWMAN 2002

Newman, M. E.: Assortative Mixing in Networks. Physical Review Letters 89 (2002) 20, id. 208701.

NEWMAN et al. 2001

Newman, M. E.; Strogatz, S. H.; Watts, D. J.: Random Graphs with Arbitrary Degree Distributions and Their Applications. Physical Review E 64 (2001) 2, id. 026118.

NIKOLOSKI et al. 2005

Nikoloski, Z.; Deo, N.; Kucera, L.: Degree-correlation of scale-free graphs. In: Felsner, S. (Ed.): Proceedings of the 2005 European Conference on Combinatorics, Graph Theory and Applications, EuroComb'05, Berlin, 05.-09.09.2005. Discrete Mathematics & Theoretical Computer Science, DMTCS Proceedings Volume AE (2005), pp. 239-244. URL: http://www.dmtcs.org/proceedings/dmAE01ind.html - Access Date: 27.02.2009.

NISSEN 1998

Nissen, M. E.: Redesigning Reengineering through Measurement-driven Inference. MIS Quarterly 22 (1998) 4, pp. 509-534.

NORDEN 1964

Norden, P. V.: Manpower utilization patterns in research and development. Ph.D. Thesis, Columbia University, 1964. New York, NY: Columbia University 1964.

NORDSIECK 1934

Nordsieck, F.: Grundlagen der Organisationslehre. Stuttgart: Poeschel 1934.

O'DONNELL & DUFFY 2005

O'Donnell, F. J.; Duffy, A. H. B.: Design Performance. London: Springer 2005. ISBN: 1-85233-889-X.

O'DONOVAN et al. 2005

O'Donovan, B. D.; Eckert, C.; Clarkson, P. J.; Browning, T. R.: Design Planning and Modelling. In: Clarkson, P. J. et al. (Eds.): Design Process Improvement - A Review of Current Practice. London: Springer 2005, pp. 60-87. ISBN: 1-85233-701-X.

O'DONOVAN 2004

O'Donovan, B. D.: Modelling and Simulation of Engineering Design Processes. Ph.D. Thesis, University of Cambridge, 2000. Cambridge: University of Cambridge 2004.

ODIORNE 1980

Odiorne, G. S.: Management by Objectives. München: Moderne Industrie 1980. ISBN: 3-478-32640-0.

OFFEN & JEFFERY 1997

Offen, R. J.; Jeffery, R.: Establishing Software Measurement Programs. IEEE Computer Society 14 (1997) 2, pp. 45-53.

OLIVER et al. 1997

Oliver, D. W.; Kelliher, T. P.; Keegan, J. G. Jr.: Engineering Complex Systems. New York: McGraw-Hill 1997. ISBN: 0-07-048188-1.

OESTERREICH et al. 2003

Oesterreich, B.; Weiss, C.; Schröder, C.; Weilkiens, T.; Lenhard, A.: Objektorientierte Geschäftsprozessmodellierung mit der UML. Heidelberg: dpunkt 2003. ISBN: 3-89864-237-2.

OVIEDO 1980

Oviedo, E. I.: Control Flow, Data Flow and Program Complexity. Proceedings of the 4th IEEE International Computer Software & Applications Conference, COMPSAC'80, Chicago, IL, 27.-31.10.1980. Los Alamitos, CA: IEEE Computer Society 1980, pp. 146-152.

SIMPSON et al. 1989

Simpson, J. A.; Weiner, E. S. C.; Murry, J. A. H. (Eds.): The Oxford English Dictionary. Oxford: Oxford University Press 1989. ISBN: 0-19-861186-2.

PADULO & ARBIB 1974

Padulo, L.; Arbib, M. A.: System Theory - A Unified State-space Approach to Continuous and Discrete Systems. Philadelphia: Saunders 1974. ISBN: 0-7216-7035-0.

PAHL et al. 2007

Pahl, G.; Beitz, W.; Feldhusen, J.; Grote, K.-H.: Konstruktionslehre. Berlin: Springer 2007. ISBN: 978-3-540-34060-7.

PAPADIMITRIOU 1994

Papadimitriou, C. H.: Computational Complexity. Reading, MA: Addison-Wesley 1994. ISBN: 0-201-53082-1.

PATZAK 1982

Patzak, G.: Systemtechnik – Planung komplexer, innovativer Systeme. Berlin: Springer 1982. ISBN: 3-540-11783-0.

PETERSON 1981

Peterson, J. L.: Petri Net Theory and the Modeling of Systems. Englewood Cliffs, NJ: Prentice Hall 1981. ISBN: 0-13-661983-5.

PFEIFFER & GEHLERT 2005

Pfeiffer, D.; Gehlert, A.: A framework for comparing conceptual models. In: Desel, J. et al. (Eds.): Proceedings of the Workshop Enterprise Modelling and Information Systems Architectures, EMISA 2005, Klagenfurt, 24.-25.10.2005. Bonn: Gesellschaft für Informatik, GI 2005, pp. 108-122. ISBN: 3-88579-404-7.

PIMMLER & EPPINGER 1994

Pimmler, T. U.; Eppinger, S. D.: Integration Analysis of Product Decompositions. In: Hight, T. K. et al. (Eds.): Proceedings of the ASME Design Technical Conferences, 6th International Conference on Design Theory and Methodolog, DTM'94, Minneapolis, MN, 11.-14.09.1994. New York, NY: ASME 1994. ISBN: 0-7918-1282-0.

PIWOWARSKI 1982

Piwowarski, P.: A Nesting Level Complexity Measure. ACM SIGPLAN Notices 19 (1982) 9, pp. 44-50.

PMI 2003

PMI: A Guide To The Project Management Body Of Knowledge. Newtown Square: Project Management Institute, PMI 2003. ISBN: 1-930699-21-2.

PONN & LINDEMANN 2008

Ponn, J.; Lindemann, U.: Konzeptentwicklung und Gestaltung technischer Produkte. Berlin: Springer 2008. ISBN: 978-3-540-68562-3.

PONN & LINDEMANN 2005

Ponn, J.; Lindemann, U.: Characterization of Design Situations and Processes and a Process Module Set for Product Development. In: Samuel, A. E. et al. (Eds.): Proceedings of the 15th International Conference on Engineering Design, ICED'05, Melbourne, Australia, 15.-18.08.2005. Barton ACT: Engineers Australia 2005. ISBN: 0-85825-788-2.

PROBST 1987

Probst, G. J. B.: Selbst-Organisation - Ordnungsprozesse in sozialen Systemen aus ganzheitlicher Sicht. Berlin: Parey 1987. ISBN: 3-489-63334-2.

PROBST 1981

Probst, G. J. B.: Kybernetische Gesetzhypothesen als Basis für Gestaltungs- und Lenkungsregeln im Management. Dissertation, Universität St. Gallen, 1981. Bern: Haupt 1981. ISBN: 3-258-03116-9.

PULLAN & BHADESHIA 2000

Pullan, W.; Bhadeshia, H. (Eds.): Structure - In Science and Art. Cambridge: Cambridge University Press 2000. ISBN: 0-521-78258-9.

PULM 2004

Pulm, U.: Eine systemtheoretische Betrachtung der Produktentwicklung. Dissertation, Technische Universität München, 2004. München: Dr. Hut 2004. ISBN: 3-89963-062-9.

RECHTIN 1991

Rechtin, E.: Systems Architecting - Creating & Building Complex Systems. Englewood Cliffs, NJ: Prentice Hall 1991. ISBN: 0-13-880345-5.

RECKER 2007

Recker, J.: A Socio-pragmatic Constructionist Framework for Understanding Quality in Process Modeling. Australasian Journal of Information Systems 14 (2007) 2, pp. 43-63.

REED 2001

Reed, W. J.: The Pareto, Zipf and other power laws. Economics Letters 74 (2001) 1, pp. 15-19.

REICHMANN & FORM 2000

Reichmann, T.; Form, S.: Balanced Chance and Risk Management. Controlling 12 (2000) 4-5, pp. 189-199.

REICHMANN & LACHNIT 1977

Reichmann, T.; Lachnit, L.: Kennzahlensysteme als Instrument zur Planung, Steuerung und Kontrolle von Unternehmungen. Maschinenbau 9 (1977) pp. 45-53 and 10 (1977) pp. 13-19.

RICHTER 2007

Richter, K.: Methoden zur Unterstützung bei der Entwicklung plattformübergreifender Benutzerschnittstellen. Dissertation, Technische Universität Darmstadt, 2007. Darmstadt: Technische Universität Darmstadt 2007.

RIND 1999

Rind, D.: Complexity and Climate. Science 284 (1999) 5411, pp. 105-107.

RITTEL & WEBBER 1973

Rittel, H.; Webber, M.: Dilemmas in a General Theory of Planning. Policy Sciences, 4 (1972) 2, pp. 155-169. ISSN: 0032-2687.

ROBERTSON & SEYMOUR 1986

Robertson, N.; Seymour, P. D.: Graph minors. II. Algorithmic aspects of tree-width. Journal of Algorithms 7 (1986) 3, pp. 309-322.

ROELOFSEN et al. 2008

Roelofsen, J.; Krehmer, H.; Lindemann, U.; Meerkamm, H.: Using the Design-Structure-Matrix for the Avoidance of Unnecessary Iterations. In: Kreimeyer, M. et al. (Eds.): Proceedings of the 10th International DSM Conference, Stockholm, 11.-12.11.2008. Munich: Hanser 2008, pp. 209-218. ISBN: 978-3-446-41825-7.

ROLÓN et al. 2006a

Rolón, E.; Ruiz, F.; García, F.; Piattini, M.: Applying Software Process Metrics in Business Process Model. RPM-AEMES 3 (2006) 2, pp. 45-61.

ROLÓN et al. 2006b

Rolón, E.; Ruiz, F.; García, F.; Piattini, M.: Evaluation Measures for Business Process Models. In: Haddad, H. M. (Ed.): Proceedings of the 21st Annual ACM Symposium on Applied Computing, SAC 2006, Dijon, France, 23.-27.04.2006. New York, NY: Association for Computing Machinery, ACM 2006, pp. 1567-1568. ISBN: 1-595-93108-2.

RUMBAUGH et al. 2005

Rumbaugh, J.; Jacobson, I.; Booch, G.: The Unified Modeling Language Reference Manual. Boston: Addison-Wesley 2005. ISBN: 0-321-24562-8.

SABBAGHIAN et al. 1998

Sabbaghian, N.; Eppinger, S. D.; Murman, E.: Product Development Process Capture and Display Using Web-Based Technologies. In: Zhou, M.-C. (Ed.): Proceedings of the 28th IEEE International Conference on Systems, Man, and Cybernetics, San Diego, CA, 11.-14.10.1998. Piscataway, NJ: IEEE Service Center, 1998, pp. 2664-2669. ISBN: 0-7803-4778-1.

SAPUAN et al. 2006

Sapuan, S. M.; Osman, M. R.; Nukman, Y.: State of the art of the concurrent engineering technique in the automotive industry. Journal of Engineering Design 17 (2006) 2, pp. 143-157.

SCHEER 1999

Scheer, A.-W.: ARIS - Business Process Modeling. Berlin: Springer 1999. ISBN: 3-540-65835-1.

SCHEER et al. 1997

Scheer, A.-W.; Nüttgens, M.; Zimmermann, V.: Objektorientierte Ereignisgesteuerte Prozeßkette (oEPK) - Methode und Anwendung. Veröffentlichungen des Instituts für Wirtschaftsinformatik, IWi, Universität des Saarlandes, IWi-Heft 141. Saarbrücken: Universität Saarbrücken 1997.

SCHLICK et al. 2008

Schlick, C. M.; Duckwitz, S.; Gärtner, T.; Schmidt, T.: A Complexity Measure for Concurrent Engineering Projects Based on the DSM. In: Kreimeyer, M. et al. (Eds.): Proceedings of the 10th International DSM Conference, Stockholm, 11.-12.11.2008. Munich: Hanser 2008, pp. 219-230. ISBN: 978-3-446-41825-7.

SCHMELZER & SESSELMANN 2006

Schmelzer, H.; Sesselmann, W.: Geschäftsprozessmanagement in der Praxis - Kunden zufrieden stellen - Produktivität steigern - Wert erhöhen. München: Hanser 2006. ISBN: 3-446-40589-5.

SCHÖN 1983

Schön, D. A.: The Reflective Practitioner - How Professionals Think in Action. New York: Basic Books 1983. ISBN: 0-465-06874-X.

SCHÜRRLE 1995

Schürrle, L.-H.: Prozessorientierte Kennzahlen als Analyseinstrument. Dissertation, Technische Universität Darmstadt, 1995. Aachen: Shaker 1995. ISBN: 3-8265-1638-9.

SHANNON & WEAVER 1998

Shannon, C. E.; Weaver, W.: The Mathematical Theory of Communication. Urbana, IL: University of Illinois Press 1998. ISBN: 0-252-72546-8.

SHAO & WANG 2003

Shao, J.; Wang, Y.: A New Measure of Software Complexity based on Cognitive Weights. Canadian Journal of Electrical and Computer Engineering 28 (2003) 2, pp. 69-74.

SIMON & MENDLING 2007

Simon, C.; Mendling, J.: Integration of Conceptual Process Models by the Example of Event-driven Process Chains. In: Oberweis, A. et al. (Eds.): Proceedings of the 8th Internationale Tagung Wirtschaftsinformatik, WI 2007, Karlsruhe, 28.02.-02.03.2007. Karlsruhe: Universitätsverlag Karlsruhe 2007, pp. 677-694. ISBN: 978-3-86644-094-4.

SIMON & SIMON 2005

Simon, D.; Simon, F.: Das wundersame Verhalten von Entwicklern beim Einsatz von Quellcode-Metriken. In: Büren, G. et al. (Eds.): Proceedings of the DASMA Software Metrik Kongress, MetriKon 2005, Kaiserslautern, 15.-16.11.2005. Aachen: Shaker 2005, pp. 263-272. ISBN: 3-8322-4615-0.

SMITH 1996

Smith, P. G.: Your product development process demands ongoing improvement. Research Technology Management 39 (1996) 2, pp. 37-44.

SOSA 2008

Sosa, M.: A Structured Approach to Predicting and Managing Technical Interactions in Software Development. Research in Engineering Design 19 (2008) 1, pp. 47-70.

SOSA et al. 2004

Sosa, M.; Eppinger, S. D.; Rowles, C. M.: The Misalignment of Product Architecture and Organizational Structure in Complex Product Development. Management Science 50 (2004) 12, pp. 1674-1689.

SPATH & WEISBECKER 2008

Spath, D.; Weisbecker, A. (Eds.): Business Process Management Tools 2008. Stuttgart: Fraunhofer IRB 2008. ISBN: 978-3-8167-7596-6.

SPATH et al. 2001

Spath, D.; Dill, C.; Scharer, M.: Vom Markt zum Markt - Produktentstehung als zyklischer Prozess. Stuttgart: LOG_X 2001. ISBN: 3-932298-09-8.

SPUR et al. 1993

Spur, G.; Mertins, K.; Jochem, R.; Warnecke, H. J.: Integrierte Unternehmensmodellierung. Berlin: Beuth 1993. ISBN: 3-410-12923-5.

STETTER 2000

Stetter, R.: Method Implementation in Integrated Product Development. Dissertation, Technische Universität München, 2000. München: Technische Universität München 2000.

STEVENS 1946

Stevens, S. S.: On the Theory of Scales of Measurement. Science 103 (1946) 2684, pp. 677-680.

STEWARD 1981

Steward, D. V.: The Design Structure System: A Method for Managing the Design of Complex Systems. IEEE Transactions on Engineering Management 28 (1981) pp. 71-74.

STROGATZ 2001

Strogatz, S. H.: Exploring complex networks. Nature 410 (2001) 6825, pp. 268-276.

SUH 1999

Suh, N. P.: A Theory of Complexity, Periodicity and the Design Axioms. Research in Engineering Design 11 (1999) 2, pp. 116-132.

SUMMERS & SHAH 2003

Summers, J.; Shah, J.: Developing Measures of Complexity for Engineering Design. In: Gupta, S. K. (Ed.): Proceedings of the ASME 2003 Design Engineering Technical Conferences and Computers and Information in Engineering Conference, DETC'03, 15th International Conference on Design Theory and Methodology, Chicago, IL, 02.-06.09.2003. New York, NY: ASME 2003. ISBN: 0-7918-3701-7.

SUPPES & ZINNES 1963

Suppes, P.; Zinnes, J.: Basic Measurement Theory. In: Luce, R.D.; Bush, R.R.; Galante, E. (Eds.): Mathematical Psychology. New York: John Wiley and Sons 1963, pp.4-76.

TITTMANN 2003

Tittmann, P.: Graphentheorie - Eine anwendungsorientierte Einführung. München: Hanser 2003. ISBN: 3-446-22343-6.

TJADEN et al. 1996

Tjaden, G. S.; Narashimhan, S.; Mitra, S.: Structural effectiveness metrics for business proceses. In: Pirkul H. et al. (Eds.): Proceedings of the 1th Conference on Information Systems and Technology, INFORMS, Washington, DC, 05.-09.05.1996. Washington, DC: INFORMS 1996, pp. 396-400.

TSAI et al. 1986

Tsai, W. T.; Lopex, M. A.; Rodriguez, V.; Volovik, D.: An approach to measuring data structure complexity. In: Proceedings of the 10th IEEE Computer Society's Annual International Computer Software and Applications Conference, COMPSAC'86, Chicago, IL, 08.-10.10.1986. New York, NY: IEEE 1986, pp. 240–246. ISBN: 0-8186-4727-2.

TUFTE 2001

Tufte, E. R.: The Visual Display of Quantitative Information. Cheshire: Graphics Press 2001. ISBN: 0-9613921-4-2.

VAJNA 2005

Vajna, S.: Workflow for Design. In: Clarkson, P. J. et al. (Eds.): Design Process Improvement - A Review of Current Practice. London: Springer 2005, pp. 366-385. ISBN: 1-85233-701-X.

VAN DER AALST & TER HOFSTEDE 2005

van der Aalst, W. M. P.; ter Hofstede, A. H. M.: YAWL: yet another workflow language. Information Systems 30 (2005) 4, pp. 245-275.

VAN DER AALST & TER HOFSTEDE 2003

van der Aalst, W. M. P.; ter Hofstede, A. H. M.; Kiepuszewski, B.; Barros, A. P.: Workflow Patterns. Distributed and Parallel Databases 14 (2003) 1, pp. 5-51.

VAN DER AALST & TER HOFSTEDE 2002

van der Aalst, W. M. P.; van Hee, K. M.: Workflow Management - Models, Methods, and Systems. Cambridge: MIT Press 2002. ISBN: 0-262-01189-1.

VAN DONGEN et al. 2006

van Dongen, B. F.; Mendling, J.; van der Aalst, W. M. P.: Structural Patterns for Soundness of Business Process Models. In: Proceedings of the 10th IEEE International Enterprise Distributed Object Computing Conference, EDOC'06, Hong Kong, 16.-20.10.2006. Los Alamitos: IEEE Computer Society 2006, pp. 116-128. ISBN: 0-7695-2558-X.

VANDERFEESTEN et al. 2008

Vanderfeesten, I.; Mendling, J.; Reijers, H. A.; van der Aalst, W.; Cardoso, J.: On a quest for good process models: The cross-connectivity metric. In: Bellahsène, Z. et al. (Eds.): Proceedings of the 20th international conference on Advanced Information System Engineering, CAiSE'08, Montpellier, 16.-20.06.2008. Lecture Notes in Computer Science, 5074. Berlin: Springer 2008, pp. 480-494. ISBN: 978-3-540-69533-2.

VANDERFEESTEN et al. 2007

Vanderfeesten, I.; Reijers, H. A.; van der Aalst, W.; Cardoso, J.; Mendling, J.: Quality Metrics for Business Process Models. In: Fischer, L. (Ed.): 2007 BPM and Workflow Handbook. Lighthouse Point, FL: Future Strategies Inc. 2007, pp. 179-190. ISBN: 978-0-9777527-1-3.

VOLKSWAGEN AG 2008

Volkswagen AG: Phaeton Bordnetz. Kassel: Forschungsverbund Fahrzeugsysteme, Universität Kassel 2008. URL: http://www.uni-kassel.de/fb16/ fsg/bilder/phaeton_bordnetz.jpg - Access Date: 01.06.2008.

VOLKSWAGEN AG 2007

Volkswagen AG: Phaeton. Wolfsburg: Volkswagen AG, Vertrieb Kundendienst, VST 1/2 Technische Kundeninformation und Dokumentenmanagement 2007.

VON BERTALANFFY 1950

von Bertalanffy, L.: An Outline of General Systems Theory. British Journal for the Philosophy of Science 1 (1950) 2, pp. 134-165.

VON BERTALANFFY 1968

von Bertalanffy, L.: General System Theory - Foundations, Development, Applications. New York, NY: Braziller 1968. ISBN: 0-8076-0453-4.

WALDMAN & SANGAL 2007

Waldman, F.; Sangal, N.: Applying DSM to Enterprise Architectures. In: Lindemann, U. et al. (Eds.): Proceedings of the 9th International DSM Conference, Munich, 16.-18.10.2008. Aachen: Shaker 2007, pp. 61-71. ISBN: 978-3-8322-6641-7.

WAND & WEBER 1990

Wand, Y.; Weber, R.: Mario Bunge's Ontology as a Formal Foundation for Information System Concepts. In: Weingartner, P. et al. (Eds.): Studies in Bunge's Treatise on Basic Philosophy of the Sciences and the Humanities. Amsterdam: Rodopi 1990, pp. 123-149. ISBN: 90-5183-187-0.

WANG 2006

Wang, Y.: Cognitive Complexity of Software and its Measurement. In: Yao, Y. (Ed.): Proceedings of the 5th IEEE International Conference on Cognitive Informatics, ICCI 2006, Beijing, China, 17.-19.07.2006. Piscataway, NJ: IEEE 2006, pp. 226-235. ISBN: 1-424-40475-4.

WASSON 2006

Wasson, C. S.: System Analysis, Design, and Development Concepts, Principles, and Practices. Hoboken, NJ: Wiley-Interscience 2006. ISBN: 0-471-39333-9.

WATTS & STROGATZ 1998

Watts, D. J.; Strogatz, S. H.: Collective Dynamics of 'Small-world' Networks. Nature 393 (1998) 6684, pp. 440-442.

PONN & LINDEMANN 2005

Weber, C.: What is "complexity"? In: Samuel, A. E. et al. (Eds.): Proceedings of the 15th International Conference on Engineering Design, ICED'05, Melbourne, Australia, 15.-18.08.2005. Barton ACT: Engineers Australia 2005. ISBN: 0-85825-788-2.

WENG & BHALLA 1999

Weng, G.; Bhalla, U. S.; Iyengar, R.: Complexity in Biological Signaling Systems. Science 284 (1999) 5411, pp. 92-96.

WEYUKER 1988

Weyuker, E. J.: Evaluating Software Complexity Measures. IEEE Transactions on Software Engineering 14 (1988) 9, pp. 1357-1365.

WHITE & MIERS 2008

White, S. A.; Miers, D.: BPMN Modeling and Reference Guide. Lighthouse Point, FL: Future Strategies 2008. ISBN: 978-0-9777527-2-0.

WHITFIELD et al. 2000

Whitfield, R. I.; Coates, G.; Duffy, A. H.; Hill, B.: Coordination Approaches and Systems- Part I: A Strategic Perspective. Research in Engineering Design 12 (2000) 1, pp. 48-60.

WIENER 1948

Wiener, N.: Cybernetics - Or the Control and Communication in the Animal and the Machine. New York: Technology Press 1948.

WOODWARD et al. 1979

Woodward, M. R.; Hennell, M. A.; Hedley, D.: A Measure of Control Flow Complexity in Program Text. IEEE Transactions on Software Engineering SE-5 (1979) 1, pp. 45-50.

WYNN et al. 2009

Wynn, D.; Nair, S.; Clarkson, P.J.: The P3 platform: An Approach and Software System for Developing Diagrammatic Model-Based Methods in Design Research. In: Proceedings of International Conference on Engineering Design - ICED 2009. Glasgow: The Design Society 2009.

WYNN 2007

Wynn, D. C.: Model-Based Approaches to Support Process Improvement in Complex Product Development. Ph.D. Thesis, University of Cambridge, 2007. Cambridge: University of Cambridge 2007.

WYNN et al. 2007

Wynn, D. C.; Eckert, C. M.; Clarkson, P. J.: Modelling Iteration in Engineering Design. In: Bocquet, J.-C. (Ed.): Proceedings of the 16th International Conference on Engineering Design, ICED'07, Paris, 28.-31.08.2007. Glasgow: The Design Society 2007. ISBN: 1-904670-02-4.

WYNN et al. 2006

Wynn, D. C.; Eckert, C. M.; Clarkson, P. J.: Applied Signposting: A Modeling Framework to Support Design Process Improvement. In: Proceedings of the ASME 2006 International Design Engineering Technical Conferences & Computers and Information in Engineering Conference, IDETC/CIE 2006, 18th International Conference on Design Theory and Methodology, DTM, Philadelphia, 10.-13.09.2006. New York, NY: ASME 2006. ISBN: 0-7918-4258-4.

YASSINE 2004

Yassine, A.: An Introduction to Modeling and Analyzing Complex Product Development Processes Using the Design Structure Matrix (DSM) Method. Italian Management Review 9 (2004) pp. 72-88.

YASSINE et al. 2003

Yassine, A.; Whitney, D.; Daleiden, S.; Lavine, J.: Connectivity maps: modeling and analysing relationships in product development processes. Journal of Engineering Design 14 (2003) 3, pp. 377-394.

YASSINE et al. 1999

Yassine A.; Falkenburg D.; Chelst K.: Engineering Design Management: An Information Structure Approach. International Journal of Production Research, 37 (1999) 13, pp. 2957-2975. ISSN: 0020-7543.

ZACHMAN 1987

Zachman, J. A.: A framework for information systems architecture. IBM Systems Journal 26 (1987) 3, pp. 276-292.

ZAKARIAN & KUSIAK 2000

Zakarian, A.; Kusiak, A.: Analysis of Process Models. IEEE Transactions on Electronics Packaging Manufacturing 23 (2000) 2, pp. 137-147.

ZIMMERMANN 2008

Zimmermann, I.: Schulungspaket Prozessmanagement. Modul 1: Grundlagen Prozessmanagement. Kissing: WEKA-Media 2008. ISBN: 978-3-8276-4206-6.

ZUR MUEHLEN 2004

zur Muehlen, M.: Workflow-based Process Controlling - Foundation, Design, and Implementation of Workflow-driven Process Information Systems. Dissertation, Universität Münster, 2002. Advances in Information Systems and Management Sciences. Berlin: Logos 2004. ISBN: 3-8325-0388-9.

ZURN 1991

Zurn, J. T.: Problem discovery function: a useful tool for assessing new product introduction. IEEE Transactions on Engineering Management 38 (1991) 2, pp. 110-119.

ZUSE 1998

Zuse, H.: A Framework of Software Measurement. Berlin: de Gruyter 1998. ISBN: 3-11-015587-7.

# 10. Appendix

In the appendix, the following topics are detailed:

| Description of section of appendix | Section (page) | Explanation in section (page) |
|---|---|---|
| Structural content of process modeling methodologies | 10.1 (pp. 272 - 286) | 2.2.3 (p. 58) |
| Conversion of a process with logic operators into matrices | 10.2 (pp. 288 - 294) | 4.4.3 (p. 114) |
| Nesting of Boolean operators | 10.3 (pp. 295 - 296) | 4.4.3 (p. 114) |
| Structural Process Architecture | 10.4 (p. 297) | 4.3 (p. 104) |
| List of structural metrics and description of their structural significance | 10.5 (pp. 298 - 388) | 5.2 (p. 143) |
| Computability of metrics | 10.6 (p.390) | 5.2.2 (p. 147) |
| Classification of metrics | 10.7 (p. 392) | 5.2.4 (p. 157) |
| S-GQM framework | 10.8 (p. 396) | 6.2 (p. 179) |
| Complete results from case study 7.2 | 10.9 (p. 398) | 7.2.3 (p. 218) |

## 10.1 Structural content of process modeling methodologies

In this section of the appendix, common process modeling methodologies are evaluated concerning their structural content. For this purpose the structural meta-model for each methodology is shown with its core domains and the relationship types linking these domains.

| Process modeling methodology | Acronym |
|---|---|
| Extended Event-driven Process Chains | eEPC |
| Object-oriented Event-driven Process Chains | oEPC |
| Business-Process Modelling (Integrierte Unternehmensmodellierung) | IUM |
| Unified Modeling Language | UML |
| Structured Analysis and Design Technique | SADT |
| Integrated Definition Method | IDEF0 / IDEF3 |
| Business Process Modeling Notation | BPMN |
| Yet Another Workflow Language | YAWL |
| Signposting | |
| Petri-Nets | |
| Process Module Methodology | PMM |
| Program Evaluation and Review Technique | PERT |
| Objektorientierte Methode für die Geschäftsprozessmodellierung und -analyse | OMEGA |

### 10.1.1 Extended Event-driven Process Chains

EPC is a semi-formal graphical notation to model workflows and processes at several levels of detail. At its core, the behavior of a process is represented concerning various economic foci, e.g., process costs or lead time. eEPC extends the basic structure of place/transition nets (like Petri nets) to illustrate business processes [SCHEER 1999]. With its well-structured representation of a process, eEPC is one of the most common modeling standards in German industry.

EPC supports six basic domains with a wide range of pre-defined relationship types. It is possible to extend the basic definitions to suit actual modeling needs. The basic notation includes AND, OR, and XOR, which branch off from the principal flow in the model.

The following MDM shows the basic domains and relationship types that are employed in eEPC models. In that manner, it is similar to the ARIS House of Business Engineering (HOBE), which also includes the control view. The control view integrates all actual elements and relations of the process into a control flow. As it does not bring in an additional domain, it is not represented in the MDM.

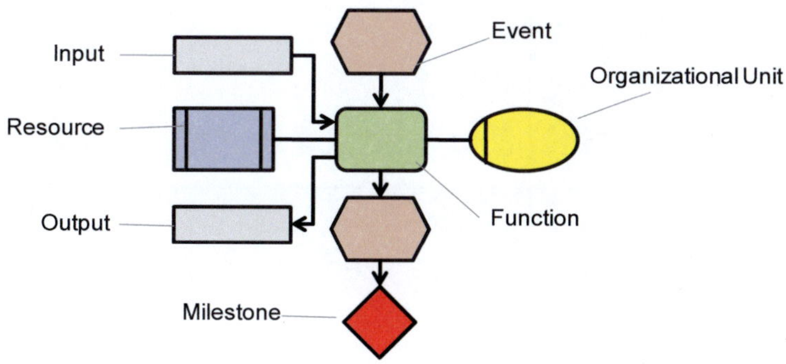

| | Function | Event | Input / Output | Org. Unit | Milestone | Resource |
|---|---|---|---|---|---|---|
| **Function** | | • Generates | • Creates<br>• Uses<br>• Produces<br>• Delivers output for | • Belongs to | • Is finished at<br>• Starts at | |
| **Event** | • Starts | | | | | |
| **Input / Output** | • Supports<br>• Is input for | | • Is created out of<br>• Is generalization of<br>• Is in conflict with | • Belongs to | • Is created at<br>• Is needed at | |
| **Org. Unit** | • Has to inform about result of<br>• Has to be informed about<br>• Acts as consultant for<br>• Is disciplinary responsible for<br>• Performs<br>• Is responsible for data processing for<br>• Takes decision about<br>• Is involved in<br>• Is produced by | | • Uses<br>• Is authorized to read<br>• Is responsible for updates of<br>• Creates<br>• Deletes<br>• Updates<br>• Reads | • Is disciplinary head of<br>• Is formed by<br>• Is functional head of<br>• Relevant / responsible for<br>• Takes role as<br>• Belongs to | | |
| **Mile-stone** | | | | | | |
| **Resource** | • Supports | | • Creates<br>• Processes | • Is finished at<br>• Starts at<br>• Is active during | | • Is predecessor of<br>• Contains |

### 10.1.2 Object-oriented Event-driven Process Chains

Object-orientated Event-driven Process Chains do not focus on the function as a central activity but on objects as discrete, distinguishable entities progressing through the process [SCHEER et al. 1997]. They integrate the concept of object-orientation into common process modeling to enable the representation of the transition of a business object. Like an object in software, a business object can be modeled with its different attributes and methods that are used to process the object. As such, the object in oEPC regroups the function view and the data view of eEPC into one singe modeling construct. The control flow, i.e., the network of relationships, is created by the messages the objects exchange while they progress from one event to the next. This means that the messages in oEPC take the place of the events in eEPC. The control flow can be split or joined using AND, OR, and XOR. The formal bipartite structure of messages and objects can be broken up by interrelating objects directly; however, in a formal process model, this is not the basic intention of the modeling scheme [SCHEER et al. 1997, p. 9].

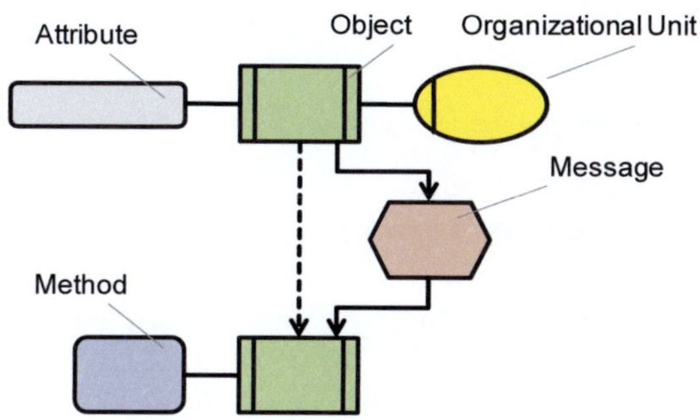


| | Object | Message | Method | Attribute (object) | Attribute (resources) |
|---|---|---|---|---|---|
| **Object** | •(has relation) | •Sends | | | |
| **Message** | •Starts | | | | |
| **Method** | •Is attribute to | | | | |
| **Attribute (object)** | •Is attribute to | | | | |
| **Attribute (resources)** | •Is attribute to | | | | |

### 10.1.3 Integrated Business-Process Modeling (IUM)

Integrated Business-Process Modeling (in German: "Integrierte Unternehmensmodellierung") regards the whole company from the perspective of the production process [SPUR et al. 1993] [MERTINS & JOCHEM 1998]. Unlike EPC, it does not have an explicit control view, but it models entities out of a basic set of objects in any sequence necessary. The entities represent the core objects, as shown below. Decisions can be modeled using AND, OR and XOR; however, these are not modeled explicitly but by representing the states the process takes in one case or the other (i.e., by case distinction). IUM is particularly centered on the "Action" as the processing task in a process, and discerns between basic "actions" as simple verbal descriptions of a task, "functions" that process inputs to outputs, and activities that specify resource and control information.

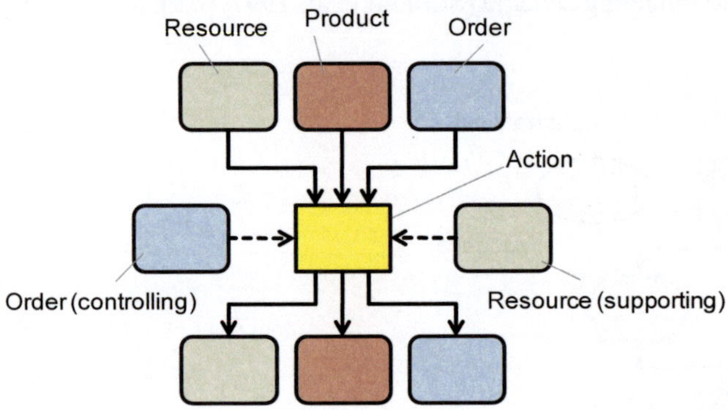


| | Product | Action | Order | Resource |
|---|---|---|---|---|
| **Product** | • Is part of<br>• Consists of | • Is processed by | • Is processed by | • Is processed by |
| **Action** | • Processes | • Leads to | • Processes | • Processes |
| **Order** | | • Controls | • Is part of<br>• Consists of | |
| **Resource** | | • Supports | | • Is part of<br>• Consists of |

### 10.1.4 Unified Modeling Language

UML is a set of notations for the design and analysis of object-oriented software [RUMBAUGH et al. 2005]. Offering a variety of models, activity diagrams have proved to be most suitable for process modeling, as they are one of the central behavioral models offered in UML 2.0 specification [ÖSTERREICH et al. 2003, p. 12] [BULLINGER & SCHREINER 2001]. UML as a highly formalized notation is especially useful, as business processes are often modeled to be later embedded into an information system, which, in turn, will be specified in UML. Activity models allow for the splitting and joining of the process flow; however, they do not necessarily represent the logic operators.

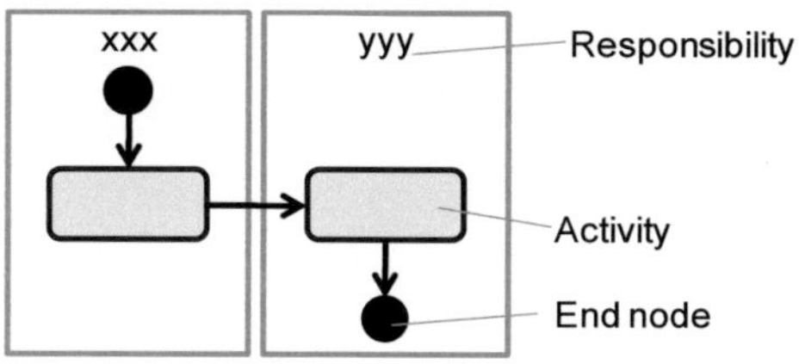


| | Initial / final node | Activity | Responsibility |
|---|---|---|---|
| **Initial / final node** | | • Starts process (initial node) | |
| **Activity** | • Ends process (final node) | • Transits into | |
| **Responsibility** | | • Is responsible for | |

### 10.1.5 Structured Analysis and Design Technique (SADT) and IDEF 0

SADT is a graphical modeling language [MARCA & MCGOWEN 1988] developed for use in the integrated computer-aided manufacturing studies of the U. S. Airforce in the 1970s. It became the basic model later used in the IDEF0 (see following section). As a very basic notation, it allows modeling the basic entities and relations to sketch the algorithm of a software program using the semantics shown in the MDM below. Although activities are related directly to one another, the relationship between them represents an input/output relationship, which is why these inputs and outputs are resolved as a separate domain. Mechanisms represent all possible resources like roles or technical equipment [BICHLMAIER 2000]. Logic operators cannot be represented.

IDEF0 is built on the functional structure of SADT (see section above) and uses the same semantics. The SADT MDM is, therefore, applicable for IDEF0 as well. Only the modeling syntax (e.g., types of forks) differs slightly from SADT from a structural point of view (see next page).

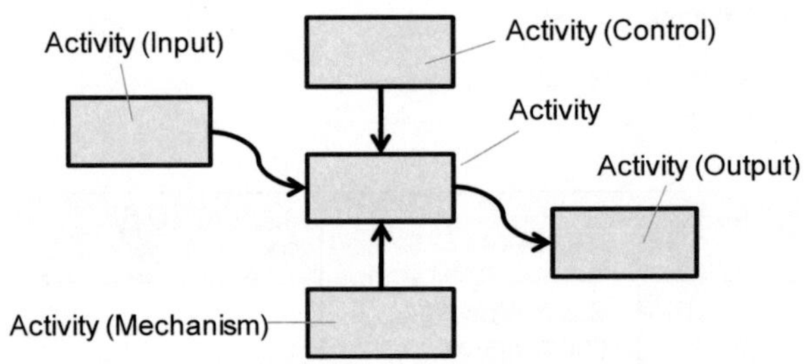


| | Control | Activity | Input / Output | Mechanism |
|---|---|---|---|---|
| **Control** | | • Controls | | |
| **Activity** | • Serves as control for | | • Has output of | |
| **Input / Output** | | • Is input for | • Output of... is input for... | |
| **Mechanism** | | • Supports | | |

### 10.1.6 Integrated Definition Method

IDEF (*Integration DEFinition*) was developed as a set of definitions and notations for various purposes of modeling [KNOWLEDGE BASED SYSTEMS 1992]. The IDEF family provides various modeling languages for software engineering; two of them are relevant for process modeling, as they are able to model sequences of activities and functions. IDEF 0, being similar to SADT, is explained on the previous page.

IDEF3 is designed as an *Integrated DEFinition for Process Description Capture Method*, complementing IDEF0 [BADICA & FOX 2005]. It provides a notation to describe process flows by modeling the relationships between actions as well as the specific states ("labels") a process undergoes. As such, it is an object-centered approach, whereas the object is part of the transformation that takes place within each task ("unit of behavior"), also referred to as an "action box". In the object-centered view, units of behavior (although shown differently in the MDM) are actually represented as attributes of the edges between the different states. IDEF3 allows the modeling of logical operators. Besides transitional relationship types, constraints can be represented. In addition to units of behavior, these can be represented by referents, which basically duplicate a unit of behavior in the model.

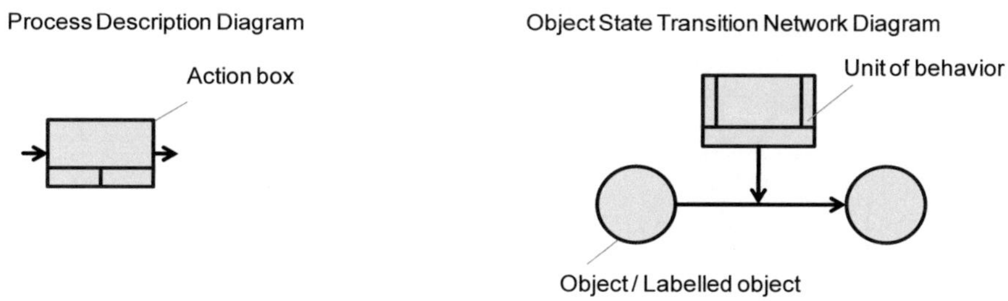


| | Unit of behavior / action box | Label | Objects / object states |
|---|---|---|---|
| **Unit of behavior/ action box** | • Precedes<br>• (others possible, too) | | |
| **Label** | | | • Relates (object) to (object) |
| **Objects/ object states** | • Uses…in transition | | • Transits into |

### 10.1.7 Business Process Modeling Notation

The Business Process Modeling Notation is a graphical notation to represent workflows and business processes [WHITE & MIERS 2008]. It was, in part, designed to allow the link to execution languages to run simulations of a process. Its primary focus, however, is an intuitive understanding of complex models.

For all domains, specific descriptions of the types of elements are available: For example, activities can be modeled as tasks, multiple instances of a task, sub processes and other constructs. Besides common relationships (sequence and message flows), conditions can also be modeled. Pools and lanes are used to represent stakeholders or applications that are active during an activity. Employing so-called gateways that represent an individual domain, the process model can be enhanced using decision logics. As in IDEF3, text annotations can also be attached to the relationships (not shown in MDM) [ALLWEYER 2008].

Currently, the Business Process Modeling Notation is available as release 1.2; an update to 2.0 exists as a proposal. The proposal is intended to extend BPMN to include the extraction of specific views onto the process, and to improve the exchange of models among different modeling tools and mappings to other process modeling schemes.

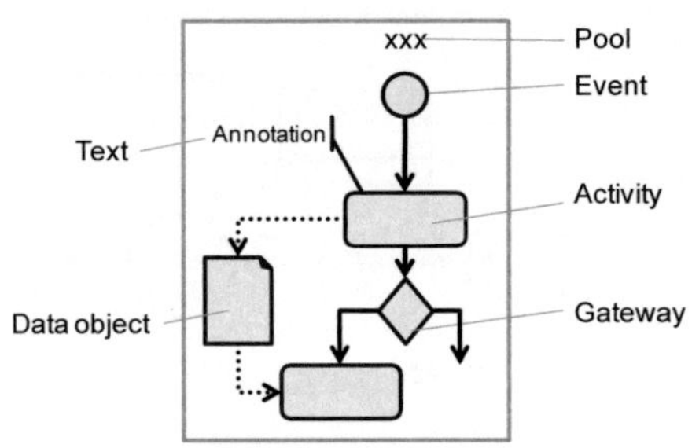


| | Activity | Event | Gateway | Data object | Text | Pool / Lane |
|---|---|---|---|---|---|---|
| **Activity** | • Precedes<br>• Transmits message to | • Ends | • Precedes<br>• Transmits message to | • Associate s with as output | | |
| **Event** | • Starts | | | | | • Occurs in |
| **Gateway** | • Precedes<br>• Transmits message to | | | | | |
| **Data object** | • Associates with as input | | | | | |
| **Text** | • Annotate | • Annotate | • Annotate | • Annotate | | • Annotate |
| **Pool / Lane** | • Executes | | • Takes decision | | | • Transmits message to |

### 10.1.8 Yet Another Workflow Language

YAWL is a modeling language to represent workflows in a way that can be executed in a software system based on Petri nets [VAN DER AALST & TER HOFSTEDE 2005]. It is based on a rather simple network of tasks that are arranged using workflow patterns and data objects that are exchanged. Decision points (AND, OR, XOR) can be modeled using conditions. The modeling method strongly relies on the introduction of attributes for further detail and decomposes the basic modeling objects, as shown in the MDM.

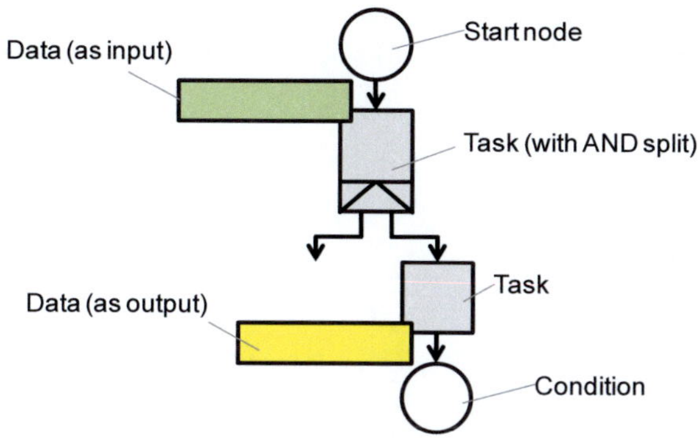


| | Start / end node | Task | Data | Condition |
|---|---|---|---|---|
| **Start / end node** | | • Starts process with | | |
| **Task** | • Ends process in | • Precedes | • Generates… as output | • Leads to |
| **Data** | | • Is input for | | |
| **Condition** | | • Is reached by | | |

### 10.1.9 Signposting

The signposting approach was primarily developed to capture expert knowledge about a process, to use this knowledge to plan and simulate processes, and to identify and assess potentials for process improvement. Therefore, the approach looks at behavior as a starting point for modeling and then deduces how high level processes arise from lower level decisions taken during runtime. At its core, the model, therefore, uses tasks that represent the activities during the process. These are attributed in particular with their inputs and outputs using parameters, the resources they require or consume, and their temporal aspects. Decisions can be represented by attributing multiple relationships to a task, where each is related to a possible scenario that actually represents the behavior at that decision point [CLARKSON & HAMILTON 2000] [WYNN et al. 2006] [WYNN 2007] [WYNN et al. 2009].

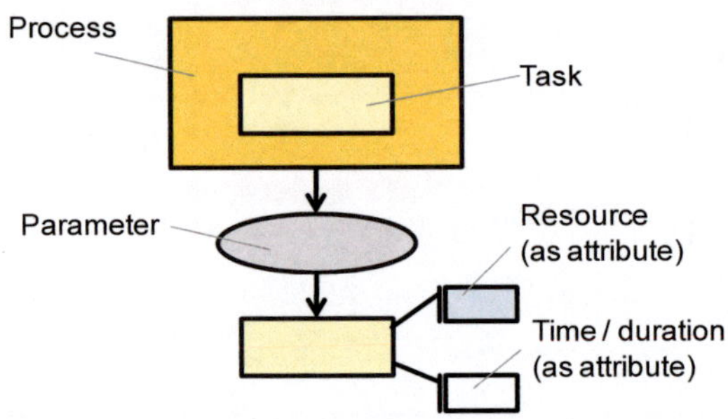


| | Task | Parameter | Resource | Process | Time / Duration |
|---|---|---|---|---|---|
| **Task** | | • Creates<br>• Changes<br>• Requires… as input<br>• Results in | • Requires | | • Has duration of |
| **Parameter** | • Informs | | | | |
| **Resource** | • Is required for<br>• Is occupied by | | | | • Is available at<br>• Is available during<br>• Is organized in |
| **Process** | • Provides detail for<br>• Encapsules | • Encapsules | | | |
| **Time / Duration** | | | | | |

### 10.1.10 Petri-Nets

Petri nets, also referred to as place/transition nets, are a modeling notation to describe discrete systems by modeling the transition of states the process undergoes. This is done by representing the process as a bipartite graph that changes between places, representing the state at a certain point of time, and transitions that represent the activities undertaken during the process. Unlike all other process modeling methodologies, Petri nets are based on a sound mathematical formalization that allows detailed simulation and analysis [PETERSON 1981].

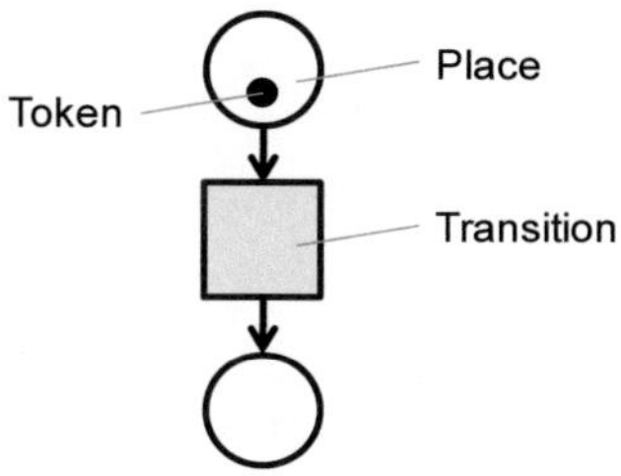


| | Transition | Place |
|---|---|---|
| **Transition** | | • Emits token to |
| **Place** | • Fires via token | |

### 10.1.11 Process Module Methodology

The Process Module Methodology decomposes the process into compact, integrated modules ("Baustein", i.e., building blocks) that are the basic entities for setting up a part of a process or a whole process to allow flexible planning. It permits a comprehensive identification of critical parts of the process. Most importantly, one module represents an individual work package that is often of a transdiciplinary character [BICHLMAIER & GRUNWALD 1999]. The modules are interrelated by the information that is transferred, and the dependencies can be of different type (e.g., general information, product data). Furthermore, the modules can be attributed with the necessary competences and tools or methods. While organizational dependencies can be modeled using the input/output information, logics and decisions cannot be represented, as the model does not focus on a behavioral view [BICHLMAIER 2000, p. 81].

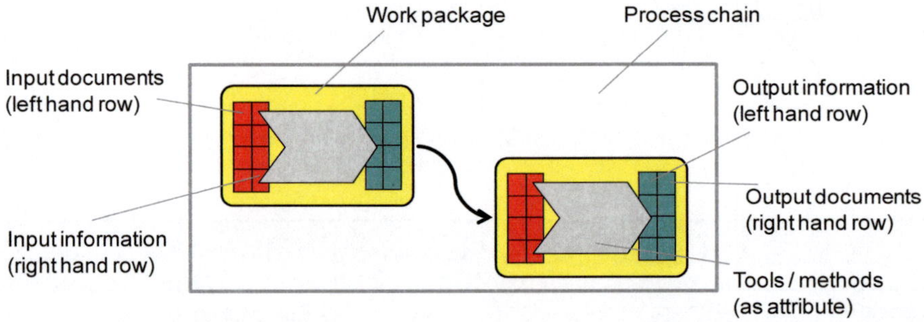


| | Work package | Information | Document | Tool / method | Process chain |
|---|---|---|---|---|---|
| **Work package** | | • Is input to | | | • Is part of |
| **Information** | • Is output of | | | | |
| **Document** | | • Represents | | | |
| **Tool / method** | • Supports<br>• Is necessary for | | | | |
| **Process chain** | • Supports<br>• Is necessary for | | | | |

### 10.1.12 Program Evaluation and Review Technique (PERT)

The Program Evaluation and Review Technique was developed in the late 1950's to plan complex engineering projects, allowing for parallelization where possible while admitting to a certain randomness of the process. Therefore, it is based on a network of tasks and their interdependencies. These tasks result in events that can synchronize multiple relationships in a process. Originally developed as an "activity on arc" network (i.e., the edges between events represent the tasks and their attributes), it was later converted to an "activity on node" network, for which it is most commonly used. Tasks within PERT are commonly used with the optimistic and pessimistic time for their execution as well as the most likely time to run the task (i.e., a basic probabilistic model is applied to estimate the expected time). Using these attributes, the critical path, i.e., the path through the process that is most likely to delay the overall process because it has no float, can be determined using the Critical Path Method (CPM) [PMI 2003]. However, CPM does not use the most likely time but the expected time to determine the critical path.

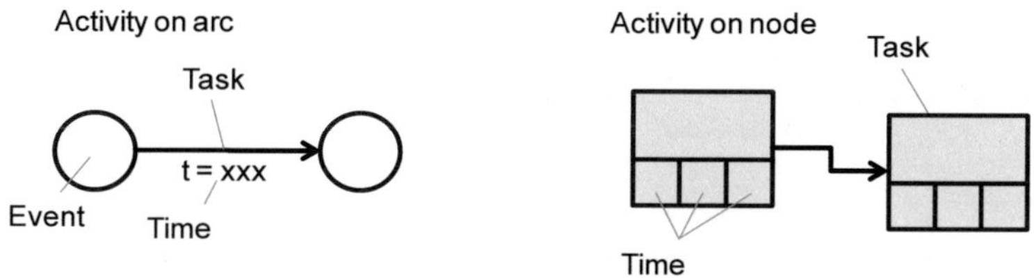


| | Task | Event | Time |
|---|---|---|---|
| **Task** | | • Results in | • Has minimum runtime of<br>• Has maximum runtime of<br>• Has most likely runtime of |
| **Event** | • Starts | | |
| **Time** | | | |

### 10.1.13 OMEGA

The OMEGA method (German acronym for "Objektorientierte Methode für die Geschäftsprozessmodellierung und -analyse": Object-oriented method for the modeling and analysis of business processes) was developed towards the end of the 1990's to overcome problems with common process modeling methodologies like SADT and similar methods: a quick understanding of the model, a systematic analysis and a means of synthesis for suggestions for improvement of the process is provided by [FAHRWINKEL 1995] [GAUSEMEIER & FINK 1999]. Besides the domains and relationships shown in the MDM below, Omega also supports the regrouping of activities into business processes to establish a control flow. This control flow can include AND, OR, and XOR. Furthermore, it is possible to describe communication relationships by attributing a supporting IT system to the edge (unlike what is shown below).

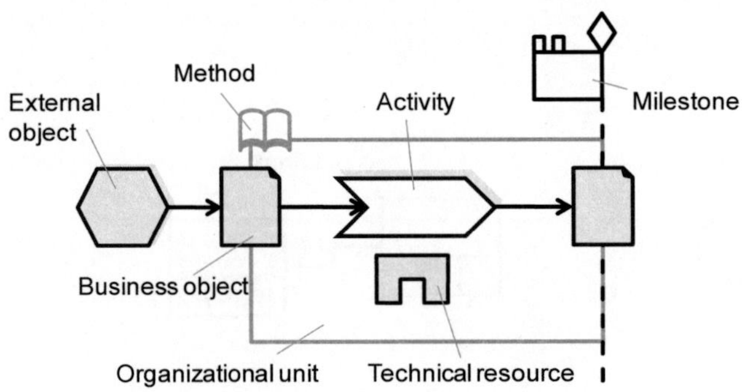

| | Activity | Business object | Milestone | Organizational unit | External object | Technical Resource | Method |
|---|---|---|---|---|---|---|---|
| Activity | | • Is output of | | | | | |
| Business object | • Is input for | | • Is necessary at | | • Receives…to finish process | • Is stored in | |
| Milestone | | | | | | | |
| Org. unit | • Executes | | • Is responsible for<br>• Assists reaching of | | | | |
| Ext. object | | • Is represented in<br>• Starts process by sending | | | | | |
| Technical Resource | • Supports | • Emits<br>• Stores<br>• Supports transmission of | | | | | |
| Method | • Supports | | | | | | |

## 10.2 Conversion of a process with logic operators

As shown in section 4.4.3, five conversions of Boolean operators into matrix-based notation are possible. In this section, they are presented using an exemplary design process for a mechatronic product, as shown in Figure 10-1, taken from [BELHE & KUSIAK 1996]. The process is modeled in IDEF3 notation and is set up as follows:

1. Prepare product specifications
2. Preliminary design
3. Evaluate cost
4. Thermal analysis
5. Electrical analysis
6. Analyze test data
7. Finalize design details

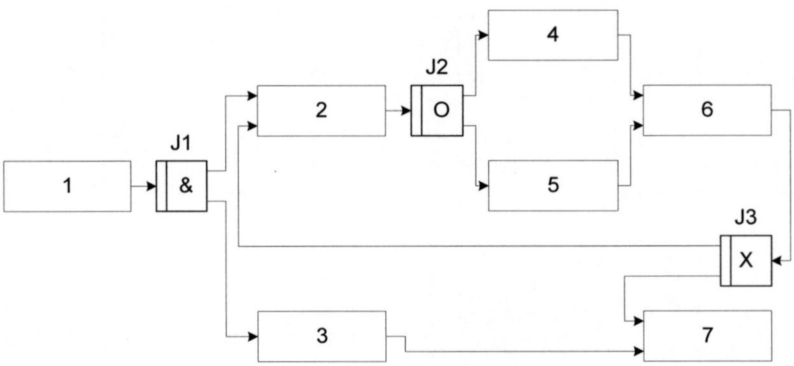


*Figure 10-1: Exemplary design process [BELHE & KUSIAK 1996]*

### 10.2.1 Rule 1: Resolve all logical connections

**Alternative 1:** $C_1 = \{(2,4), (6,7)\}$

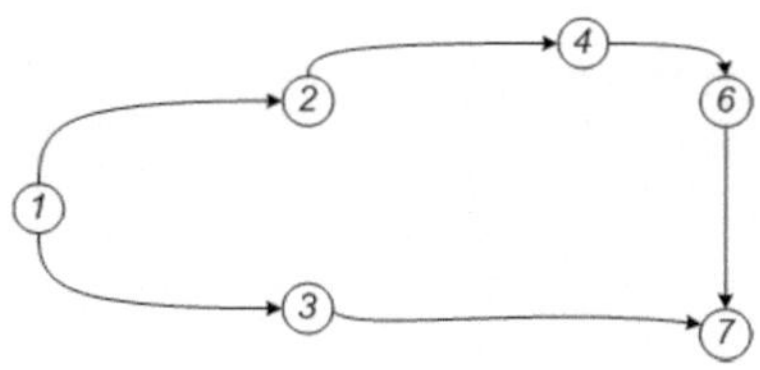


| | 1 | 2 | 3 | 4 | 5 | 6 | 7 |
|---|---|---|---|---|---|---|---|
| 1 | | X | X | | | | |
| 2 | | | | X | | | |
| 3 | | | | | | | X |
| 4 | | | | | | X | |
| 5 | | | | | | | |
| 6 | | | | | | | X |
| 7 | | | | | | | |

*Figure 10-2: Conversion according to rule 1, alternative result 1*

**Alternative 2:** $C_2 = \{(2,5), (6,7)\}$

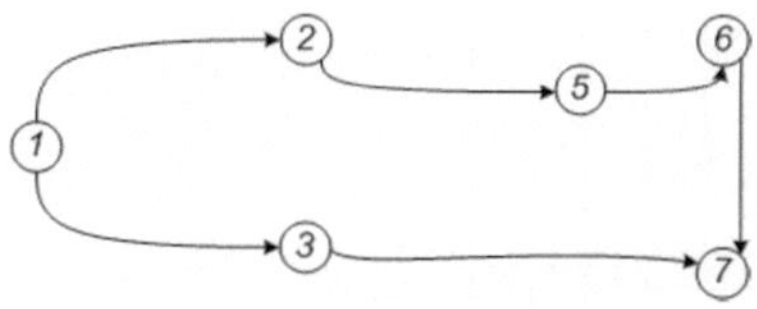


| | 1 | 2 | 3 | 4 | 5 | 6 | 7 |
|---|---|---|---|---|---|---|---|
| 1 | | X | X | | | | |
| 2 | | | | | X | | |
| 3 | | | | | | | X |
| 4 | | | | | | | |
| 5 | | | | | | X | |
| 6 | | | | | | | X |
| 7 | | | | | | | |

*Figure 10-3: Conversion according to rule 1, alternative result 2*

**Alternative 3:** $C_3 = \{(2,4), (2,5), (6,7)\}$

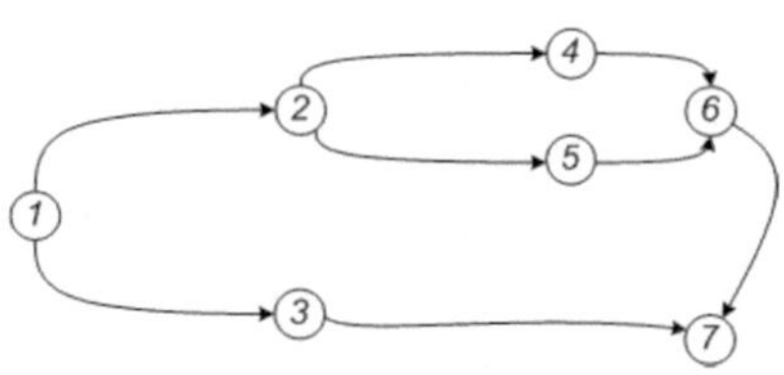


| | 1 | 2 | 3 | 4 | 5 | 6 | 7 |
|---|---|---|---|---|---|---|---|
| 1 | | X | X | | | | |
| 2 | | | | X | X | | |
| 3 | | | | | | | X |
| 4 | | | | | | X | |
| 5 | | | | | | X | |
| 6 | | | | | | | X |
| 7 | | | | | | | |

*Figure 10-4: Conversion according to rule 1, alternative result 3*

**Alternative 4:** $C_4 = \{(2,4), (6,2), (6,7)\}$

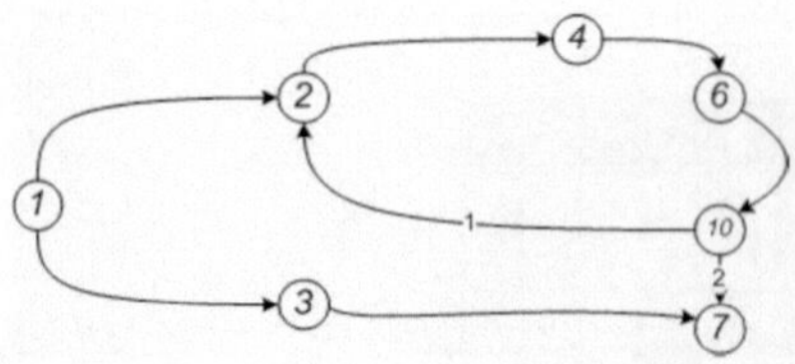


| | 1 | 2 | 3 | 4 | 5 | 6 | 7 |
|---|---|---|---|---|---|---|---|
| 1 | | X | X | | | | |
| 2 | | | | X | | | |
| 3 | | | | | | | X |
| 4 | | | | | | X | |
| 5 | | | | | | | |
| 6 | | X | | | | | X |
| 7 | | | | | | | |

*Figure 10-5: Conversion according to rule 1, alternative result 4*

**Alternative 5:** $C_5 = \{(2,5), (6,2), (6,7)\}$

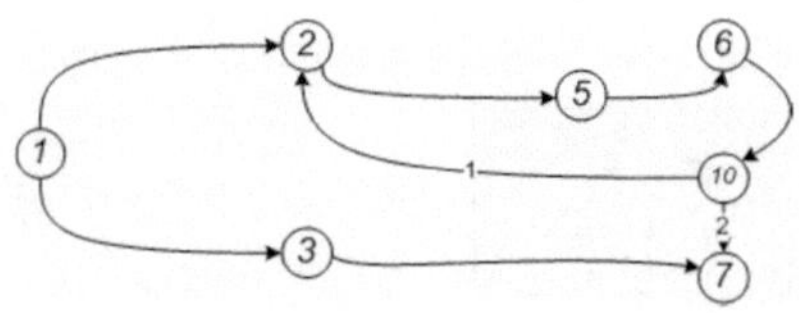


| | 1 | 2 | 3 | 4 | 5 | 6 | 7 |
|---|---|---|---|---|---|---|---|
| 1 | | X | X | | | | |
| 2 | | | | | X | | |
| 3 | | | | | | | X |
| 4 | | | | | | | |
| 5 | | | | | | X | |
| 6 | | X | | | | | X |
| 7 | | | | | | | |

*Figure 10-6: Conversion according to rule 1, alternative result 5*

**Alternative 6:** $C_6 = \{(2,4), (2,5), (6,2), (6,7)\}$

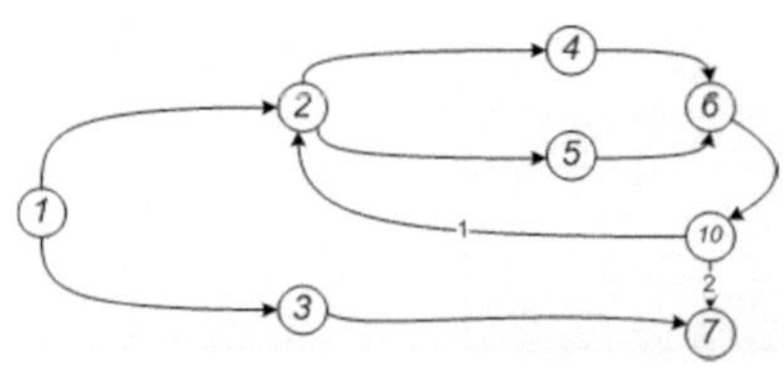


| | 1 | 2 | 3 | 4 | 5 | 6 | 7 |
|---|---|---|---|---|---|---|---|
| 1 | | X | X | | | | |
| 2 | | | | X | X | | |
| 3 | | | | | | | X |
| 4 | | | | | | X | |
| 5 | | | | | | X | |
| 6 | | X | | | | | X |
| 7 | | | | | | | |

*Figure 10-7: Conversion according to rule 1, alternative result 6*

### 10.2.2 Rule 2: Neglect the operators

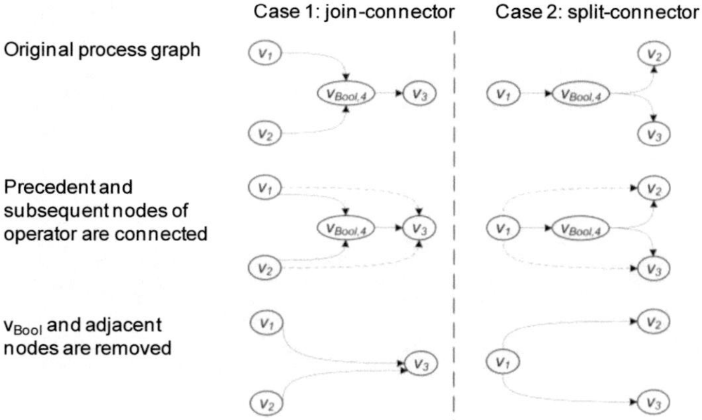


*Figure 10-8: Conversion according to rule 2 (two possible cases)*

Figure 10-9 shows the conversion: First, the new edges are added (dashed lines), then the nodes originating from logic operators are removed as well as all adjacent edges thereof (nodes 8-10 and adjacent edges).

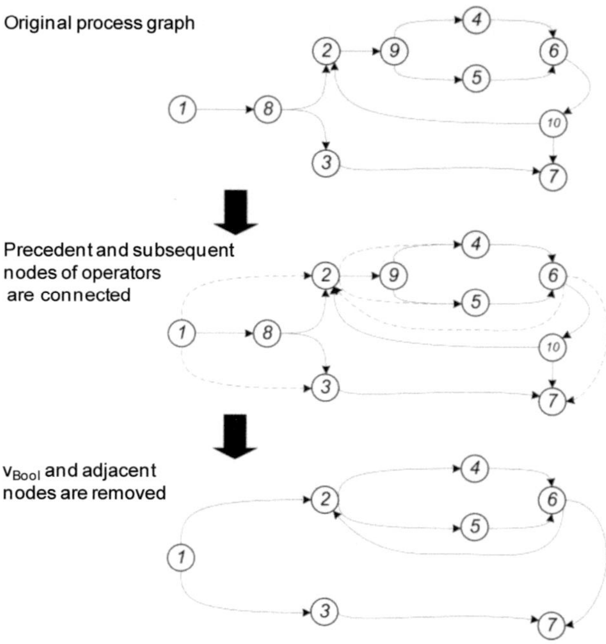


*Figure 10-9: Conversion according to rule 2: exemplary process graph*

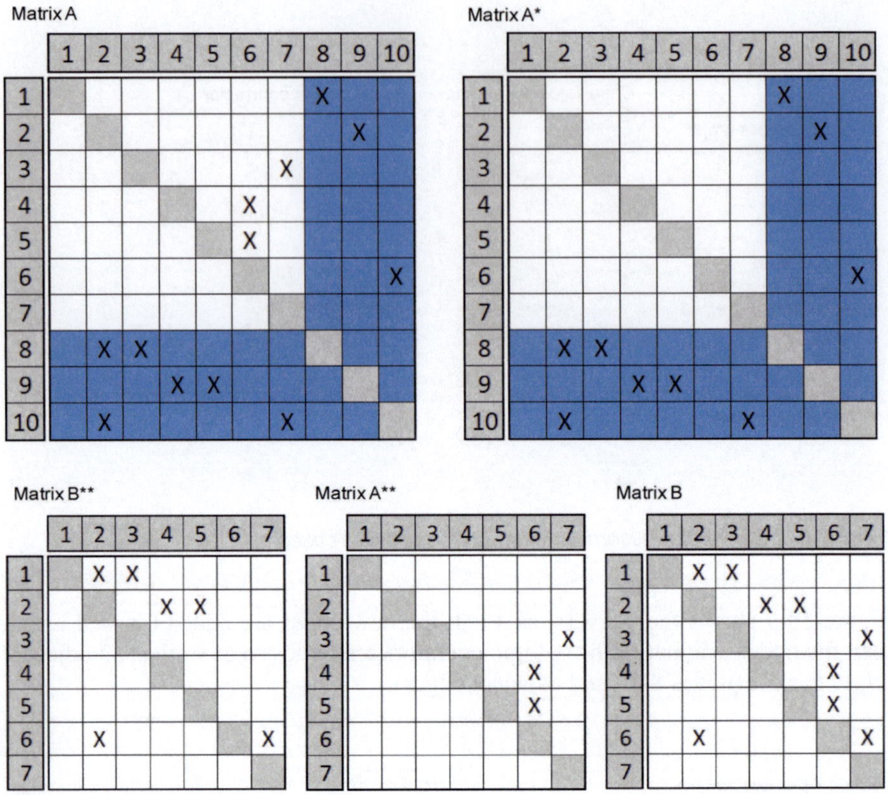


*Figure 10-10: Matrix B as resulting description of rule 2 (lower right) with intermediate matrices from algorithm*

### 10.2.3 Rule 3: Translate operators into probabilities

There are no probabilities given in the example of [BELHE & KUSIAK 1996]; therefore, the following values are used to illustrate rule 3:

OR-connector: p=0.8 for each outdoing edge (i.e., it is possible that both edges eventuate)

XOR-connector: p=0.5 for each outgoing edge

AND-connector: p=1 for each outgoing edge

| | 1 | 2 | 3 | 4 | 5 | 6 | 7 |
|---|---|---|---|---|---|---|---|
| 1 | | 1 | 1 | | | | |
| 2 | | | | 0.8 | 0.8 | | |
| 3 | | | | | | | 1 |
| 4 | | | | | | 1 | |
| 5 | | | | | | 1 | |
| 6 | | 0.5 | | | | | 0.5 |
| 7 | | | | | | | |

*Figure 10-11: Conversion according to rule 3*

### 10.2.4 Rule 4: Logic operators as additional entities

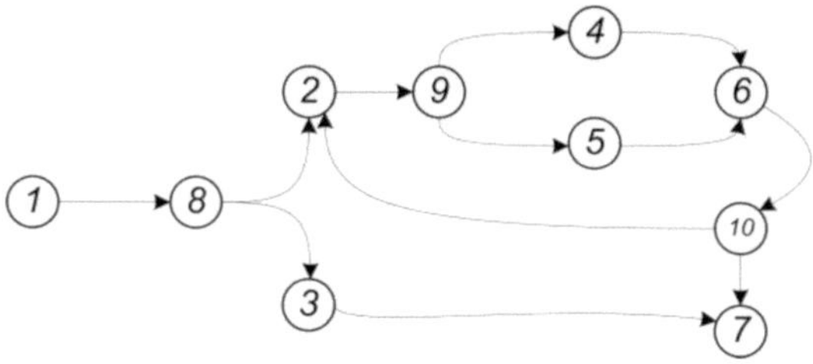


| | 1 | 2 | 3 | 4 | 5 | 6 | 7 | 8 | 9 | 10 |
|---|---|---|---|---|---|---|---|---|---|---|
| 1 | | | | | | | | X | | |
| 2 | | | | | | | | | X | |
| 3 | | | | | | | X | | | |
| 4 | | | | | | X | | | | |
| 5 | | | | | | X | | | | |
| 6 | | | | | | | | | | X |
| 7 | | | | | | | | | | |
| 8 | | X | X | | | | | | | |
| 9 | | | | X | X | | | | | |
| 10 | | X | | | | | X | | | |

*Figure 10-12: Conversion according to rule 4*

Nodes $v_8, v_9, v_{10}$ represent the logic operators that were converted into regular nodes.

### 10.2.5 Rule 5: Logic operators as additional entities with their characteristics

| | 1 | 2 | 3 | 4 | 5 | 6 | 7 | J1 | J2 | J3 | AND | OR | XOR |
|---|---|---|---|---|---|---|---|---|---|---|---|---|---|
| 1 | | | | | | | | 1 | | | | | |
| 2 | | | | | | | | | 1 | | | | |
| 3 | | | | | | | 1 | | | | | | |
| 4 | | | | | | 1 | | | | | | | |
| 5 | | | | | | 1 | | | | | | | |
| 6 | | | | | | | | | | 1 | | | |
| 7 | | | | | | | | | | | | | |
| Ja | | 1 | 1 | | | | | | | | 1 | | |
| J2 | | | | 1 | 1 | | | | | | | 1 | |
| J3 | | 1 | | | | | 1 | | | | | | 1 |
| AND | | | | | | | | | | | | | |
| OR | | | | | | | | | | | | | |
| XOR | | | | | | | | | | | | | |

*Figure 10-13: Multiple-Domain Matrix according to rule 5*

## 10.3 Nesting of Boolean operators

Calculation of the activity of task 1

Connector_2 is first calculated

$$A_{connector\_2} = \sum_{i=1}^{n} EF_i = (1+1) = 2$$

The result is then inserted into the calculation of connector_1

$$A_{connector\_1} = \frac{1}{n}\sum_{i=1}^{n} EF_i = \frac{1+1+2}{3} = 1{,}33$$

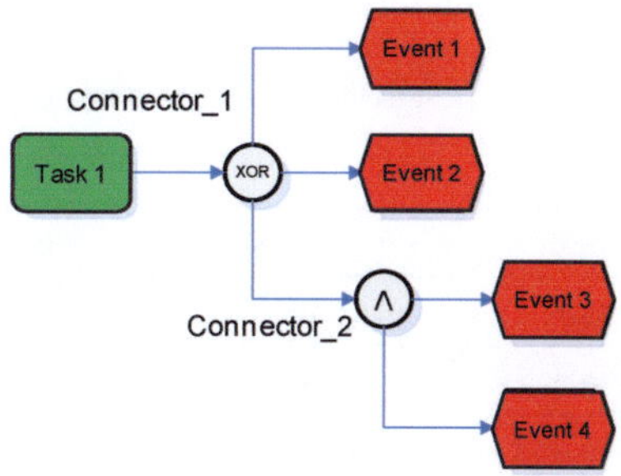


*Figure 10-14: Direct calculation of nested operators (not using weights)*

While the example in Figure 10-14 does not use the weights but shows how a successive operator following a split is integrated into its predecessor, the example in Figure 10-15 shows how weights are used for a more complex setting of nested Boolean operators.

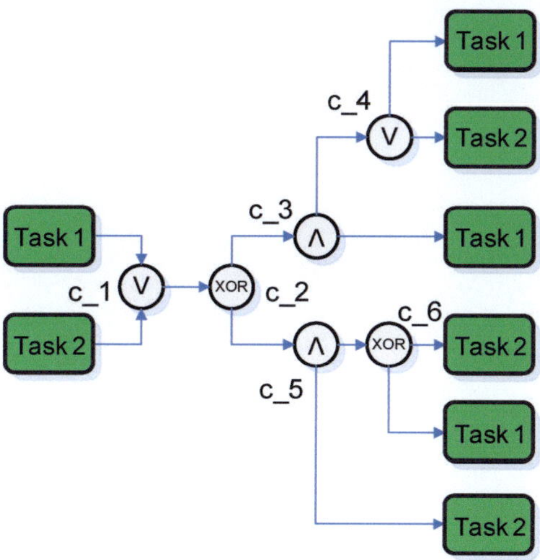


*Figure 10-15: Weights of connectors to calculate the activity and passivity of the involved tasks*

Table 10-1 lists the weights of each connector and the activity and passivity of the tasks in the example. For example, the activity of task 1 calculates as 2.16, as starting from the farthest reachable node, c_4 calculates as 1.33. Connector c_3 then combines with 1.33 and an additional edge as an AND-split to a value of 2.33. This value is then inputted to calculate c_3, also using the weight of c_5, which calculates as 2. The XOR combines thus one outgoing edge weighted as 2.33 and one as 2; assuming the equal structural relevance, the weight of c_2 thus calculates as 2.16. As no previous splits exist, this value is the activity of task 1.

*Table 10-1: Weights, activities and passivities of all entities in the example from Figure 10-15*

| task | activity | passivity |
|---|---|---|
| task 1 | 2.16 | 0 |
| task 2 | 2.16 | 0 |
| task 3 | 0 | 1.33 |
| task 4 | 0 | 1.33 |
| task 5 | 0 | 1.33 |
| task 6 | 0 | 1.33 |
| task 7 | 0 | 1.33 |
| task 8 | 0 | 1.33 |

| connector | weight |
|---|---|
| c_1 | -2.16 |
| c_2 | 2.16 |
| c_3 | 2.33 |
| c_4 | 1.33 |
| c_5 | 2 |
| c_6 | 1 |
| | |
| | |

## 10.4 The complete Structural Process Architecture

| | Task | Artifact | Event | Organizational unit | Resource | Time | Product attribute |
|---|---|---|---|---|---|---|---|
| **Task** | • **precedes temporally**<br>• precedes logically | • **has output of**<br>• processes<br>• controls<br>• changes | • **generates**<br>• processes<br>• ends in | • **belongs to** | • **requires**<br>• processes | • **is active at**<br>• is finished at<br>• has duration of<br>• starts at | • **processes** |
| **Artifact** | • **is input for**<br>• supports<br>• controls<br>• starts | • **transits into**<br>• is used to create<br>• is in conflict with | • **sends**<br>• is created at<br>• is needed at | • **belongs to** | • **is processed by**<br>• is stored in | • **is created at**<br>• is needed at | • **represents**<br>• describes (part of) |
| **Event** | • **starts**<br>• controls | • **produces** | | • **occurs in** | | | |
| **Org. unit** | • **is responsible for**<br>• is necessary for<br>• supports<br>• is involved in<br>• takes decision about<br>• has to be informed about<br>• processes data for | • **has responsibility for**<br>• uses<br>• creates / deletes / updates<br>• is authorized to read | | • **communicates with**<br>• belongs to<br>• is formed by<br>• is head of<br>• is responsible for<br>• takes role as | • **uses** | • **is responsible for**<br>• assists reaching of<br>• is finished at<br>• starts at<br>• is active during | • **generates**<br>• is responsible for<br>• consults about<br>• has knowledge about |
| **Resource** | • **is required for**<br>• supports<br>• is occupied by | • **processes**<br>• creates<br>• emits<br>• stores<br>• transmits | | • **transmits message to** | • **transmits message to** | • **is available at**<br>• is available during | • |
| **Time** | | • **is attribute to** | | | | | |
| **Product attribute** | | • **starts process by sending** | | | | • **is defined at**<br>• is needed at | |

## 10.5 List of structural metrics

Each metric as listed in

Table ***10-2*** is detailed in the following section. The description is completed by the list of possible structural significance for the domains and principal relationship types as provided by the Structural Process Architecture. This significance is intended to facilitate the interpretation of the results of each metric. It does not indicate the only possibilities of interpretation and is not intended as a complete checklist of possible meanings.

*Table 10-2: Available metrics for structural analysis (as compiled in this research)*

| **Size and density** |
|---|
| Number of domains |
| Number of nodes |
| Number of edges |
| Number of classes |
| Number of interfaces between domains |
| Number of edges per node |
| Relational density |
| Number of unconnected nodes |
| **Adjacency** |
| Activity / Passivity |
| Degree correlation (nodes) |
| Degree correlation (edges) |
| Degree distribution |
| Fan criticality |
| Synchronization points / distribution points |
| Number of independent sets |
| **Attainability** |
| Number of reachable nodes |
| Reachability of a node |
| **Closeness** |
| Proximity |
| Relative centrality (based on betweenness) |
| **Connectivity** |
| Node connectivity |
| Edge connectivity |
| **Paths** |
| Number of paths |
| Path length |
| Weight of an edge |
| Centrality of path (based on centrality) |
| Centrality of path (based on degree) |
| Degree of progressive oscillation |

| **Hierarchies** |
|---|
| Height of hierarchy |
| Width of hierarchy |
| Tree criticality |
| Snowball factor |
| Forerun factor |
| Tree-robustness |
| Maximum nesting depth |
| **Clustering** |
| Number of cliques |
| Cluster-coefficient (local) |
| Cluster-coefficient (global) |
| Module quality 1 (flow of information) |
| Module quality 2 (compactness) |
| **Cycles** |
| Number of cycles |
| number of cycles per node |
| Number of cycles per edge |
| Number of feedbacks |
| Activation of cycle |
| Number of starting points for iterations |
| Iterative oscillation |
| **Several domains** |
| Bipartite density |
| Number of organizational interfaces |
| **Cognition** |
| Cognitive weight |
| Degree of non-planarity |
| **Boolean Operators** |
| McCabe Cyclomatic Number |
| Control-Flow Complexity |
| Log-based Complexity |

### 10.5.1 Number of domains

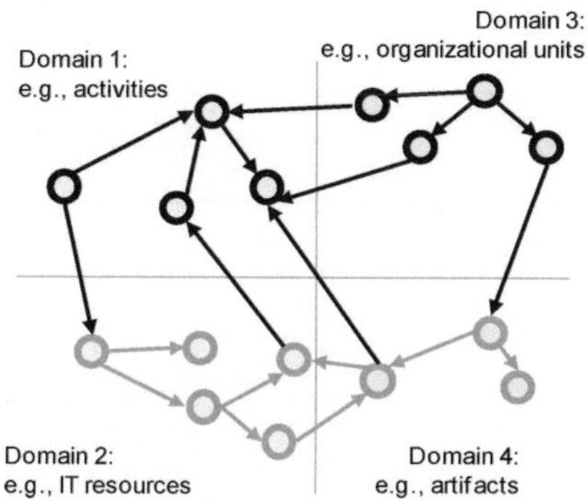


| Process | Number of domains |
|---|---|
| Process 1 | 4 |
| Process 2 | ... |
| ... | |

**Definition**

- Number of different domains within the network

**Structural significance**

- Evaluation of the multi-factedness of the network
- Number of possible views and stakeholders in a process

**Representation**

- Metric for each process

**Literature**

[GRUHN & LAUE 2006b]

### 10.5.2 Number of nodes

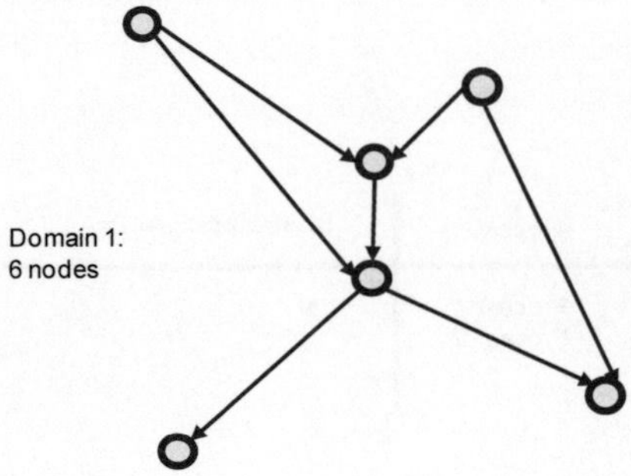


| Domain | Number of nodes per domain |
|---|---|
| Domain 1 | 6 |
| Domain 2 | ... |
| Domain 3 | |
| ... | |

**Definition**

- Number of nodes per domain
- Can be detailed to count start- and end-nodes explicitly (2 start-nodes and 2 end-nodes are found in the example)

**Structural significance**

- Size of the network
- Assessment basis to put other metrics into perspective

**Representation**

- Metric per domain

**Literature**

[AZUMA & MOLE 1994], [BROWNING 2002], [GRUHN & LAUE 2006b]

## 10.5.3 Number of edges

Domain 1:
8 edges

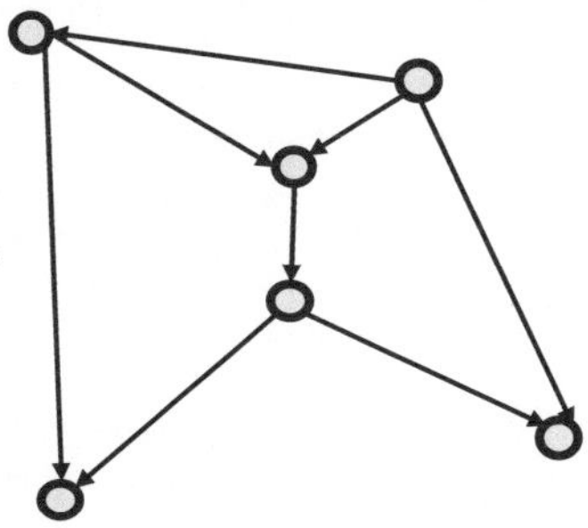

| Domain | Number of edges per domain |
|---|---|
| Domain 1 | 8 |
| Domain 2 | ... |
| Domain 3 | |
| ... | |

**Definition**

- Number of edges within a single domain
- Differentiation of directed and undirected edges possible

**Structural significance**

- Determination of the level of interaction within a domain
- Estimation of the number of interfaces and communication activity

**Representation**

- Metric per domain

**Literature**

[BROWNING 2001b]

## Structural significance

| | Number of domains | Number of nodes | Number of edges |
|---|---|---|---|
| Task | • Level of decomposition within tasks (e.g., tasks, work packages) | • Number of tasks in process (number of instances) | • Extent of network of tasks<br>• Number of interfaces among tasks |
| Artifact | • Level of decomposition within artifacts (e.g., documents, files) | • Number of artifacts in process (number of instances) | • Extent of network of artifacts<br>• Number of interfaces among artifacts |
| Org. unit | • Level of decomposition within organization units (e.g., division, teams) | • Number of organizational units in process (number of instances) | • Extent of network of organizational units<br>• Number of interfaces among organizational units |
| Time | • Level of decomposition within time (e.g., phases, milestones) | • Number of points in time during process (number of instances) | • Extent of network of points in time<br>• Number of interfaces among points in time |
| Event | • Level of decomposition within events (e.g. states, messages) | • Number of events in process (number of instances) | • Extent of network of events<br>• Number of interfaces among events |
| Resource | • Level of decomposition within resources (e.g., IT system, machine) | • Number of resources in process (number of instances) | • Extent of network of resources<br>• Number of interfaces among resources |

### 10.5.4 Number of classes

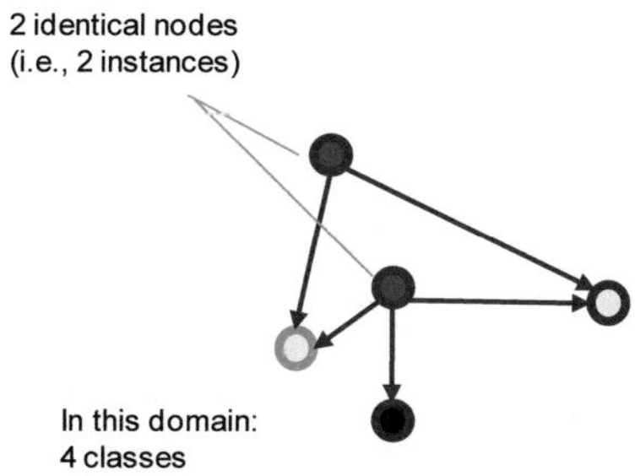


| Domain | Number of classes |
|---|---|
| Domain 1 | 4 |
| Domain 2 | ... |
| Domain 3 | |
| ... | |

**Definition**

- Number of unique nodes per domain, (i.e., number of nodes that do not bear the same name, as opposed to total number of nodes)

**Structural significance**

- Evaluation of the diversity of the network
- Relativization of node count when using object-oriented models (i.e., when nodes are instantiated several times)

**Representation**

- Metric per domain

**Literature**

[GRUHN & LAUE 2006b], [HENRY et al. 1981]

### 10.5.5 Number of interfaces between domains

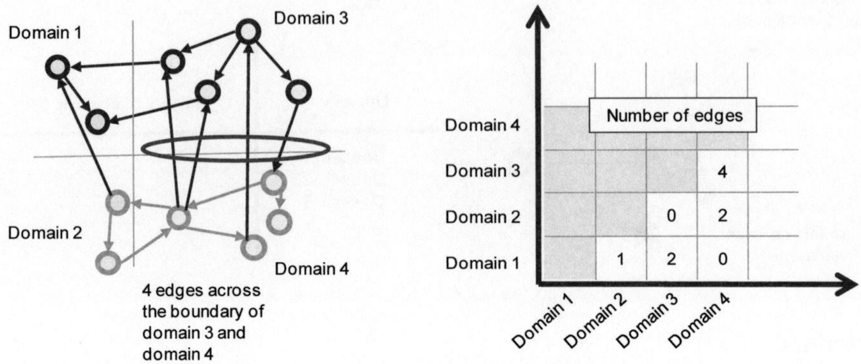


**Definition**

- Number of edges between each pair of domains
- Differentiation of directed and undirected edges possible

**Structural significance**

- Determination of the level of interaction between each pair of domains
- Evaluation of the size of the interface between two domains

**Representation**

- Metric for each pair of domains

**Literature**

[BROWNING 2002]

## Structural significance

| | Number of classes | Number of interfaces between domains |
|---|---|---|
| Task | • Number of unique tasks in process | • Extent of interfaces of tasks to another domain (e.g., to resources that support the tasks)<br>• Number of interfaces to another domain |
| Artifact | • Number of unique artifacts in process | • Extent of interfaces of artifacts to another domain (e.g., to resources that process the artifacts)<br>• Number of interfaces to another domain |
| Org. unit | • Number of unique organizational units in process | • Extent of interfaces of organizational units to another domain (e.g., to tasks that the organizational units are responsible for)<br>• Number of interfaces to another domain |
| Time | • Number of unique points in time during process | • Extent of interfaces of points in time to another domain (e.g., to tasks that are finished at the points in time)<br>• Number of interfaces to another domain |
| Event | • Number of unique events in process | • Extent of interfaces of events to another domain (e.g., to tasks that produce an event)<br>• Number of interfaces to another domain |
| Resource | • Number of unique resources in process | • Extent of interfaces of resources to another domain (e.g., to tasks that are supported by a resource)<br>• Number of interfaces to another domain |

### 10.5.6 Number of edges per node

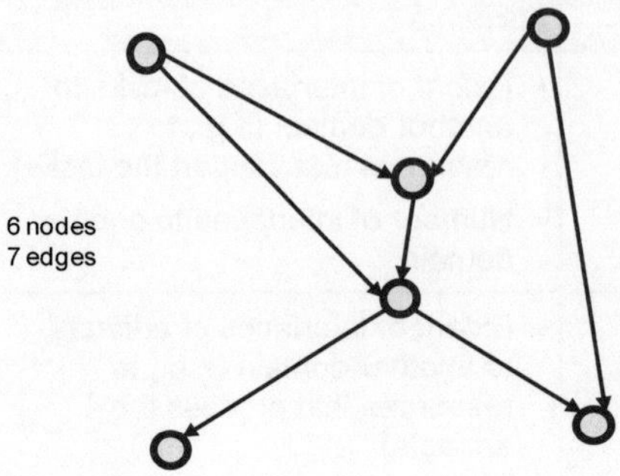


| Domain | Number of edges per node |
|---|---|
| Domain 1 | 1.1667 |
| Domain 2 | ... |
| Domain 3 | |
| ... | |

**Definition**

- Quotient of the number of edges and the number of nodes
- For each domain; also possible for the complete process

**Structural significance**

- Evaluation of the density of networking within the process
- Description of the level of cross-linking within the network

**Representation**

- Metric per domain

**Literature**

[BROWNING 2002]

### 10.5.7 Relational density

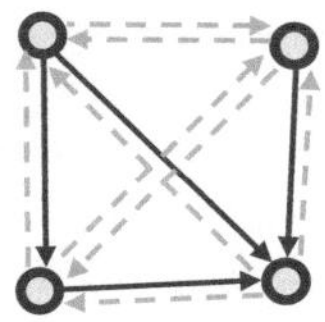

4 existing edges out of 12 possible edges (directed case, possible other edges dashed)

| Domain | Relational density |
|---|---|
| Domain 1 | 0.333 |
| Domain 2 | ... |
| Domain 3 | |
| ... | |

**Definition**

- Quotient of the number of edges in a domain and the number of possible edges
- For each domain; also possible for the complete process

**Structural significance**

- Evaluation of the density and intensity of networking in the process
- Intensity of cross-linking

**Representation**

- Metric per domain

**Literature**

[VANDERFEESTEN et al. 2007], [MENDLING 2008]

## Structural significance

| | Number of edges per node | Relational density |
|---|---|---|
| Task | • Mean degree of dependency of one task on another task<br>• Degree of proximity of task network to an ideal sequence (smaller numbers) | • Extent to which the potential for networking is exhausted<br>• Degree to which Concurrent Engineering is implemented<br>• Mean degree to which the process can be decomposed into meaningful modules to raise transparency |
| Artifact | • Mean degree of dependency of one artifact on another artifact<br>• Degree of proximity of artifact network to an ideal lifecycle of growing maturity (smaller numbers) | • Extent of clear arrangement of artifacts<br>• Degree of comprehensibility of relations among artifacts<br>• Degree to which Concurrent Engineering is reflected by harmonized documents |
| Org. unit | • Mean degree of dependency of one organizational unit on another organizational unit<br>• Density of social network | • Extent to which the potential for a social network is exhausted<br>• Degree of lone fighting in the process<br>• Estimation of potential for process improvement through social and organizational measures |
| Time | • Mean degree of dependency of one point in time on another point in time<br>• Proximity to ideal sequence | • Degree of linearity of the process<br>• Extent to which the potential for synchronization of points in time is exhausted |
| Event | • Mean degree of dependency of one event on another event<br>• Estimation of effort necessary for transferring the process to an idea sequence | • Degree of linearity of the process<br>• Extent to which the potential for synchronization of events is exhausted |
| Resource | • Mean number of interfaces between two resources<br>• Mean degree of consistency of transfer of artifacts among resources | • Extent to which isolated resources dominate the process<br>• Estimation of potential for implementing better interfaces between resources |

## 10.5.8 Number of unconnected nodes

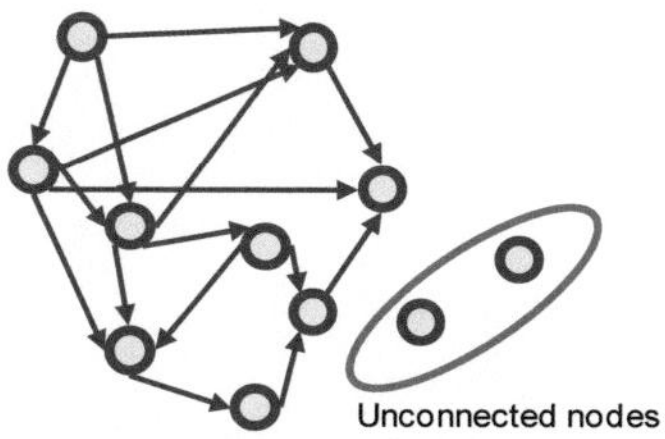


| Domain | Number of un-connected elements |
|---|---|
| Domain 1 | 2 |
| Domain 2 | ... |
| Domain 3 | |
| ... | |

**Definition**

- Number of nodes which are not connected to the graph

**Structural significance**

- Number of independent entities
- Identification of possible modeling errors

**Representation**

- Number of unconnected nodes per domain

**Literature**

[MAURER 2007, p. 209]

## Structural significance

<table>
<tr><th></th><th>Number of unconnected nodes</th></tr>
<tr><td>Task</td><td rowspan="6">• Independent node (task, artifact)<br>• Missing relationship in process<br>• Missing relationship in model<br>• Wrong level of detail of model<br>• Wrong system border (node is not part of model)</td></tr>
<tr><td>Artifact</td></tr>
<tr><td>Org. unit</td></tr>
<tr><td>Time</td></tr>
<tr><td>Event</td></tr>
<tr><td>Resource</td></tr>
</table>

### 10.5.9 Activity / Passivity

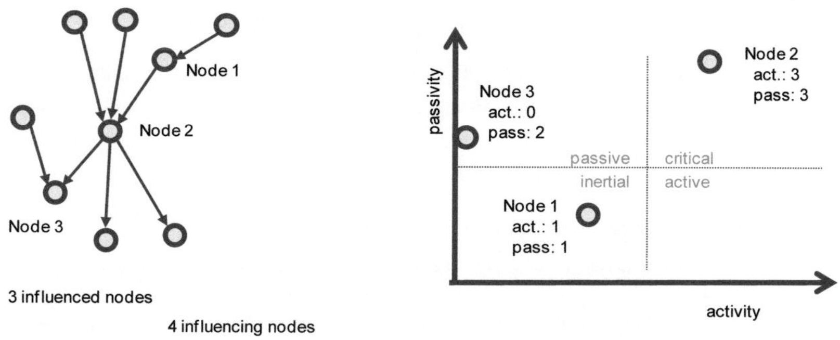


**Definition**

- Number of outgoing edges (= activity) or number of incident edges (= passivity)
- Also applicable for logic operators (see section 4.4.3)

**Structural significance**

- Intensity of changes that a node exerts on or receives from its immediate neighbors
- Quick identification of nodes that are highly relevant for the process
- Degree of homogeneity of network
- Importance of node for local process (immediate environment)

**Representation**

- Portfolio containing all existing nodes

**Literature**

[LINDEMANN 2007, p. 256], [DAENZER & HUBER 2002]

## Structural significance

| | Activity / passivity |
|---|---|
| Task | • Degree of influence exercised /received by a task<br>• Degree of importance of a task in its immediate environment / degree of independence of a task from its immediate environment<br>• Extent of effort to coordinate with neighboring tasks (degree of communication)<br>• Risk to distribute / be impacted by errors<br>• Importance of decision point (only for logic operators) |
| Artifact | • Degree of change impact of / on an artifact (number of changes absorbed / sent)<br>• Degree of (in)dependence on other artifacts: Identification of possible partial results that can be generated independently<br>• Impact of artifacts for individual decision points (only for logic operators) |
| Org. unit | • Degree of direct control and authority / of dependency on superiors<br>• Extent of direct integration of an organizational unit into the process organization |
| Time | • Degree of control of individual points in time / degree of dependency on other points in time<br>• Degree to which a point in time serves as a synchronization point / as a buffer for delays<br>• Impact of points in time on decision points in the process (only for logic operators) |
| Event | • Degree of direct impact of one event on the next immediate event(s) / of dependency on the previous immediate event(s)<br>• Degree of control over the quality of one event<br>• Risk of distributing errors at one event<br>• Impact of events on decision points in the process (only for logic operators) |
| Resource | • Importance of single resources for neighboring resources<br>• Potential for consistency that is possible with a resource<br>• Degree of openness of a resource for flexible use in the process |

### 10.5.10 Degree correlation (nodes)

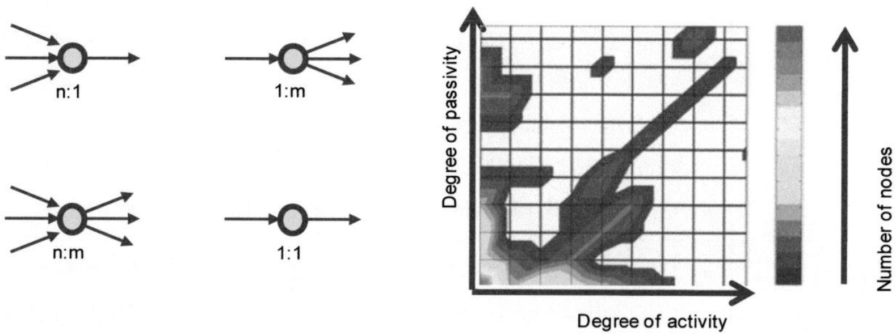


**Definition**

- Occurrence of a correlation of the degrees of incident and outgoing edges for all nodes (i.e., occurrence of each possible pair of activity and passivity)

**Structural significance**

- Degree to which a node impacts (or is impacted by) the process
- Tendency to which the process relies on individual nodes to coordinate the overall structure

**Representation**

- Plot of the occurrence of each pair of activity and passivity

**Literature**

[AHN et al. 2007], [NIKOLOSKI et al. 2005]

## Structural significance

| | **Degree correlation (nodes)** |
|---|---|
| Task | • Degree of the possibility that tasks are more interdependent than they appear to be explicitly (because of indirect relationships, weak relationships, or missing relationships in the model)<br>• Degree to which the overall process is laced by critical tasks that ensure the overall connectivity of the process<br>• Level of risk of distribution of errors of the overall process |
| Artifact | • Degree of the possibility that artifacts are not consistent among themselves because of relationships that were not respected (indirect relationships, weak relationships, or missing relationships in the model)<br>• Degree to which the overall process is dependent on artifacts that serve as communication hubs |
| Org. unit | • Degree to which the potential for networking among organizational units is capitalized (measure for possible social relationships that are not modeled)<br>• Level of the risk that the process depends on a few central organizational units |
| Time | • Degree of possible delays from unexpected reasons (because of indirect relationships, weak relationships, or missing relationships in the model)<br>• Level of the risk that the process depends on a few critical points in time that, if delayed, delay the overall process noticeably |
| Event | • Degree to which an event is possibly more dependent on other events than expected (because of indirect relationships, weak relationships, or missing relationships in the model)<br>• Degree of controllability of the overall process by only a few selected states |
| Resource | • Measure for the possible need of more direct interfaces among resources<br>• Level of the risk that consistent information transfer between the resources is focused on few interfaces |

### 10.5.11 Degree correlation (edges)

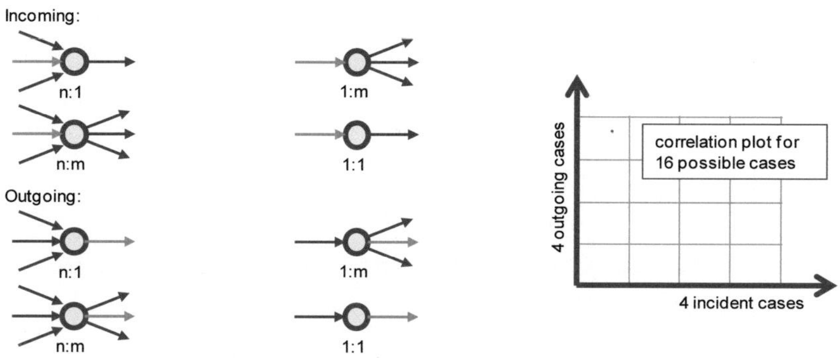


**Definition**

- Occurrence of a correlation of the degrees of incident and outgoing edges for each edge (i.e., occurrence of each possible pair of activity and passivity ordered to each individual edge, classified by incident and outgoing edges)

**Structural significance**

- Degree to which an edge impacts (or is impacted by) the network
- Degree of the process to be dependent on individual paths or on a networked structure
- Tendency of how the transmission of information is handled (integration or distribution)
- Identification of nodes that play a central role in integrating or distributing information
- Determination of critical edges that rely on other edges to be fully operational

**Representation**

- Plot of the occurrence of each pair of activity and passivity for the sixteen possible cases of correlations

## Structural significance

| | **Degree correlation (edges)** |
|---|---|
| Task | • Degree of the importance of a single interface (as transition between two tasks) for the overall process<br>• Measure for the tendency of the process to rely on single or on combined information transfer |
| Artifact | • Degree to which the overall process relies on the networking among individual artifacts<br>• Measure for the tendency of spreading documents in a central homogeneous manner |
| Org. unit | • Degree of the social network to be spread out as a chain or a homogeneous network |
| Time | • Measure for the degree of linearity of the process with few kickbacks |
| Event | • Measure for the homogeneity of progress in the process across a line of states or across many states simultaneously |
| Resource | • Degree of risk that individual interfaces rely on the transfer across other interfaces, which can interrupt the overall process if disrupted |

## 10.5.12 Degree distribution

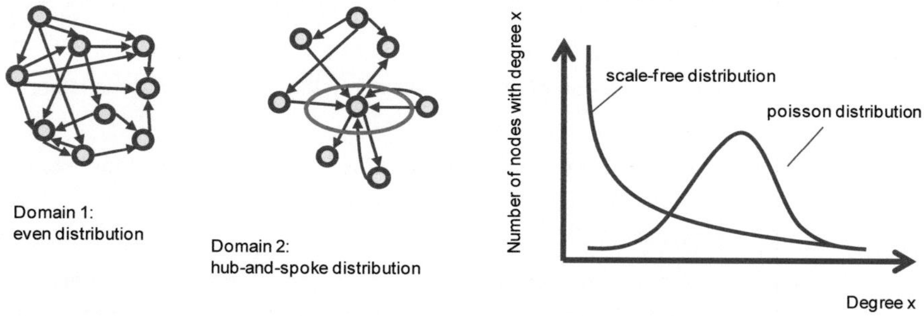


**Definition**

- Distribution of number of nodes with identical activity or passivity

**Structural significance**

- Homogeneity of the process
- Sensitivity of network to the malfunction or drop-out of individual nodes
- Identification of critical nodes that can cause a failure of the overall process
- Identification of hubs in the process
- Identification of nodes that are little integrated and possibly of little importance

**Representation**

- Plot of number of nodes with identical activity or with identical passivity

**Literature**

[ALBERT & BARABASI 2002]

## Structural significance

| | **Degree distribution** |
|---|---|
| Task | • Degree of randomness in process (the more the degrees are distributed as an exponential function, the more the process is not random)<br>• Degree for the extent of failure of tasks that is necessary for overall process to fail (i.e., fall apart or be seriously hindered in its connectivity)<br>  o Measure for the risk of success of a deliberate attack against central tasks (e.g., sabotage) to cause the overall network to fail<br>  o Measure for the risk that random failures of tasks (e.g., by producing errors) do not corrupt the overall network<br>• Degree to which the process is controlled by central tasks (i.e., all tasks can be reached across a very short path) |
| Artifact | • Degree of randomness of the consistency among documents<br>• Degree of risk caused by wrong or faulty documents<br>• Degree of risk caused by data loss or by the loss of knowledge |
| Org. unit | • Degree of organization in the social network<br>• Degree to which the process depends on core organizational units that need to be particularly protected |
| Time | • Tendency of the process to rely on (few) critical points in time that control the process<br>• Level of the risk that the overall process is delayed if single points in time are not reached in time<br>• Degree to which the process is laced with bottlenecks |
| Event | • Degree of organized progress of the process<br>• Level of risk of the overall process to depend on (few) critical events that control the process<br>• Degree for the controllability of the overall process using few events |
| Resource | • Degree of homogeneity of the interfaces in the resource system<br>• Degree of dependency of the overall process on (few) critical resources<br>• Level of risk of the overall process to fail if single resources fail |

## 10.5.13 Fan criticality

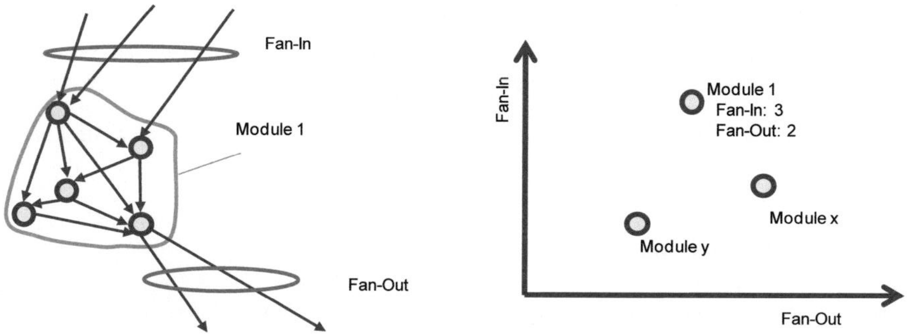


### Definition

- Activity (= Fan-Out) / passivity (= Fan-In) for a module (number of outgoing or incident edges for a module)
- Also possible for logic operators

### Structural significance

- Similar to the metric "activity / passivity" but for modules
- Only sensible for the evaluation of existing modules (= pre-defined groups of elements)
- Comparison of modules concerning their susceptibility to changes and/or their impact on the overall network

### Representation

- Portfolio of Fan-In and Fan-Out containing all modules

### Literature

[GRUHN & LAUE 2006b] , [DAENZER & HUBER 2002]

## Structural significance

| | Fan criticality |
|---|---|
| Task | • Degree of influence of a module (i.e., a defined set of tasks) on its immediate environment (active fan criticality)<br>• Degree of influence from the immediate environment on a module<br>• Level of risk of possible errors generated in a module<br>• Extent of interaction necessary with immediate environment (communication) |
| Artifact | • Degree of impact of changes made to a (pre-defined) set of artifacts on their immediate surroundings<br>• Degree of dependence of a set of artifacts on their immediate surroundings<br>• Level to which a set of artifacts can be generated independently from the process |
| Org. unit | • Number of communication channels available to a body of organizational units (e.g., a department) to send (active) or receive (passive) information<br>• Degree to which a body of organizational units is embedded into the process via its immediate environment |
| Time | • Level of independence of a set of points in time (e.g., a phase) from the overall schedule<br>• Degree of influence of a set of points in time to the schedule (active fan criticality)<br>• Degree to which a set of points in time serves as a buffer (passive fan criticality) |
| Event | • Level of integration of a set of events into the overall process<br>• Degree of independence for reaching a set of events from the overall process<br>• Level of influence of a set of events on the overall process |
| Resource | • Degree of openness of an encapsulated family of resources (e.g., a group of IT systems) to the overall process<br>• Level of risk of an encapsulated family of resources to be a bottleneck<br>• Level of decoupling of an encapsulated family of resources from the overall process |

### 10.5.14 Synchronization points / distribution points

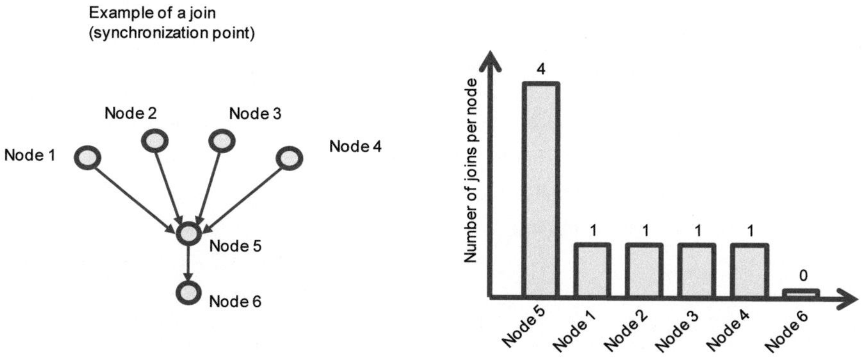


**Definition**

- Number of AND-joins (merging busses) as synchronization points concerning the flow of information
- Number of AND-splits (distributing busses) concerning the flow of information
- Only nodes with many more incident than outgoing edges are regarded (for synch. points, outgoing edges for distribution points)
- Also possible for logic operators

**Structural significance**

- Identification of critical coordination points

**Representation**

- Pareto distribution of incoming or outgoing edges for relevant nodes (with "high" degree)

**Literature**

[GRUHN & LAUE 2006b]

## Structural significance

Compare also the interpretation guideline for Activity (Synchronization points) / Passivity (Distribution points)

| | Synchronization points | Distribution points |
|---|---|---|
| Task | • Degree to which a task serves as a buffer within the overall process by collecting inputs before continuing | • Degree to which a task possibly causes delays within the overall process by withholding outputs that are necessary for other tasks to continue |
| Artifact | • Degree to which an artifact serves as a buffer within the overall process by collecting inputs before continuing | • Degree to which an artifact possibly causes delays within the overall process by being the backbone, from which other artifacts are generated |
| Org. unit | • Degree to which an organizational unit serves as an information sink within the overall process by collecting information and know-how | • Degree to which an organizational unit possibly controls the overall social net through spreading or withholding information |
| Time | • Degree to which a point in time serves as a buffer within the overall process by depending on other points in time | • Degree to which a point in time possibly causes delays within the overall process by allowing other points in time to be reached only after one point has been visited. |
| Event | • Degree to which an event serves as a buffer within the overall process by waiting for other events before continuing | • Degree to which an event possibly causes delays within the overall process by obstructing other events from being reached |
| Resource | • Degree to which a resource serves as an information sink within the overall process by integrating inputs | • Degree to which a resource possibly is a bottleneck within the overall process, through which many other nodes need to pass |

### 10.5.15 Number of independent sets

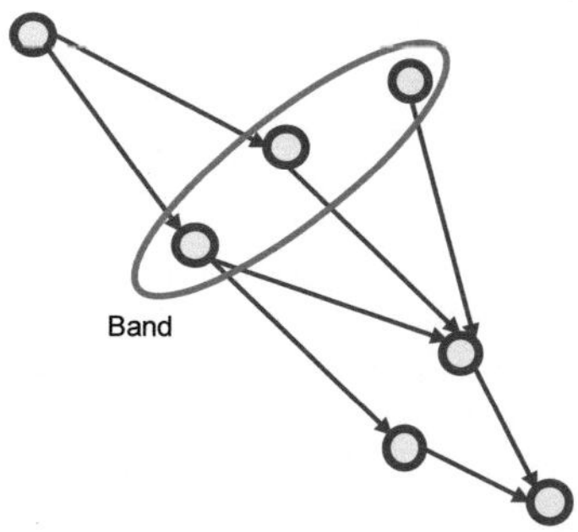


| Domain | Number of independent sets |
|---|---|
| Domain 1 | 4 |
| Domain 2 | ... |
| Domain 3 | |
| ... | |

**Definition**

- Number of bands in a banded DSM

**Structural significance**

- Groups of nodes that can be executed independently of each other
- Possibility of parallelization of process for each bands

**Representation**

- Metric per domain

**Literature**

[MAURER 2007, p. 226], [YASSINE 2004], [BROWNING 2001]

## Structural significance

| | Number of independent sets |
|---|---|
| Task | • Potential for parallelization and concurrent engineering<br>• Degree to which a process necessitates synchronization |
| Artifact | • Degree to which the artifacts are interdependent |
| Org. unit | • Potential to generate teams that operate independently towards a common goal |
| Time | • Indicator of the creation of phases |
| Event | • Robustness of the state of the process |
| Resource | • Potential to integrate resources into a common workflow that is largely independent from the rest of the resources |

## 10.5.16 Number of reachable nodes

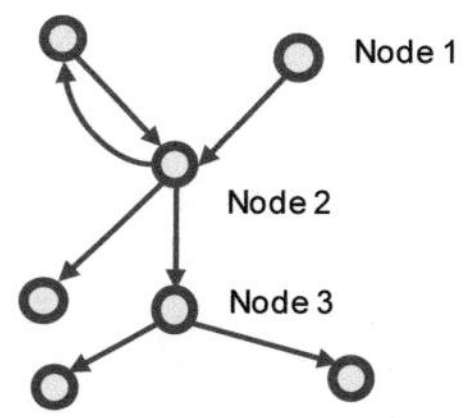


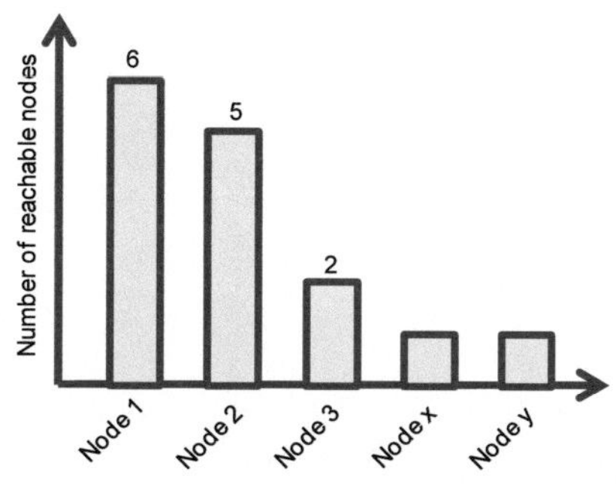


### Definition

- Number of nodes to be reached from starting node (i.e., number of all nodes within the tree of each root node)
- Can be normalized to the total number of nodes within the graph

### Structural significance

- Degree of influence of a node onto the overall network
- Extent of the downstream hierarchy
- Influence is not weighted according to distance (as opposed to hierarchies)

### Representation

- Pareto distribution of all nodes ordered by the number of passively reachable nodes
- Also possible as a plot of reachabilities and number of reachable nodes (see Metric "Reachability of a node")

### Literature

[MAURER 2007, p. 202]

### 10.5.17 Reachability of a node

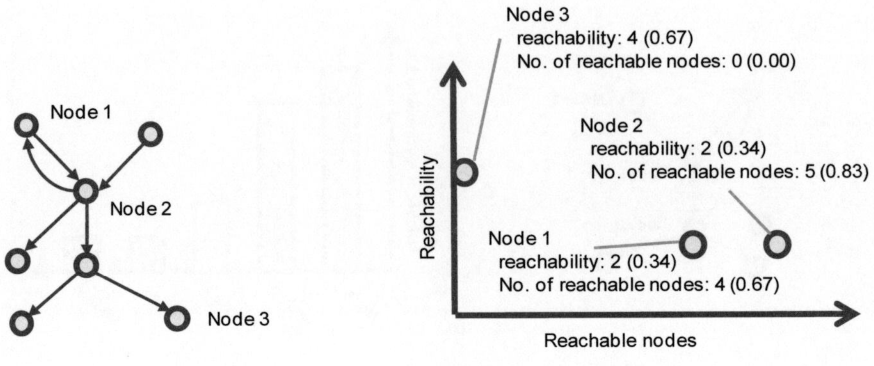


(in brackets: normalized to total number of nodes-1)

**Definition**

- Number of possible starting nodes to reach a designated node
- Can be normalized for the total number of nodes within the graph

**Structural significance**

- Influence of the overall process on a node
- Influence is not weighted according to distance (as opposed to hierarchies)

**Representation**

- Pareto distribution of nodes ordered by the number of passively reachable nodes (see metric "Number of reachable nodes")
- Also possible as a plot of reachabilities and number of reachable nodes

**Literature**

[MAURER 2007, p. 202]

## Structural significance

| | Number of reachable nodes | Reachability of a node |
|---|---|---|
| Task | • Degree to which a task is actively in control of the subsequent process<br>• Level of risk of a task to supply subsequent tasks with errors<br>• Relevance of a task for the subsequent process | • Degree to which a task serves as an information sink<br>• Level of risk of a task to be based on wrong assumptions<br>• Degree to which a task depends on previously compiled information |
| Artifact | • Extent of necessary consistency with subsequent artifacts<br>• Level of impact of an artifact on the overall, subsequent process<br>• Level of risk of an artifact to transmit errors into the subsequent process | • Extent of necessary consistency with previous artifacts<br>• Level of impact of the overall, previous process onto an artifact<br>• Level of risk of an artifact to work with erroneous information |
| Org. unit | • Level of the potential to communicate within the social network<br>• Degree of control exercised by an organizational unit over other units | • Visibility of an organizational unit within the organizational setup<br>• Extent to which information is likely to arrive at an organizational unit |
| Time | • Level of relevance of a point in time for all other subsequent points in time<br>• Degree of control of a point in time over the subsequent schedule<br>• Level of importance of a point in time for its integration into overall schedule<br>• Level of risk of a point in time to delay the overall subsequent process | • Degree of control of the schedule previous to a point in time over its timeliness<br>• Level of importance of a point in time for its integration into the overall schedule<br>• Degree to which a point in time serves as a buffer for the overall process |
| Event | • Level of relevance of an event for all subsequent events<br>• Degree of control of an event for all subsequent events<br>• Level of risk of an event to delay the overall, subsequent process | • Level of relevance of an event for all previous events in the overall process<br>• Degree of control of the process previous to an event over its timeliness<br>• Degree to which an event serves as a buffer for the overall, previous process |
| Resource | • Extent of possible consistency<br>• Degree to which a resource can be used to spread information efficiently | • Level of potential for consistency among all artifacts in the process<br>• Openness to deal with various inputs |

### 10.5.18 Proximity

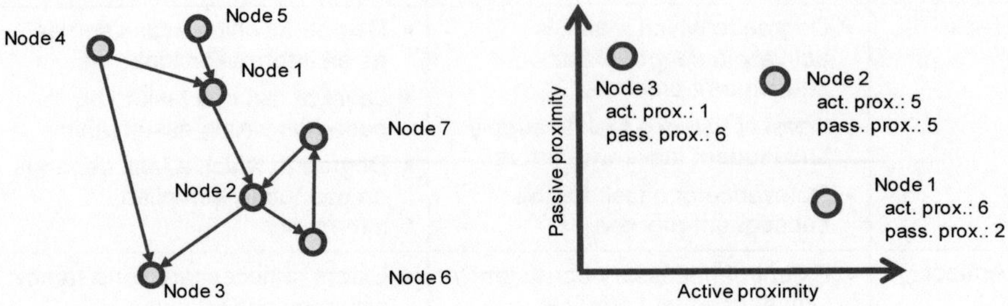


**Definition**

- Total length of all paths that cross a node
- Differentiation of active and passive path length possible
- Also referred to as closeness

**Structural significance**

- Compactness of a process from a node's point of view
- Estimation of velocity of reaching other entities in the process
- Degree of immediacy of influence on or by other nodes

**Representation**

- Portfolio according to incident and outgoing path length for all nodes
- Distribution of absolute proximity (active and passive) for all nodes

**Literature**

[MAURER 2007, p. 205]

## Structural significance

| | **Proximity** |
|---|---|
| Task | • Level of importance of a task as a broker for information within the overall process<br>• Degree of integration of a task into the overall process<br>• Degree of tangibility of long-range impacts of a task on the overall process |
| Artifact | • Level of importance of an artifact to serve as a central information sink that needs to be documented well<br>• Degree of integration of an artifact into the overall process<br>• Degree of tangibility of long-range interactions of an artifact with all other artifacts |
| Org. unit | • Extent of potential of an organizational unit to work as a central hub of information transfer and opinion formation<br>• Extent of potential of an organizational unit to communicate with all other organization units in the process |
| Time | • Level of importance of timeliness at a point in time to ensure the timeliness of the overall process<br>• Extent of integration of a point in time into the overall schedule |
| Event | • Degree of influence of an event on the overall process<br>• Visibility of an event within the overall process as a means of control<br>• Extent of integration of an event into the overall process |
| Resource | • Level of potential of a resource to serve as a central hub for information transfers and information processing<br>• Level of accessibility of resource from other resources throughout the overall process |

### 10.5.19 Relative centrality (based on between-ness)

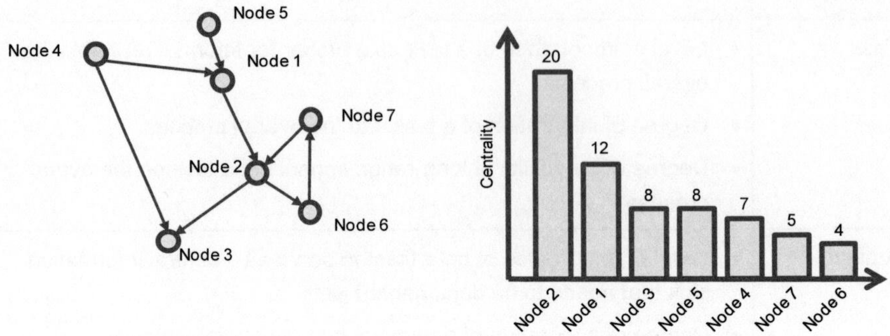


**Definition**

- Number of shortest paths between any two nodes that cross a designated node
- Normalization possible to (n-1)*(n-2) as maximum number of possible paths per node

**Structural significance**

- Degree of the communication activity in the process
- Degree of integration of an entity into the process (e.g., in terms of opinion making)
- Potential of an entity to influence the process
- Identification of hubs in the overall process concerning their role as distributors of information

**Representation**

- Pareto distribution of all nodes and their centralities

**Literature**

[MAURER 2007, p. 144], [FREEMAN 1978]

## Structural significance

| | Relative centrality (based on between-ness) |
|---|---|
| Task | • Level of importance of a task for the overall process<br>• Degree of a task to serve as a broker for information and to form opinions<br>• Level of risk of a task to be susceptible to changes<br>• Level of risk of a task to seriously hinder the overall process in case of failure<br>• Degree to which a task should be detailed when planning a process |
| Artifact | • Degree to which an artifact serves as a central information hub<br>• Level of risk that is caused by an error in an artifact<br>• Level of risk of an artifact to be susceptible to changes<br>• Degree to which an artifact possibly serves as a central point of reference |
| Org. unit | • Degree to which an organizational unit has the role of a central coordinator<br>• Degree to which an organizational unit works as a hub for information transfer<br>• Degree to which an organizational unit possibly serves as a repository of knowledge |
| Time | • Degree to which a point in time plays a central role in the overall schedule<br>• Level of risk of a point in time to be susceptible to changes and even iterations<br>• Level of importance of timeliness of a point in time for the overall timeliness |
| Event | • Degree to which an event plays a central role in reaching the overall process result<br>• Degree to which the progress of a process reflects in an event |
| Resource | • Degree to which a resource purposefully serves as a hub for information exchange<br>• Level of accessibility of a resource from the overall process |

### 10.5.20 Node connectivity

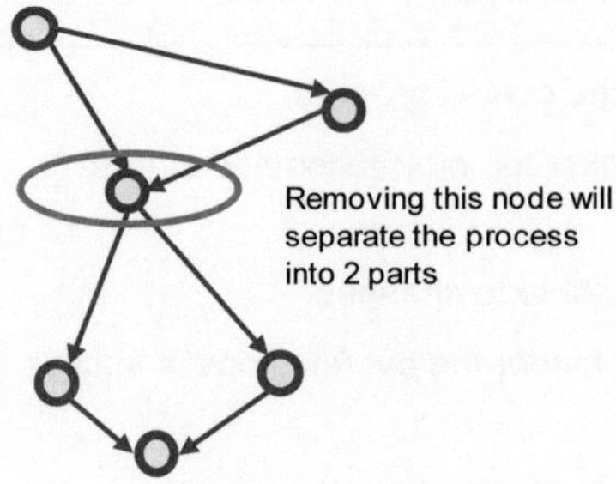


| Domains | Number of resulting networks | Necessary number of nodes to be removed |
|---|---|---|
| Domain 1 | 2 | 1 |
| | … | … |
| | n | |
| Domain 2 | 2 | |
| … | … | |

**Definition**

- Minimum number of nodes that need to be removed to separate the network into disjoint networks
- Can purposefully be related to the size of the networks that remain after separation

**Structural significance**

- Evaluation of the robustness of the process against single entities dropping out
- Tendency of the process to keep its overall integrity in case a node fails
- Identification of nodes that are critical to the coherence of the overall process

**Representation**

- Number of removed nodes necessary per domain arranged according to number of resulting disjoint networks

**Literature**

[GROSS & YELLEN 2005, p. 175]

## 10.5.21 Edge connectivity

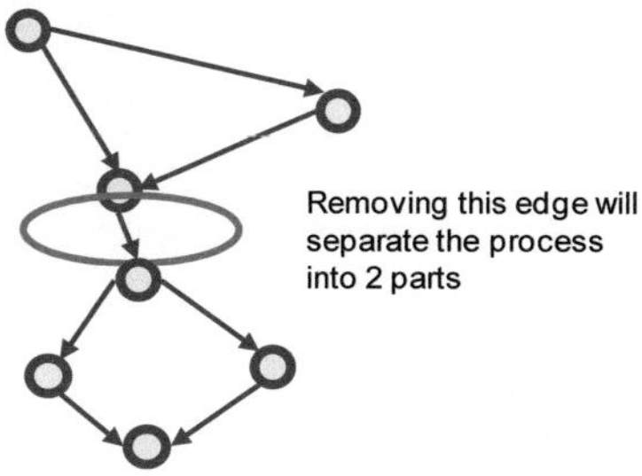


| Domains | Number of resulting networks | Necessary number of edges to be removed |
|---|---|---|
| Domain 1 | 2 | 1 |
| | … | … |
| | n | |
| Domain 2 | 2 | |
| … | … | |

**Definition**

- Minimum number of edges to be removed to separate the network into two / three /… disjoint networks
- Can purposefully be related to the size of the networks that remain after separation

**Structural significance**

- Evaluation of the robustness of the process against single edges dropping out
- Tendency of the network to keep its overall integrity in case an edge fails
- Identification of nodes that are critical to the coherence of the overall process

**Representation**

- Number of removed edges necessary per domain arranged according to number of resulting disjoint networks

**Literature**

[GROSS & YELLEN 2005, p. 175]

## Structural significance

| | Node connectivity (/edge connectivity) |
|---|---|
| Task | • Degree of integrity of the overall process based on individual tasks (/on individual communication channels)<br>• Level of risk of the overall process to fall apart if individual tasks (/individual communication channels) fail<br>• Number of tasks (/communication channels) that need to be removed to split the process into two or more independent processes<br>• Extent of independent sub-processes within the overall process, consisting of groups of tasks that are only connected to the process via one or a few connecting tasks (/communication channels)<br>• Level of flexibility to process parts of the overall process relatively independently |
| Artifact | • Extent of independency of a group of artifacts that are only connected to the overall process via transferring artifacts (/communication channels) |
| Org. unit | • Degree of integrity of the social network<br>• Level of dependency of the social network on individual organizational units (/communication channels) that ensure the overall collaboration of the process<br>• Level of risk of the overall process to fall apart in case of failure of individual organizational units (/communication channels) |
| Time | • Extent of bottlenecks in the process that synchronize and spread the process flow (/that channel the information flow)<br>• Level of risk of delays due to individual points in time not being reached<br>• Level of potential to generate results of parts of the process independently from the rest of the overall process |
| Event | • Degree of independence of the events within the overall process<br>• Level of potential to achieve certain events of the process independently from the rest of the overall process<br>• (/Level of risk of individual transitions in the overall process) |
| Resource | • Degree of integration of chain of resources<br>• Level of risk of obtaining isolated applications in case individual resources (/interfaces) fail |

### 10.5.22 Number of paths

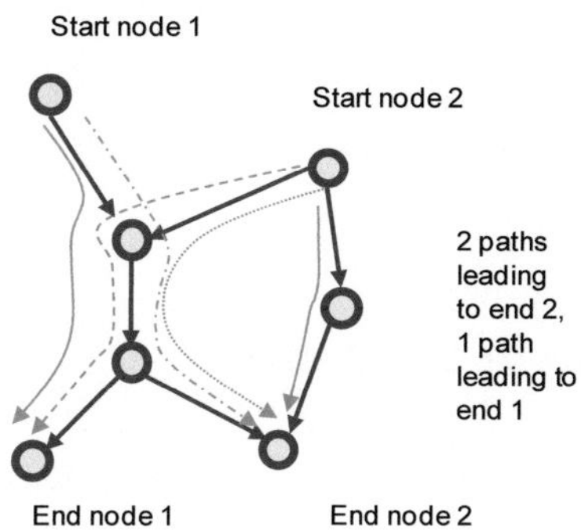


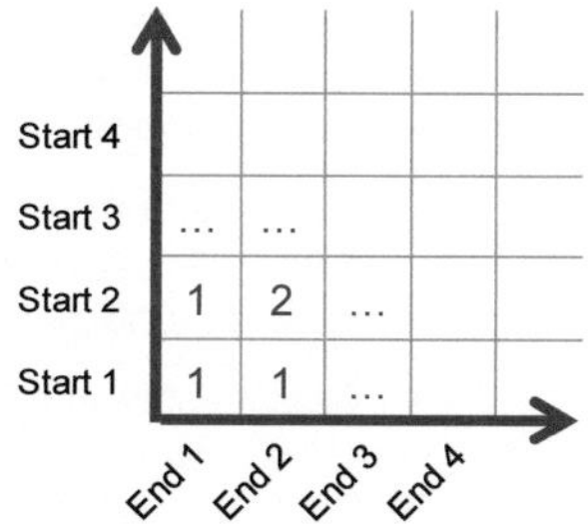


**Definition**

- Number of all possible paths per pair of start- / end-nodes
- Purposeful mostly for dedicated pairs of start and end nodes

**Structural significance**

- Evaluation of redundant pathways through the process
- Determination of clarity of processing of process
- Determination of critical start- and end-nodes

**Representation**

- Metric for each pair of nodes

**Literature**

[McCabe 1976]

## Structural significance

| | **Number of paths** |
|---|---|
| Task | • Degree of directedness of a process (unambiguous processing of tasks)<br>• Extent of redundancies within the process<br>• Degree of parallelism in processing of tasks |
| Artifact | • Level of ambiguity in processing of artifacts<br>• Degree of possible parallelism in generation of artifacts |
| Org. unit | • Degree of variety of communication paths<br>• Level of flexibility of communication |
| Time | • Level of synchronism of processing of process<br>• Extent of necessary coordination |
| Event | • Degree of parallelization of process<br>• Degree of variety of control of process |
| Resource | • Level of flexibility in resource landscape to provide the consistency and transfer of artifacts |

### 10.5.23 Path length

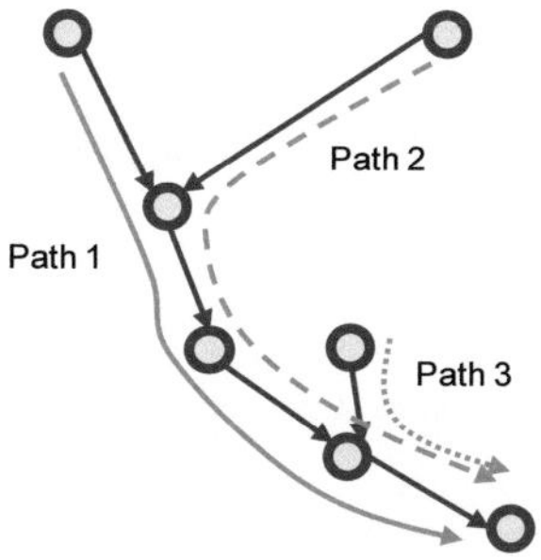


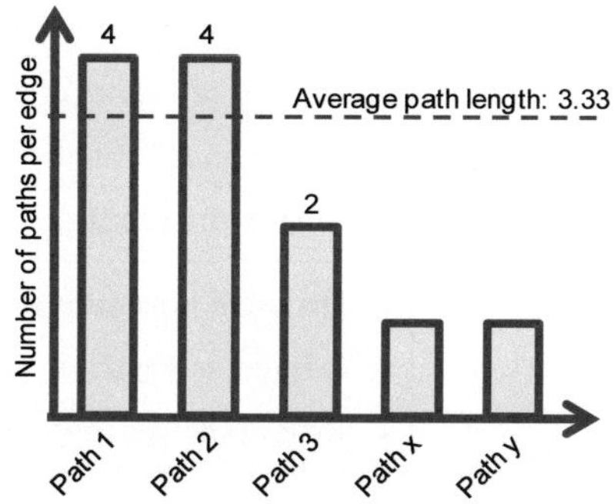


**Definition:**

- Number of edges between start-node and end-node (for a path)
- Also possible as minimum path length for two nodes
- Also possible as average path length for all paths across the overall network

**Possible meaning:**

- Difficulty to reach another designated node within the network
- Description of the size of the network

**Representation:**

- Pareto distribution
- Metric per domain (for average path length)

**Literature**

[NEWMAN 2003a]

## Structural significance

| | **Path length** |
|---|---|
| Task | • Distance between two tasks<br>• Degree of impact of one task on another task<br>• Minimum effort to synchronize two tasks (min. path length)<br>• Size of the process (average path length) |
| Artifact | • Distance between two artifacts<br>• Degree of impact of one artifact on another artifact<br>• Size of the process (average path length) |
| Org. unit | • Distance between two organizational units<br>• Effort for two organizational units to communicate |
| Time | • Possible number of phases between two points in time<br>• Duration between two points in time |
| Event | • Degree of impact of one event on another event |
| Resource | • Number of intermediate interfaces between two resources |

## 10.5.24 Weight of an edge

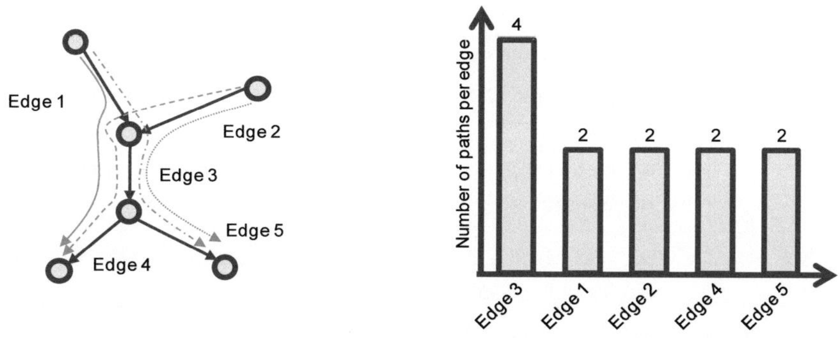


**Definition**

- Number of shortest paths that follow a designated edge

**Structural significance**

- Importance of an edge for the overall network - e.g., in terms of communication channels
- Identification of critical edges for the function of the overall process

**Representation**

- Pareto distribution of weight for all edges

**Literature**

[HENRY et al. 1981]

## Structural significance

| | **Weight of an edge** |
|---|---|
| Task | • Level of importance of an information transfer or interface between two tasks for the overall process<br>• Degree to which a transfer of information is central to overall process<br>• Degree to which a transfer of information is part of a principal process path |
| Artifact | • Extent to which a transition between two artifacts is part of central genesis of knowledge in the overall process<br>• Degree to which a transition between two artifacts is central to a process<br>• Extent of the spreading of artifacts throughout the process<br>• Level of the risk of spreading erroneous artifacts throughout the process |
| Org. unit | • Level of communication between two organizational units for the overall process<br>• Degree of centrality of an interface between two organizational units |
| Time | • Degree of centrality of a transition from one point in time to the next for the progress of the overall process<br>• Extent of progress made within the overall process at the transition between two points in time |
| Event | • Degree of centrality of a transition between two events for the overall process<br>• Level of potential to determine central events and interfaces to control the overall progress of the process |
| Resource | • Degree of importance of an interface between two resources for the integrity of information transfer within the overall process |

### 10.5.25 Centrality of path (based on centrality)

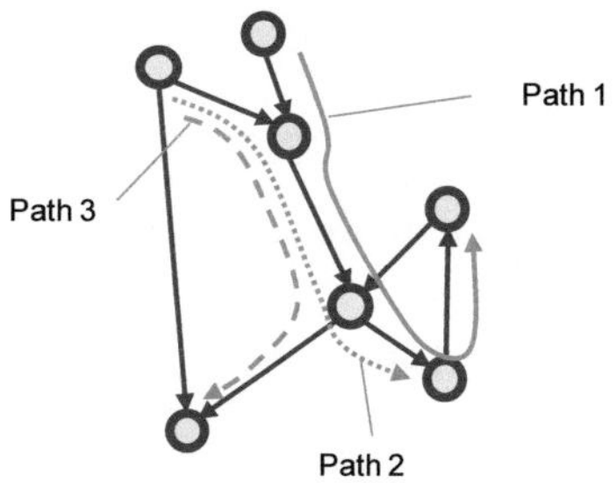


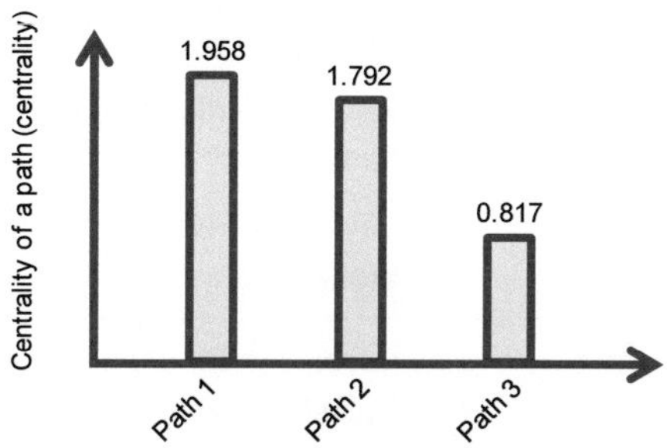


**Definition**

- Sum of all centralities of all nodes along a specific path
- Normalization of product for length of path and normalization basis of centrality

**Structural significance**

- Evaluation of relevance of an individual path for the overall process in terms of its connection to the overall process
- Identification of critical transitions and pathways through the process
- Degree to which an individual path is connected to the overall network

**Representation**

- Pareto distribution of all centralities for all relevant paths

**Literature**

[LOCH et al. 2003], [FREEMAN 1978]

## Structural significance

| | Centrality of a path (based on centrality) |
|---|---|
| Task | • Degree of independence of a sequence of tasks from the rest of the overall process<br>• Degree of influence taken by the overall process on a sequence of tasks<br>• Level of relevance of the principal process path for the overall process |
| Artifact | • Level of independence of principal path from other sources across overall process<br>• Level of relevance of a sequence of artifacts for the overall process |
| Org. unit | • Degree of over-determination of a communication path or a chain of command across overall organizational setup<br>• Extent of external influences on communication path<br>• Level of relevance of a communication path for the overall process |
| Time | • Degree of synchronization of a central path necessary with overall process<br>• Degree of possible controllability of a sequence of points in time because of their independence from the overall schedule<br>• Degree of other-directedness of schedule<br>• Level of relevance of the principal process path for the overall process |
| Event | • Degree of independence of a sequence of events from the overall process<br>• Level of controllability of a sequence of events<br>• Degree of synchronization necessary<br>• Level of relevance of the principal process path for the overall process |
| Resource | • Degree of openness of a series of resources towards other systems<br>• Extent of necessary interfaces of a series of resources with the overall process |

## 10.5.26 Centrality of path (based on degree)

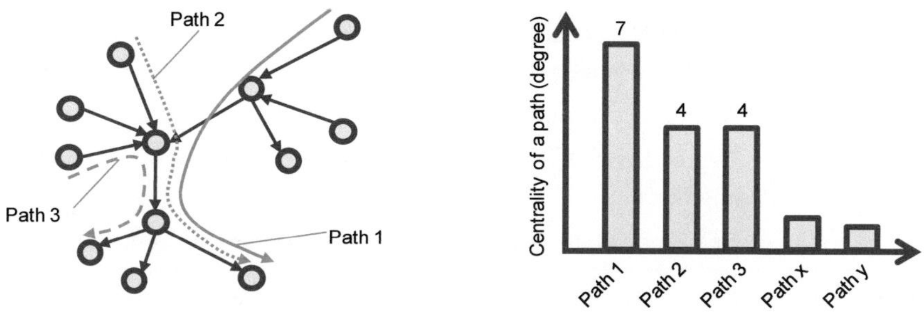


**Definition**

- Sum of all degrees of nodes along a specific path (excluding edges of path)
- Is equivalent to activity / passivity of a path
- Differentiation by active and passive degree possible

**Structural significance**

- Evaluation of relevance of an individual path for the overall process in terms of processing and distribution of information
- Identification of critical transitions and pathways through the process

**Representation**

- Pareto distribution of centralities for all relevant paths

**Literature**

[MAURER 2007, p. 144], [FREEMAN 1978]

## Structural significance

| | **Centrality of a path (based on degree)** |
|---|---|
| Task | • Number of interfaces to neighboring tasks<br>• Degree of independence of a sequence of tasks from its direct interfaces to the process<br>• Degree of influence taken by interfacing tasks<br>• Degree to which the sequence of tasks is influenced by its direct interfaces<br>• Relevance of the principal process path for the overall process |
| Artifact | • Number of interfaces to other documents in the process<br>• Extent of difficulty to achieve consistent documentation or a process<br>• Degree of independence of a sequence of documents from the direct interfaces to other documents in the process |
| Org. unit | • Number of interfaces of a communication path or a chain of command to other organizational units<br>• Extent of over-determination of a communication path |
| Time | • Number of synchronization points within the schedule<br>• Degree of other-directedness of schedule |
| Event | • Number of interfaces of series of events from other events within the overall process<br>• Extent of necessary synchronization with other events in the process<br>• Degree of other-directedness of schedule |
| Resource | • Number of interfaces of a series of resources to overall process<br>• Extent of integration into overall resource landscape |

## 10.5.27 Degree of progressive oscillation

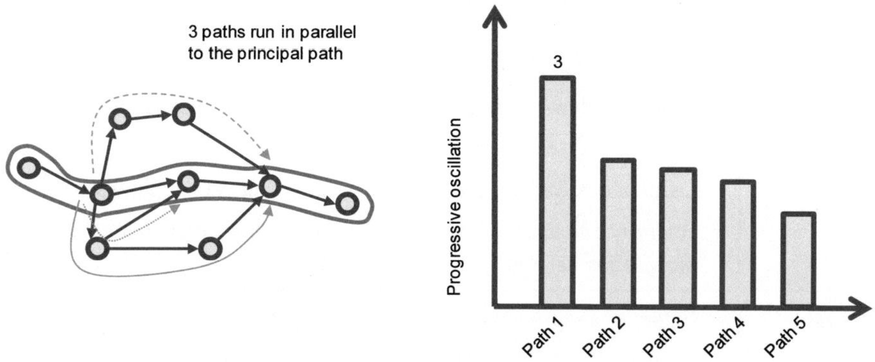


**Definition**

- Sum of length of all paths that run parallel to a designated path, starting and ending on that path

**Structural significance**

- Evaluation of the impact of supporting processes for an individual pathway
- Determination of the degree to which a path depends on supporting processes
- Identification of over-determined paths

**Representation**

- Pareto distribution of progressive oscillation for all relevant paths

**Literature**

[PONN & LINDEMANN 2005]

## Structural significance

| | **Progressive oscillation** |
|---|---|
| Task | • Number of supporting processes (as series of tasks) that contribute to direct progress of the principal process<br>• Level of independence of a principal path from supporting tasks<br>• Level of over-determination of principal path<br>• Level of risk of delays because of delays outside the principal path<br>• Degree of flexibility to execute principal process despite obstacles along principal path |
| Artifact | • Extent of supporting documentation to generate principal artifacts<br>• Level of risk of principal path to generate inconsistent documentation<br>• Degree of independence from other artifacts |
| Org. unit | • Extent of external influences on a communication path or a chain of command<br>• Level of risk of unwanted influences on a communication channel<br>• Level of risk of divergence of opinion building along a communication path |
| Time | • Extent of influence of external points in time on principal sequence<br>• Level of risk of delays outside the sequence of points in time delaying the principal sequence |
| Event | • Degree of dependence on events outside principal process<br>• Number of states of supporting processes that can impact the principal sequence of events |
| Resource | • Degree of openness of a set of resources<br>• Degree of flexibility to replace principal chain of resources with other resources |

### 10.5.28 Height of hierarchy

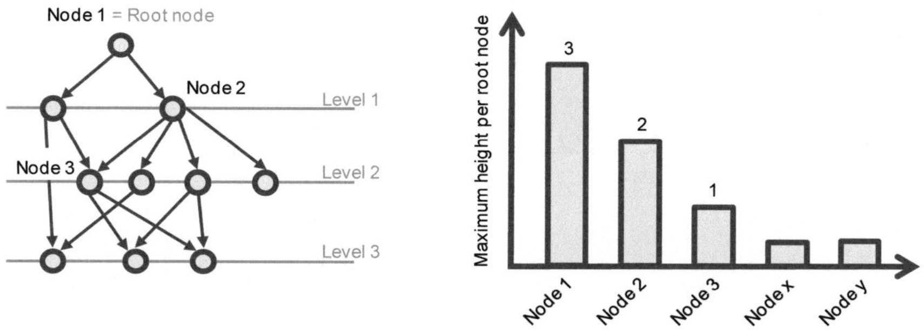


**Definition**

- Number of levels of a tree
- Hierarchy is computed level by level

**Structural significance**

- Evaluation of intensity of the distribution of information or errors
- Possible as impact on other nodes (active root node) or as feed (passive root node)
- Evaluation of secondary effects of changes to a node

**Representation**

- Pareto distribution of height of each hierarchy for all root nodes

**Literature**

[MAURER 2007, p. 218], [ROBERTSON & SEYMOUR 1986], [HARRISON & MAGEL 1981], [PIWOWARSKI 1982]

### 10.5.29 Width of hierarchy

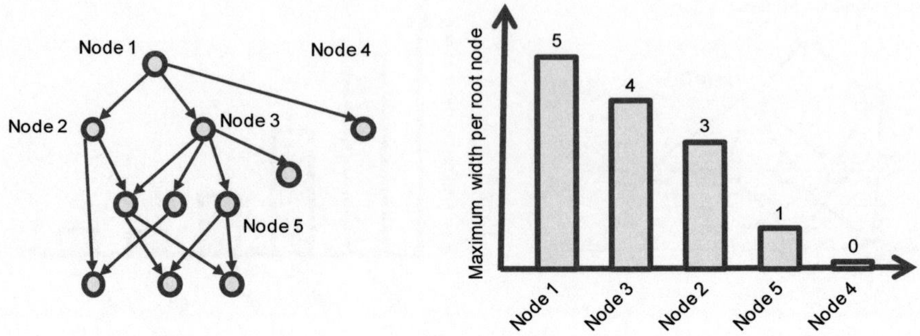


**Definition**

- Number of all end nodes (per level) of a tree
- If a node is accessed by two or more levels, the lowest level is counted
- Hierarchy is computed level by level

**Structural significance**

- Evaluation of velocity of distribution of information or errors (per level)
- Possible as impact on other nodes (active root node) or as feed (passive root node)
- Evaluation of secondary effects of changes to a node

**Representation**

- Pareto distribution per root node

**Literature**

[MAURER 2007, p. 218], [ROBERTSON & SEYMOUR 1986]

### 10.5.30 Tree criticality

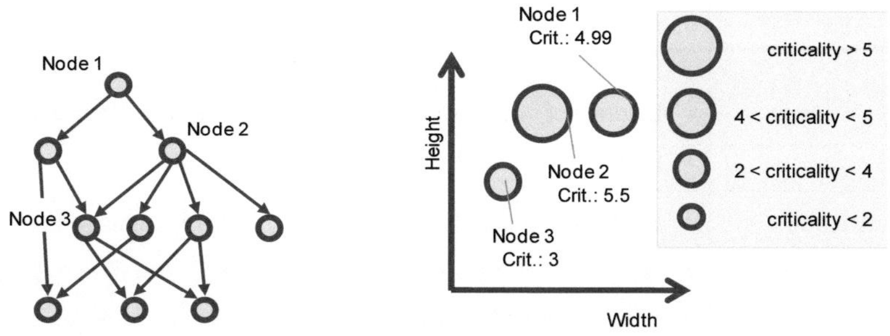


**Definition**

- Surface of a tree (width, height), weighted according to level (active root node = snowball factor, passive root node = forerun factor)
- Hierarchy is computed level by level
- Weight of level can be set individually (e.g., as inverse of distance to root node)

**Structural significance**

- Measure for the distribution of information and errors within the process
- Analysis for nodes that are robust against the propagation of errors
- Analysis of nodes that are central distributors of information
- Calculation is possible for consequences (→ active root node of hierarchy) or forerun (→ passive root node of hierarchy)

**Representation**

- Metric per root node
- Pareto distribution of height, width, and criticality for all nodes

## Structural significance

| | **Tree criticality (including width and height of a hierarchy)** |
|---|---|
| Task | • Degree of a task to rapidly spread information and errors throughout the process<br>• Degree of susceptibility for information within the process<br>• Level of risk of a task to be impacted by errors |
| Artifact | • Extent necessity to adjust other artifacts to achieve consistency<br>• Degree of artifact to rapidly spread out information to other artifacts<br>• Level of risk of errors in an artifact to spread to other artifacts rapidly |
| Org. unit | • Extent of potential of an organizational unit to rapidly address other organizational units<br>• Extent of visibility of an organizational unit for other organizational units<br>• Degree of accessibility of know-how in the organizational setup<br>• Degree of potential to rapidly spread out information from one organizational unit to others<br>• Extent of influence of an organizational unit |
| Time | • Level of risk of a point in time to spread a delay across the overall process<br>• Degree of susceptibility of a point in time to be impacted by delays across the overall process<br>• Level of importance of a point in time for the information distribution across the overall process |
| Event | • Degree of influence of an event exercised on other events in the process<br>• Level of impact received from other events in the process<br>• Degree to which an event controls the subsequent process<br>• Level of risk of delays of an event because of delays in the previous process |
| Resource | • Level of potential of a resource to allow for the consistent exchange and processing of artifacts in the overall process<br>• Degree of the ability of a resource to rapidly distribute information to other resources<br>• Degree to which a resource is well integrated into the overall network of resources |

## 10.5.31 Snowball factor

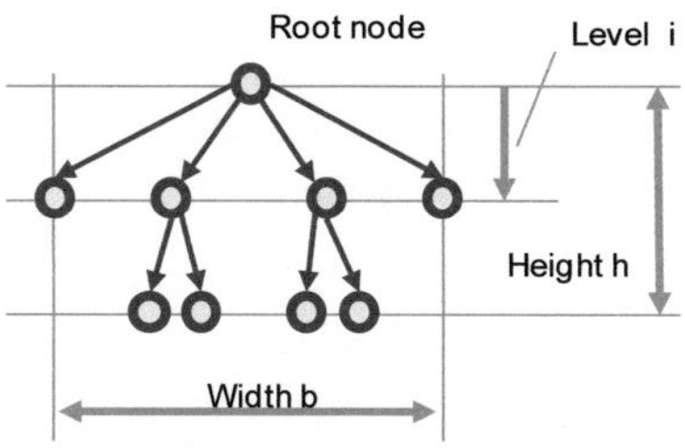


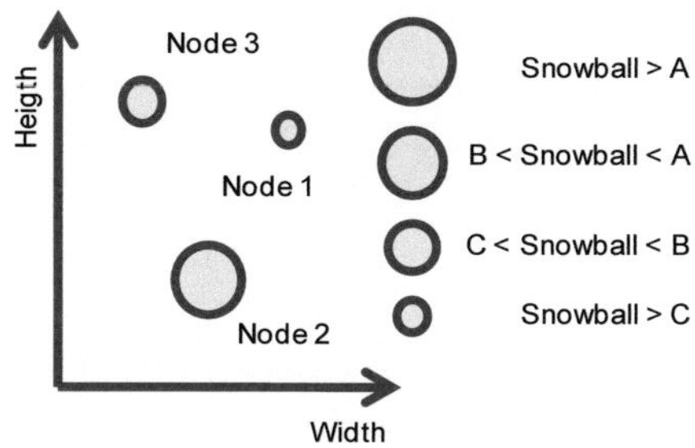


**Definition**

- Sum of product of width (per level) and height, each level weighted according to inverse of shortest path length to root node
- Hierarchy is computed level by level

**Structural significance**

- Measure for the spreading of information and error

**Representation**

- Portfolio for height, width, and snowball factor of all nodes
- Also possible as distribution of snowball factor for all root nodes (see metric "Forerun factor")

**Literature**

[LOCH et al. 2003]

## Structural significance

| | Snowball factor |
|---|---|
| Task | • Weighted degree of influence of a task on subsequent tasks<br>• Degree of importance of a task for the subsequent process<br>• Level of rapid information sharing with subsequent process<br>• Level of effort to synchronize subsequent tasks to achieve consistency<br>• Level of risk of rapidly spreading errors from task to subsequent process |
| Artifact | • Weighted degree of impact of changing an artifact<br>• Level of effort to achieve consistent documentation<br>• Level of importance of the quality of an artifact to influence the outcome of the overall process<br>• Level of risk of errors contained in an artifact to propagate to other artifacts |
| Org. unit | • Weighted degree of potential of the ability of an organizational unit to communicate within an organizational set<br>• Extent of networking within social network to enable quick distribution of information<br>• Degree of possible influence on other organizational units within organizational setup |
| Time | • Weighted level of risk of a delay at a point of time to delay the subsequent process |
| Event | • Weighted degree of influence of an event on the subsequent process<br>• Level of risk of an event to delay the subsequent process<br>• Level of importance of the quality of an event to influence the outcome of the overall process |
| Resource | • Weighted degree of potential to forward information consistently to other resources<br>• Degree of openness to transmit information to other resources<br>• Level of potential of a resource to serve as an information hub |

## 10.5.32 Forerun factor

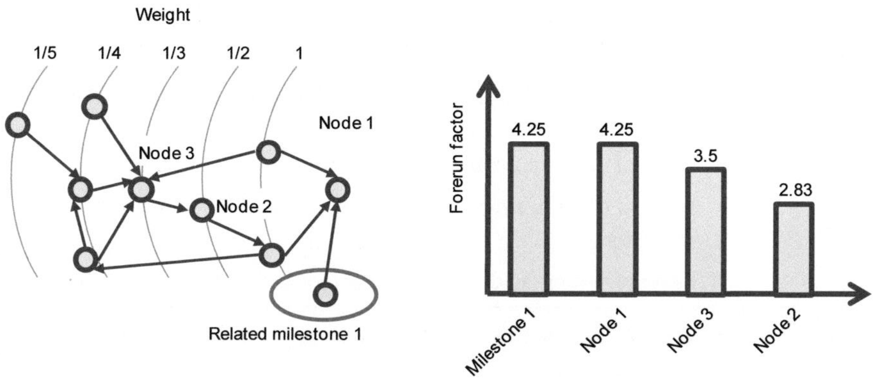


**Definition**

- Number of nodes that lead to a milestone or decision point weighted by closeness to this milestone
- Hierarchy is computed level by level

**Structural significance**

- Extent of preparations that have to be arranged previous to a milestone
- Robustness of an entity towards incoming information and/or errors

**Representation**

- Portfolio for height, width, and forerun factor of all nodes (see metric "Snowball factor")
- Also possible as distribution of forerun factor for all root nodes

**Literature**

[BADKE-SCHAUB & GEHRLICHER 2003]

## Structural significance

| | Forerun factor |
|---|---|
| Task | • Weighted degree of dependency of a task on previous tasks<br>• Degree of importance of a task to compile information from the previous process<br>• Level of rapid information access from previous process<br>• Level of effort to synchronize with previous tasks to achieve consistency<br>• Level of risk of rapidly spreading errors to influence task |
| Artifact | • Weighted degree of impact of changing the previous process into an artifact<br>• Level of effort necessary to achieve documentation consistent with previous artifacts<br>• Level of dependence of the quality of an artifact on previous artifacts<br>• Level of risk of errors to be contained in an artifact<br>• Level of importance of information search to generate an artifact |
| Org. unit | • Weighted degree of visibility of an organizational unit within the organizational setup<br>• Potential of an organizational unit to receive information<br>• Extent of networking within social network to enable quick reception of information<br>• Degree of possible impact by other organizational units within organizational setup |
| Time | • Weighted level of risk of a delay at a point of time because of delays during the previous process |
| Event | • Weighted degree of impact of previous process on an event<br>• Level of risk of an event to be delayed because of delays in the previous process<br>• Level of dependence of the quality of an event on the outcome of the previous process |
| Resource | • Weighted degree of potential to generate information consistently to other resources<br>• Degree of openness to receive information from other resources<br>• Level of potential of a resource to serve as an information hub |

## 10.5.33 Tree-robustness

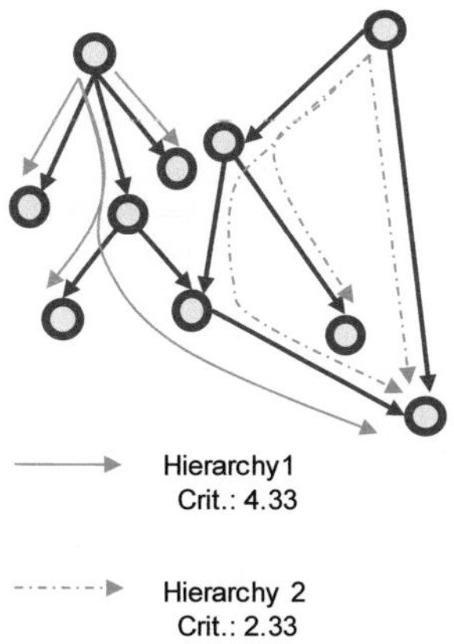


| Domains | Tree robustness per domain |
|---|---|
| Domain 1<br>Domain 2<br>... | 0.3<br>... |

**Definition**

- Quotient of number of all trees and sum of all tree criticalities
- Hierarchies are computed level by level

**Structural significance**

- Degree to which the network is interspersed with trees
- Robustness of a complete process concerning the distribution or reception of information and/or errors
- Evaluation of the overall process for its robustness against rapid propagation of errors

**Representation**

- Metric per domain

### 10.5.34 Maximum nesting depth

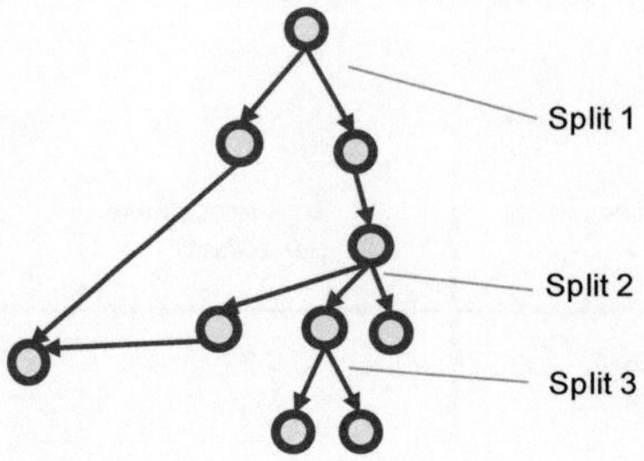


| Domains | Maximum nesting depth |
|---|---|
| Domain 1 | 3 |
| Domain 2 | ... |
| Domain 3 | |
| ... | |

**Definition**

- Number of splits in a process
- Can only be calculated for process with one starting node

**Structural significance**

- Difficulty of understanding the process model
- Estimation of well-structuredness of process if compared to maximum nesting depth for splits leading towards single end-node of process (process is well structured if difference of the two is zero)

**Representation**

- Metric per domain

**Literature**

[GRUHN & LAUE 2006a], [HARRISON & MAGEL 1981], [PIWOWARSKI 1982]

## Structural significance

| | **Tree robustness** | **Maximum nesting depth** |
|---|---|---|
| Task | • Degree of an overall process to rapidly spread out information among all tasks<br>• Degree of susceptibility of an overall process to propagate errors across all tasks | • Level of subdivision of a process into possible sub-processes (for a well-structured process) (for starting node)<br>• Number of decision points in a process (if decision points are not modeled explicitly) after starting node |
| Artifact | • Degree of (mutual) dependency of all artifacts in a process among each other<br>• Level of risk of a process to contain inconsistent artifacts<br>• Degree of rapid distribution of information across artifacts<br>• Level of risk of spreading errors rapidly among artifacts | • Number of subsequent phases of generating artifacts in a process<br>• Level of potential for clusters of artifacts that can be regrouped<br>• Level of hierarchization of artifacts |
| Org. unit | • Level of potential for communication among all organizational units in organizational setup<br>• Degree of potential to rapidly spread out information within the overall process | • Number of levels of hierarchy below an organizational unit |
| Time | • Degree of susceptibility of the schedule for the overall process to be dependent on single points in time | • Level of potential to create phases (i.e. pre-defined modules of points in time) based on well-structured groups of points in time |
| Event | • Degree to which the overall process is controlled by individual events<br>• Risk of the overall process to be delayed because of problems at individual events | • Level of organization among events<br>• Level of potential to regroup a number of states into a module to simplify process |
| Resource | • Level of potential for the continuous and consistent exchange of information among all resources<br>• Level of potential for the rapid sharing of information among artifacts | • Level of flexibility to circumnavigate individual resources in case of their failure |

### 10.5.35 Number of cliques

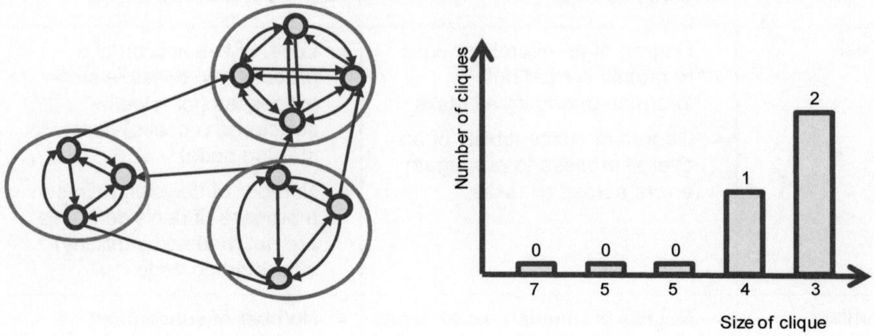


**Definition**

- Number of complete clusters within the network

**Structural significance**

- Identification of closely connected groups that involve a lot of communication
- Degree to which the network is characterized by completely connected elements

**Representation**

- Distribution according to number of connected nodes forming a clique

**Literature**

[GROSS & YELLEN 2005, p. 43]

## Structural significance

| | **Number of cliques** |
|---|---|
| Task | • Degree of compilation of discipline-spanning knowledge in collaboration among several tasks<br>• Level of quality of a process model, where single cliques could be modeled as individual tasks at a higher more rudimentary level of detail<br>• Level of complexity of the object processed in the process<br>• Level of potential that the use of process simulation can improve the schedule (simulation of best navigation of iterations) |
| Artifact | • Level of mutual dependency on other artifacts that are jointly generated during the process<br>• Level of potential of possible integration of a clique of artifacts into one overall artifact |
| Org. unit | • Degree of necessity of work groups that are located in one space<br>• Number of possible meetings per time frame necessary to efficiently synchronize organizational units<br>• Level of potential to introduce communities of practice to share knowledge across the borders of an organizational unit |
| Time | • Degree of non-linearity of the schedule<br>• Level of risk of delays in the overall process<br>• Level of potential that the use of process simulation can improve the schedule (simulation of best navigation of iterations) |
| Event | • Degree of non-linearity of the schedule<br>• Level of risk of not reaching a desired process outcome in a direct run |
| Resource | • Number of highly compatible resources |

## 10.5.36 Cluster-coefficient (local)

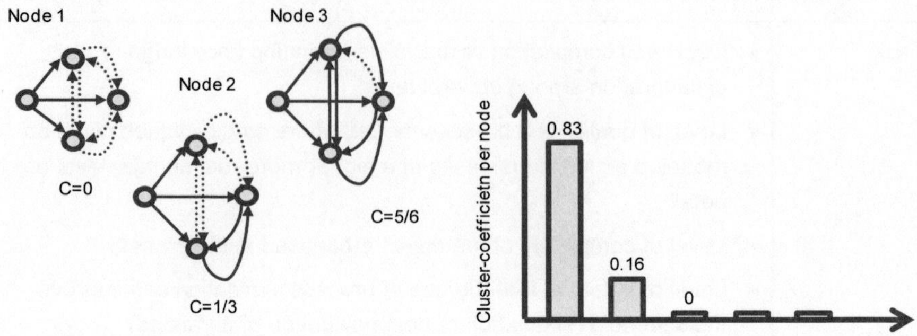


Cluster-coefficients for three adjacent nodes

**Definition**

- Quotient of number of existing edges between nodes adjacent to a node and number of possible edges
- Purposeful only for nodes with active degree > 1
- Commonly used for active node; can also be computed for passive root nodes and for non-directed graphs

**Structural significance**

- Evaluation of tendency of individual nodes to be part of a cluster
- Identification of nodes that are not fully involved in cluster
- Identification of possible synchronization / distribution nodes that do not exhaust their options

**Representation**

- Pareto distribution of cluster-coefficient for all nodes

**Literature**

[NEWMAN 2003a], [WATTS & STROGATZ 1998]

## Structural significance

| | Cluster-coefficient (local) |
|---|---|
| Task | • Level of interdependence of a task on implicit or "weak" dependencies to be consistent<br>• Level of risk of unexpected changes to a task during the process<br>• Level of risk to induce small iterations among neighboring tasks in a process<br>• Level of tendency of a task to cause non-linearity in the process schedule |
| Artifact | • Level of potential for consistency among artifacts<br>• Level of risk of a lack of consistency induced by an artifact<br>• Level of tendency of an artifact to cause non-linearity in the process schedule |
| Org. unit | • Extent of possible social contacts in an organizational setup<br>• Level of potential for networking within social network<br>• Degree of necessity to closely involve the partners of an organizational unit among each other |
| Time | • Level of risk of a point in time to be involved in a small iteration<br>• Level of tendency of a point in time to cause non-linearity in the process schedule |
| Event | • Level of risk of an event not to reach the desired state in the planned manner |
| Resource | • Approximation of the necessary interfaces of a system to allow for efficient transmission of information |

### 10.5.37 Cluster-coefficient (global)

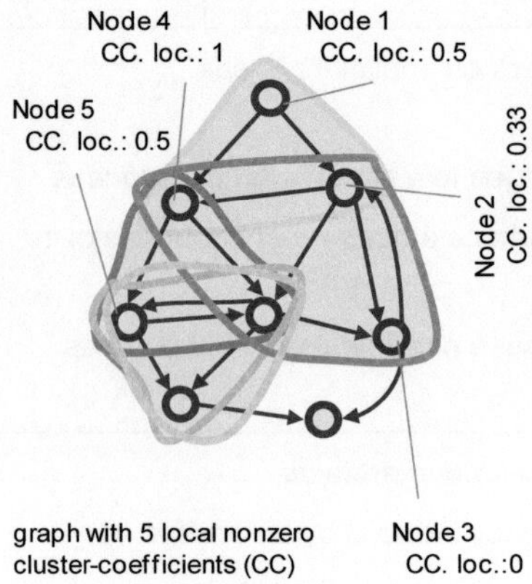


| Domains | Cluster-coefficient per domain |
|---|---|
| Domain 1 | 0.29 |
| Domain 2 | ... |
| ... | |

**Definition**

- Sum of all local cluster-coefficients divided by total number of nodes that have an active degree > 1
- Commonly used for active node; can also be computed for passive root nodes and for non-directed graphs

**Structural significance**

- Evaluation of tendency of individual nodes to be part of a cluster (esp. for social networks)
- Comparison of clustering of different networks
- Degree to which the overall process utilizes its possibilities

**Representation**

- Metric per domain

**Literature**

[NEWMAN 2003a], [WATTS & STROGATZ 1998]

## Structural significance

| | Cluster-coefficient (global) |
|---|---|
| Task | • Level of potential for linearity of the overall process<br>• Level of risk of unexpected changes to occur<br>• Level of risk of iterations in the overall process<br>• Level of risk of possible conflicts among the tasks<br>• Degree of uncertainty in the process |
| Artifact | • Degree of mutual dependencies among artifacts<br>• Level of risk of unexpected changes to occur<br>• Level of risk of possible conflicts in the documentation |
| Org. unit | • Degree of necessity of the process to be supported by a dense social network<br>• Level of possible short communication channels outside the organizational setup to quickly overcome unexpected problems |
| Time | • Level of risk of the process to be delayed because of unexpected radiation of a delay at one point in time to another point in time<br>• Degree of possible ambiguity of the process schedule<br>• Level of risk of possible conflicts in the process schedule |
| Event | • Level of risk of events to occur unexpectedly |
| Resource | • Level of potential implementation of consistent interfaces |

### 10.5.38 Module quality 1 (flow of information)

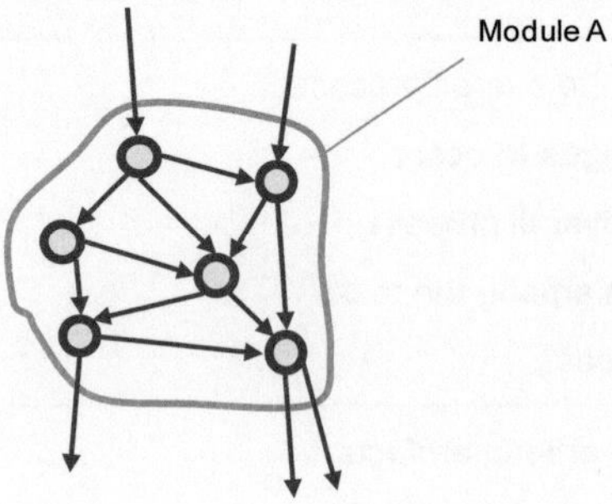


| Domains | Module | Module quality 1 |
|---|---|---|
| Domain 1 | A | 360 |
| | B | ... |
| | C | |
| Domain 2 | ... | |
| ... | | |

**Definition**

- Product of number of edges that cross the border of the module and number of edges within the module

**Structural significance**

- Degree of completeness of a module concerning complete clusters
- Only useful for the evaluation of existing modules (pre-defined groups)
- Comparison of modules concerning their interaction with their environment
- Indicator to possibly reduce complexity by introducing more modules with a reduced number of interfaces
- Means of quality to generate modules out of clusters

**Representation**

- Metric per module per domain

**Literature**

[HENRY et al. 1981]

### 10.5.39 Module quality 2 (compactness)

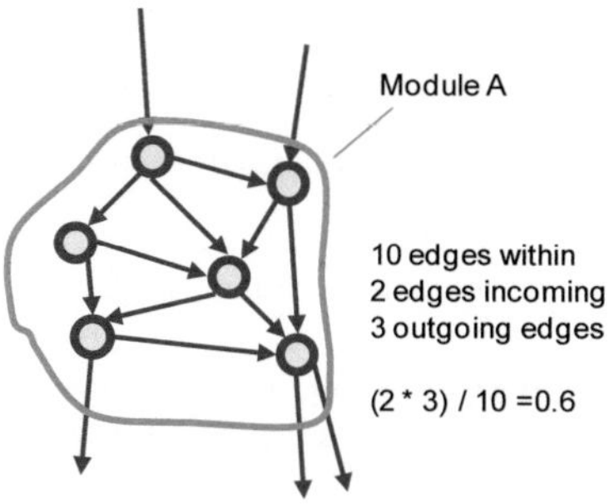


| Domains | Module | Module quality 2 |
|---|---|---|
| Domain 1 | A | 0.6 |
| | B | ... |
| | C | |
| Domain 2 | ... | |
| ... | | |

**Definition**

- Product of number of edges that cross the border of the module and number of edges within module

**Structural significance**

- Degree of closeness of a module
- Only useful to evaluate existing modules (pre-defined groups)
- Comparison of modules concerning their interaction with their environment
- Indicator to possibly reduce complexity by introducing more modules with a reduced number of interfaces
- Means of quality to generate modules out of clusters

**Representation**

- Metric per module per domain

**Literature**

[HENRY et al. 1981]

## Structural significance

| | **Module quality 1 and 2** |
|---|---|
| Task | • Degree of closeness of a module (as a pre-defined set of tasks)<br>• Level of quality of the model using well-defined modularization criteria for sub-processes<br>• Degree of distinctness of a module to reduce the number of prevailing interfaces |
| Artifact | • Degree of closeness of a module of artifacts (as a pre-defined set of artifacts)<br>• Level of potential to process a module of artifacts independently from the overall process |
| Org. unit | • Level of quality of the setup of teams and departments as modules of organizational units |
| Time | • Degree of closeness of a module of points in time (as a pre-defined set of tasks, e.g., a phase) to allow for independent processing of the process |
| Event | • Degree of independence of a module of events (as a pre-defined set of events) from external influences |
| Resource | • Level of quality of embedding a module of resources into the overall resource landscape |

## 10.5.40 Number of cycles

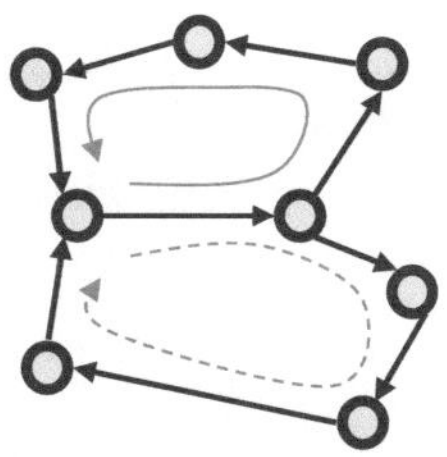

2 cycles with length 5

| Domain | Number of cycles per domain |
|---|---|
| Domain 1 | 2 |
| Domain 2 | ... |
| Domain 3 | |
| ... | |

**Definition**

- Number of paths with identical starting- and end-node
- Can purposefully be combined with length of cycles and the occurrence of specific nodes and edges

**Structural significance**

- Evaluation of the overall level of uncertainty of the process
- Determination of the degree of possible rework during process execution

**Representation**

- Metric per domain
- Pareto distribution of occurrence of cycles and their length

**Literature**

[BADKE-SCHAUB & GEHRLICHER 2003]

## Structural significance

| | Number of cycles |
|---|---|
| Task | • Degree of non-linearity of the process<br>• Number of possible iterations in the process<br>• Extent of necessity to allocate mixed workgroups that closely synchronize iterations<br>• Extent of necessity of a central instance to coordinate the overall process<br>• Level or risk of unexpected iterations to cause delays<br>• Extent of generation of knowledge in the process in an interdisciplinary context |
| Artifact | • Degree of focusing on centralized documents that are reworked and completed during the overall process<br>• Extent of risk of errors in individual documents to propagate throughout the process |
| Org. unit | • Extent of closeness of the social network<br>• Extent of implicit control within the social network in a way that information is transmitted back to the sender across a different path<br>• Level of risk of the occurrence of Chinese whisper phenomena in the process |
| Time | • Level of risk of unexpected delays in the schedule |
| Event | • Degree of non-linearity of the process |
| Resource | • Degree of continuity of the chain of resources |

### 10.5.41 Number of cycles per node

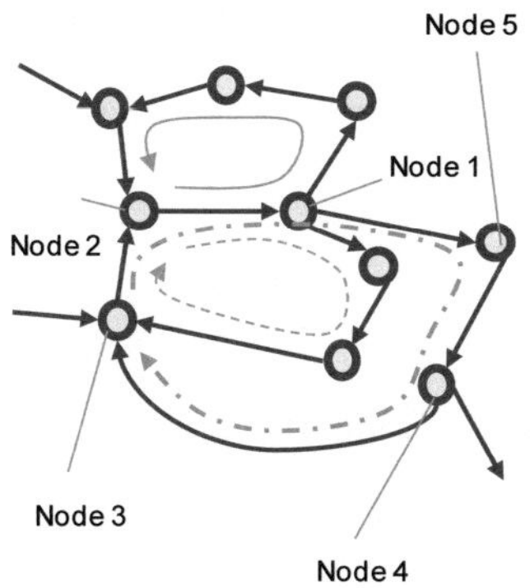


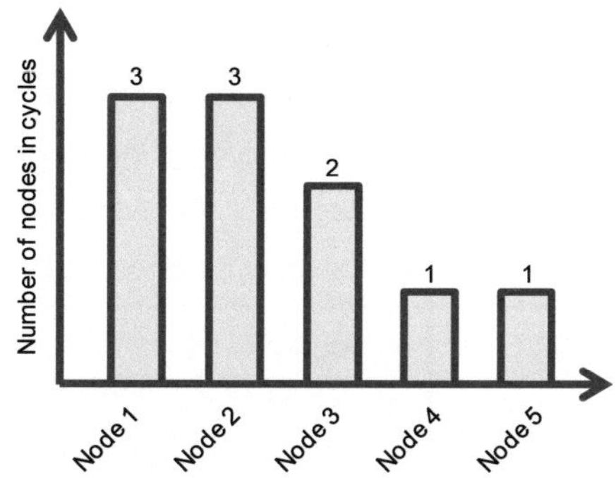


**Definition**

- Occurrence of a node in all cycles

**Structural significance**

- Determination of core entities of a process that help cope with uncertainty in the process
- Determination of entities that have an important impact on the generation of knowledge during the overall process

**Representation**

- Pareto distribution of occurrence of each node in cycles

**Literature**

[MAURER 2007, p. 236]

## Structural significance

| | Number of cycles per node |
|---|---|
| Task | • Degree of importance of a task towards the generation of knowledge (as possible core competency)<br>• Extent of insecurity processed in a task<br>• Degree of coordination exercised by a task<br>• Extent of interfaces that run through a task |
| Artifact | • Degree of importance of an artifact to document the progress of concretization in the process<br>• Extent of risk of errors in individual documents to propagate throughout the process<br>• Degree of synchronization of information that is carried out via a document<br>• Extent to which a document serves as an interface to assure consistency of information |
| Org. unit | • Degree to which communication revolves around an organizational unit |
| Time | • Degree of necessity of buffering of a point in time to avoid delays<br>• Level of importance of a point in time for the timeliness of the overall process |
| Event | • Extent of insecurity that results in an event<br>• Level of risk of not reaching an event because of unexpected results underway |
| Resource | • Degree of importance of a resource to ensure the data transfer among various workgroups |

## 10.5.42 Number of cycles per edge

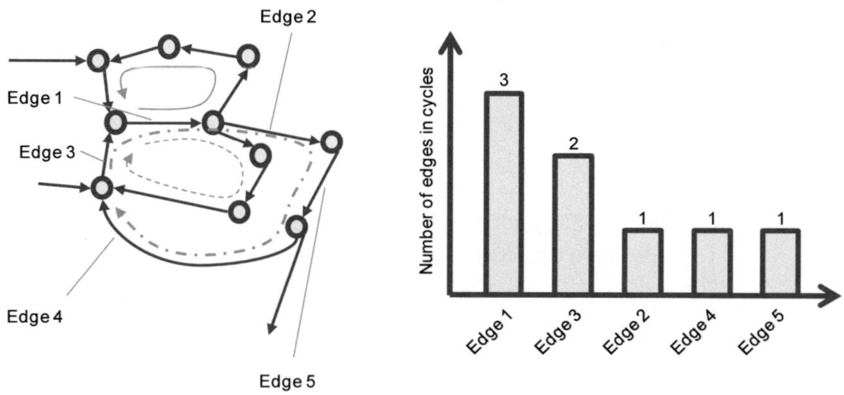


### Definition

- Occurrence of an edge in cycles

### Structural significance

- Dependencies that are highly relevant to coping with uncertainty in the process
- Identification of possible drivers for handling uncertainty

### Representation

- Pareto distribution of occurrence of edges in cycles

### Literature

[MAURER 2007, p. 236]

## Structural significance

| | Number of cycles per edge |
|---|---|
| Task | • Degree of importance of a communication channel between two tasks towards the generation of knowledge<br>• Extent of insecurity that is communicated between two tasks<br>• Degree of coordination exercised by a communication channel |
| Artifact | • Degree of importance of an interface between two artifacts to document the progress of concretization in the process<br>• Extent of maturity generated with the transition of an artifact into another artifact<br>• Degree of synchronization of information that is carried via an interface<br>• Extent to which a document serves as an interface to assure consistency of information |
| Org. unit | • Degree to which communication revolves around an two organizational units |
| Time | • Degree of necessity of buffering between two points in time to avoid delays<br>• Level of risk associated to a structural bottleneck in the schedule |
| Event | • Extent of insecurity that is processed from one event to the next<br>• Level of risk of not reaching an event because of unexpected results underway |
| Resource | • Degree of importance of a resource to ensure the data transfer among various workgroups |

## 10.5.43 Number of feedbacks

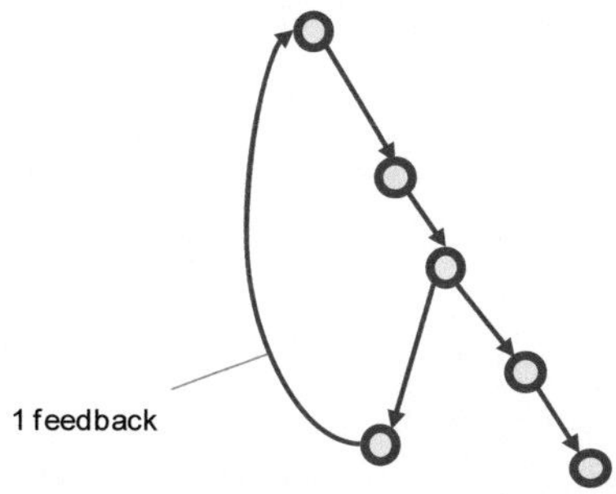


| Domains | Number of feedbacks per domain |
|---|---|
| Domain 1 | 1 |
| Domain 2 | ... |
| Domain 3 | |
| ... | |

**Definition**

- Number of edges that impede the ideal triangularization of a DSM
- Computation is nondeterministic as there is no unique form of triangularization; the minimum number of edges that impede ideal triangularization cannot always be computed

**Structural significance**

- Evaluation of the degree of uncertainty in the process
- Determination of the degree of deviation from an ideal sequence

**Representation**

- Metric per domain

**Literature**

[BROWNING 2001a]

### 10.5.44 Activation of cycle

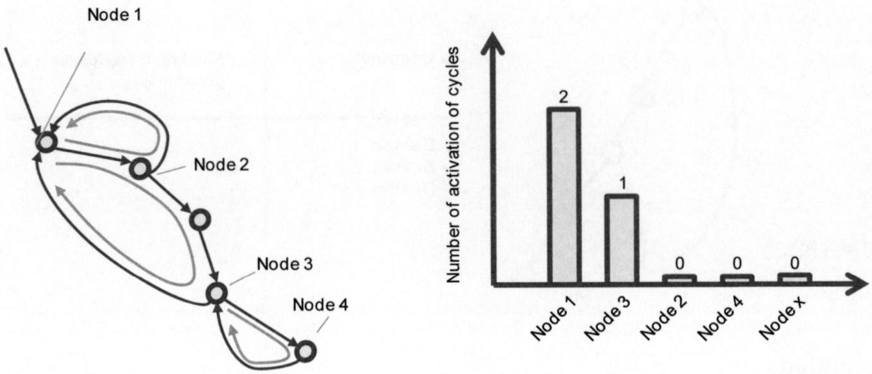


**Definition**

- Number of nodes that are the first ones in a cycle (in a triangularized DSM)
- Computation is nondeterministic as there is no unique form of triangularization; the minimum number of edges that impede ideal triangularization cannot always be computed

**Structural significance**

- Nodes that are relevant for handling uncertainty
- Identification of nodes that possibly lay the groundwork in one or more iteration(s)

**Representation**

- Pareto distribution of occurrence of activating nodes in cycles

## Structural significance

| | Number of feedbacks | Activation of cycle |
|---|---|---|
| Task | • Degree of non-linearity of the process<br>• Extent of necessity of small, tightly cross-linked work groups<br>• Level of risk of the overall process to be delayed by (possibly unexpected) rework<br>• Extent of collaboration in an interdisciplinary context or based on strong division of labor | • Degree of a task to have a preparatory effect in the generation of knowledge in the process<br>• Level of risk of a task to lead to rework in case of insufficient quality of the results of that task<br>• Extent of knowledge necessary to process a task |
| Artifact | • Degree of focusing on (few or many) artifacts that control the process flow as transition points in iterations | • Level of involvement of an artifact to document information that is at the heart of an iteration<br>• Degree of informational value of a artifact<br>• Level of risk of errors contained in an artifact |
| Org. unit | • (not applicable) | • (not applicable) |
| Time | • Level of risk of delays in the overall process | • Extent of necessary planning to ensure robustness of a point in time against possible delays through rework<br>• Extent of necessary investment at a point of time to ensure high quality of the outcome and to reduce the chance of starting an iteration |
| Event | • Degree of non-linearity of the concretization of results in the process | • Extent of necessary preparations to be done for an event<br>• Extent of insecurity present at an event<br>• Extent of necessity of quality control at an event |
| Resource | • (not applicable) | • Level of quality necessary for the results of a resource |

## 10.5.45 Number of starting points for iterations

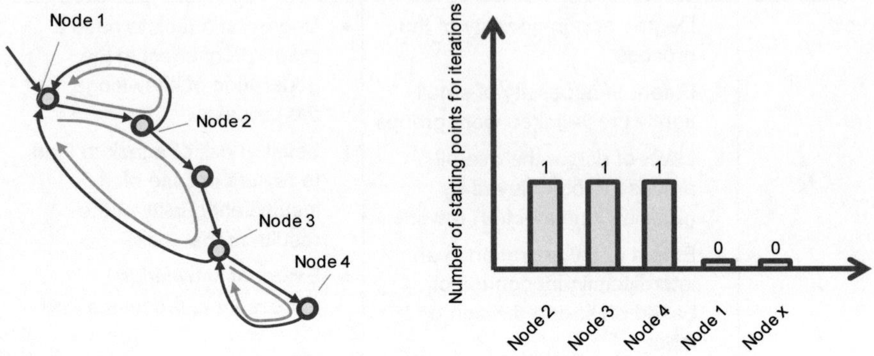


**Definition**

- Number of nodes that start iterations (nodes that are starting nodes of edges that impede the ideal triangularization of the DSM)
- Computation is nondeterministic as there is no unique form of triangularization; the minimum number of edges that impede ideal triangularization cannot always be computed

**Structural significance**

- Criticality of an entity to start an iteration
- Determination of possible decision points that can cause iterations
- Determination of entities where uncertainty in the process is handled

**Representation**

- Pareto distribution of outgoing edges that are starting points for iterations and number of initiated iterations

**Literature**

[LUKAS et al. 2007]

## Structural significance

| | Number of starting points for iterations |
|---|---|
| Task | • Degree of a task to evaluate the generation of knowledge in the process and to initialize possible rework<br>• Level of risk of a task to lead to rework in case of insufficient quality of the input and results of that task<br>• Extent of knowledge necessary to process a task (that should be made available early in the process)<br>• Level of risk associated to errors that are overlooked at a task |
| Artifact | • Level of involvement of an artifact to transport information that is at the heart of an iteration<br>• Degree of informational value of a artifact to prepare the decision for rework<br>• Level of risk of errors contained in an artifact to cause problems at a later stage |
| Org. unit | • Potential of an organizational unit to influence iterations before they take place |
| Time | • Extent of necessary planning to ensure robustness of a point in time against possible delays through rework<br>• Extent of knowledge necessary at a point of time to assess the quality of the input<br>• Extent of potential to install point in time as milestone |
| Event | • Extent of an event to serve as a decision point to control the flow of the process<br>• Extent of necessity of quality control at an event<br>• Extent of potential to install event as milestone |
| Resource | • Level of quality necessary for the results of a resource |

### 10.5.46 Iterative oscillation

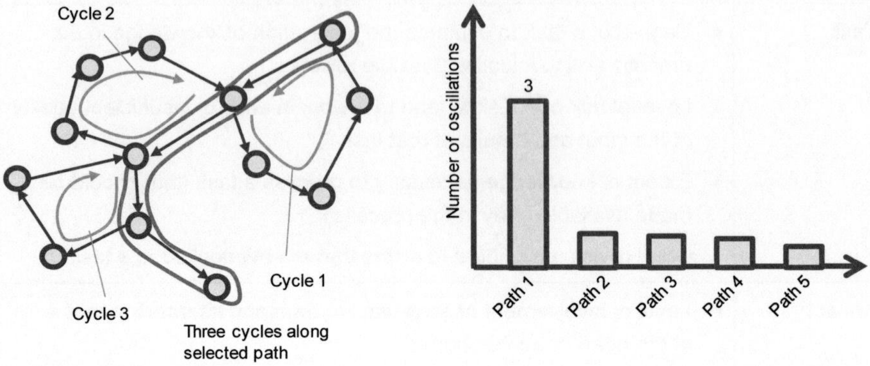


**Definition**

- Sum of length of all cycles that share at least one edge with a selected path

**Structural significance**

- Degree to which a path interacts with other nodes based on uncertainty within the process
- Determination of pathways through the network that are highly susceptable to changes long the way

**Representation**

- Pareto distribution of number of cycles per path

**Literature**

[LOCH et al. 2003]

## Structural significance

| | **Iterative oscillation** |
|---|---|
| Task | • Extent of effort necessary outside the principal process flow to execute the process<br>• Degree of linearity of the process |
| Artifact | • Degree of linear progress in the process<br>• Extent of forecast reliability of a process to be represented as a simple Gantt chart<br>• Extent to which a series of artifacts are dependent on supporting artifacts |
| Org. unit | • Level of risk of possible influences exercised on a communication path outside the intended or official communication path |
| Time | • Extent of forecast reliability of a process to be represented as a simple Gantt chart |
| Event | • Extent of forecast reliability of a process to be represented as a simple Gantt chart |
| Resource | • Extent of possible support to be accessed by main chain or resources |

## 10.5.47 Bipartite density

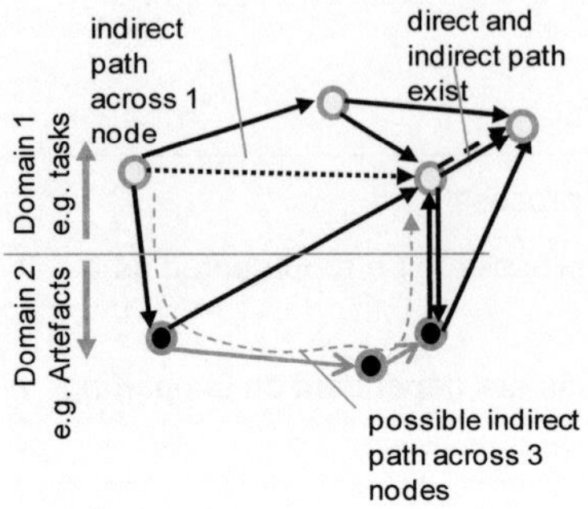


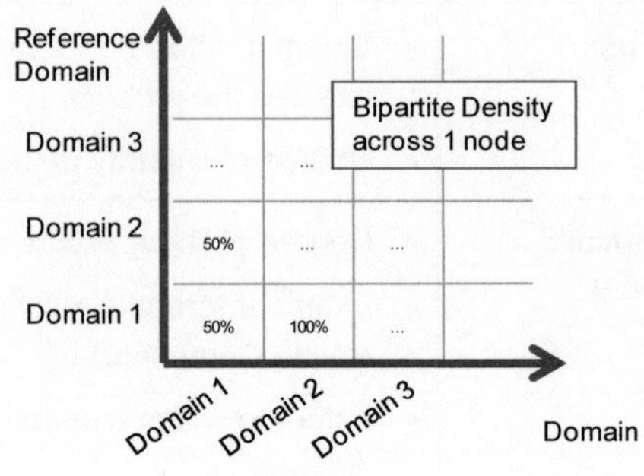


**Definition**

- Percentage of existing implicit relations (within the same or via a different domain) in relation to the number of possible relations
- Can be calculated across one or several level (i.e., across one or more nodes): an implicit path equals a shortest path to a reachable node that is not directly connected; the path length of that reachability serves as a parameter to the determination of implicit paths
- Within a domain roughly similar to cluster coefficient (local)

**Structural significance**

- Comparison of alignment with other domains
- Analysis of appropriateness of direct relations
- Assessment of modeling accuracy (direct dependencies that should be modeled as indirect dependencies and vice versa)

**Representation**

- Metric for each domain related to another domain

**Literature**

[VANDERFEESTEN et al. 2007], [MAURER et al. 2006]

## Structural significance

Only applicable for a domain (left column) in relation to a reference domain (not specified).

If a domain is assessed with a view to a second reference domain, each metric represents:

- the degree to which a network of entities (e.g., tasks) depends on supporting entities (e.g., resources) in the domain of reference to be processed.
- the degree to which a network of entities (e.g., tasks) is aligned with the needs imposed by the domain of reference (e.g., product attributes).

| | **Bipartite density** |
|---|---|
| Task | • Extent of implicit communication among task that are not explicitly in place in process<br>• Extent of meetings and other means of synchronization necessary in the process<br>• Extent of correct modeling of the process (in terms of implicit relationships that possibly exist but were not modeled) |
| Artifact | • Extent of implicit transitions among task that drive the process and advance the maturity of the artifacts in the process<br>• Extent of synchronization necessary to achieve consistent documentation<br>• Extent of correct modeling of the process (in terms of implicit relationships that possibly exist but were not modeled) |
| Org. unit | • Density of the social network via indirect relationships<br>• Extent of possible shortcuts to support quick distribution of information<br>• Extent of meetings and other means of synchronization necessary in the process<br>• Extent of correct modeling of the process (in terms of implicit relationships that possibly exist but were not modeled) |
| Time | • Extent of implicit transitions among points in time that drive the process<br>• Degree of attention that has to be paid to other points in time when planning the schedule of the process<br>• Extent of correct modeling of the process (in terms of implicit relationships that possibly exist but were not modeled) |
| Event | • Extent of implicit transitions among events that drive the process<br>• Extent of correct modeling of the process (in terms of implicit relationships that possibly exist but were not modeled) |
| Resource | • Extent of possibly purposeful interfaces among resources that should be implemented to facilitate the process and to support consistency among the processed artifacts<br>• Extent of correct modeling of the process (in terms of implicit relationships that possibly exist but were not modeled) |

### 10.5.48 Number of organizational interfaces

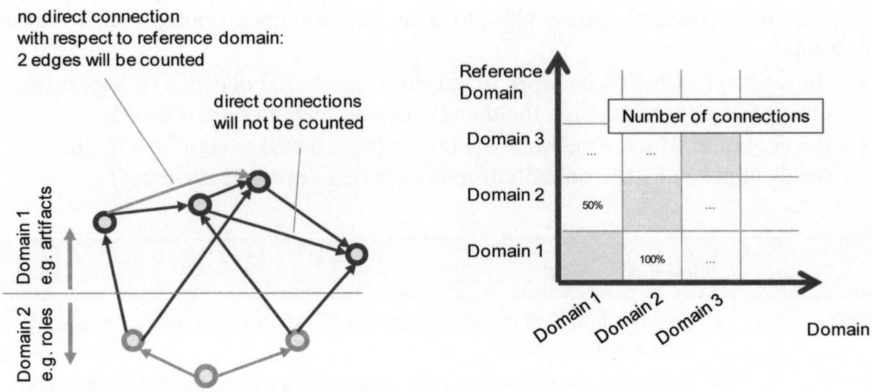


**Definition**

- Number of edges within one domain that link two nodes which are not attributed to the same node in a different domain (= reference domain)

**Structural significance**

- Analysis of the effort taken for a transition between two nodes because the node of reference is changed (e.g., different responsibility, different format, different media, different model)
- Identification of those transitions that demand special interfaces
- Comparable to attribution of two domains via a swimlane-model

**Representation**

- Metric per domain

**Literature**

[ANDERL & TRIPPNER 2000, p. 11], [BECKER et al. 2005, p. 123]

## Structural significance

| | Number of organizational interfaces (with view to reference domain) |
|---|---|
| Task | • Extent of effort necessary at the interface between two tasks with view to the transfers necessary via a supporting domain of reference (e.g., how many resources are applied to support the interface) |
| Artifact | • Extent of effort necessary at the transition between two artifacts with view to the transfers necessary via a supporting domain of reference (e.g., through how many organizational units an artifact is transferred to make a transition to the next artifact) |
| Org. unit | • Extent of effort necessary at the interface between two organizational units with view to the transfers necessary via a supporting domain of reference (e.g., how many artifacts are necessary for one organizational unit to communicate with another organizational unit) |
| Time | • Extent of effort necessary at the transition between two points in time with view to the transfers necessary via a supporting domain of reference (e.g., how many resources support the process) |
| Event | • Extent of effort necessary at the transition between two events with view to the transfers necessary via a supporting domain of reference (e.g., through how many resources an event is transferred to the next event) |
| Resource | • Extent of effort necessary at the interface between two resources with view to the transfers necessary via a supporting domain of reference (e.g., how many documents are in between two resources) |

## 10.5.49 Cognitive weight

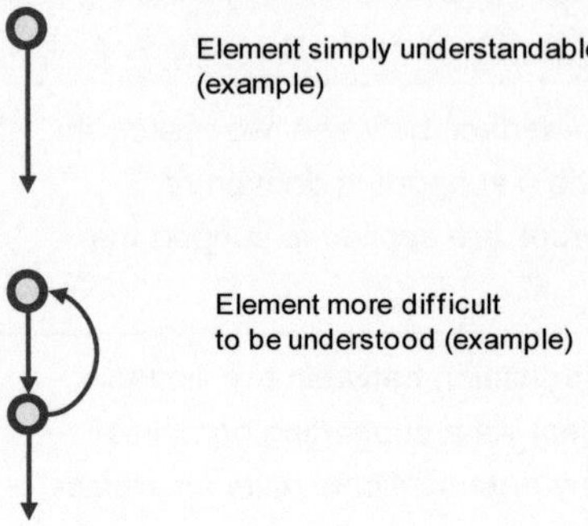


| Domains | Cognitive weight |
|---|---|
| Domain 1 | xx |
| Domain 2 | yy |
| Domain 3 | zz |
| ... | ... |

**Definition**

- Based on the attribution of empirically founded characteristic values of typical constellations of edges and nodes
- Summation of all cognitive weights

**Structural significance**

- Description of the human ability to grasp individual parts of the process as well as its global structure

**Representation**

- Metric per domain

**Literature**

[SHAO & WANG 2003], [MCQUAID 1997], [WANG 2006]

### 10.5.50 Degree of non-planarity

at least 1 edge needs to be removed to obtain a planar graph

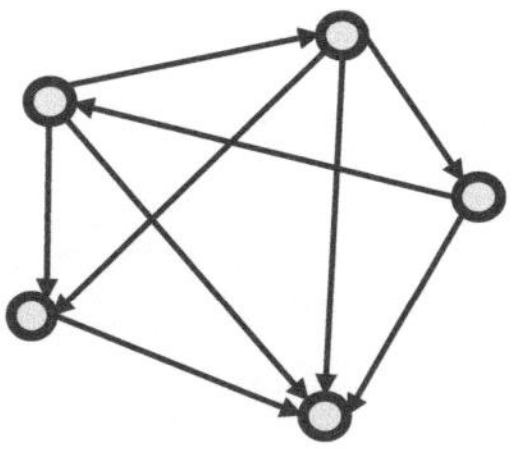

| Domain | Degree of non-planarity |
|---|---|
| Domain 1 | 1 |
| Domain 2 | ... |
| Domain 3 | |
| ... | |

**Definition**

- Minimum number of edges that have to be removed to obtain a planar graph
- Computation is non-deterministic

**Structural significance**

- Possibility to measure the clarity and transparency of the process
- Evaluation of the understandability of the process
- Determination of the ascertainability of the network model
- Description of the transparency of the process model

**Representation**

- Metric per domain

**Literature**

[KORTLER et al. 2009]

## Structural significance

| | **Cognitive weight / degree of non-planarity** |
|---|---|
| Task | • Degree of comprehensibility of the arrangement of tasks to form the overall process<br>• Level of chance to comprehend the role of a task in the context of the overall process<br>• Level of the chance to identify a possibly erroneous task in the context of the overall process |
| Artifact | • Degree of comprehensibility of the arrangement of artifacts<br>• Degree of clear arrangement of the landscape of artifacts<br>• Level of chance to comprehend the importance of an artifact for the overall process<br>• Level of the chance to identify a possibly erroneous artifact in the context of the overall process |
| Org. unit | • Degree of comprehensibility of the social network<br>• Level of chance to comprehend the importance of relevant organizational units for the overall process<br>• Level of the chance to identify core personnel |
| Time | • Degree of comprehensibility of the interaction of points in time and their impact on planning<br>• Level of the chance to locate a point in time with respect to all its dependencies<br>• Level of risk of not integrating all dependencies into the planning of the schedule suitably |
| Event | • Degree of comprehensibility of the transition of events into each other<br>• Level of risk of not integrating all transitions into the planning of the schedule |
| Resource | • Degree of comprehensibility of the cross-linking of resources among each other<br>• Degree of clear arrangement of the landscape of resources<br>• Level of the chance to identify a possibly useful resource and access it |

### 10.5.51 McCabe Cyclomatic Number

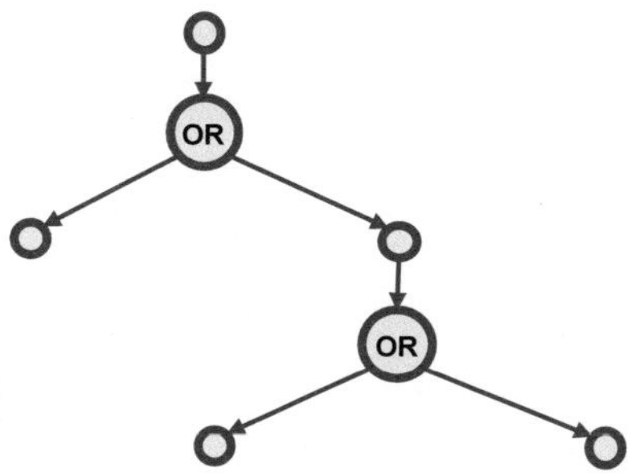


| Domains | Cyclomatic number |
|---|---|
| Domain 1 | 3 |
| Domain 2 | ... |
| Domain 3 | |
| ... | |

**Definition**

- Difference of number of edges and number of nodes (excluding logical split connectors) plus two minus
- Only applicable for processes with one initial node (i.e., root node of the process)
- Only useful with Boolean operators to represent decision points
- Adaptation for bipartite process graphs (e.g., EPC) necessary

**Structural significance**

- Number of possible paths in a control flow
- Number of binary decisions in control flow

**Representation**

- Metric per domain

**Literature**

[McCabe 1976], [Cardoso 2006]

### 10.5.52 Control-Flow Complexity

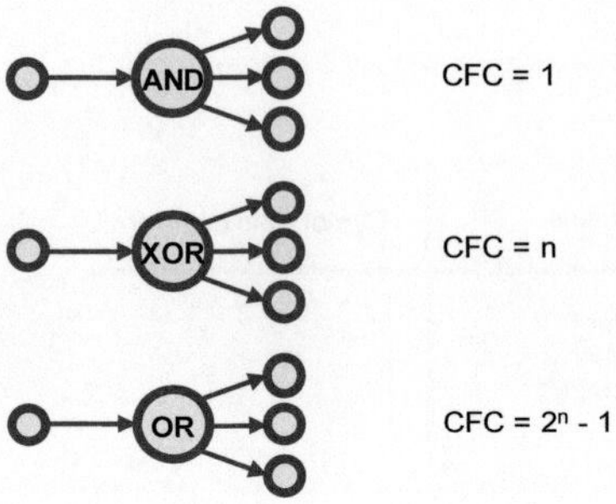


| Domains | Control-flow Complexity |
|---|---|
| Domain 1 | x |
| Domain 2 | ... |
| Domain 3 | |
| ... | |

**Definition**

- Sum of all possible constellations of outgoing edges from logic operators (splits)

**Structural significance**

- Number of all possible decisions in a process
- Impact of a single decision (similar to activity of a split)

**Representation**

- For individual nodes: Pareto-distribution of Control-flow Complexity for all nodes
- For overall process: Metric per domain

**Literature**

[CARDOSO 2005a], [GRUHN & LAUE 2006a]

## Structural significance

| | Cyclomatic number | Control-flow complexity |
|---|---|---|
| Task | • Extent of moving targets in the process<br>• Degree of flexibility built into the process<br>• Degree of plurality of scenarios of process execution | • Number of possible self-contained modules of tasks (pre-defined process flows through a part of the process) that can be regrouped independently from decision points<br>• Degree of adaptivity of the process |
| Artifact | • Extent of fluctuating artifacts within the process<br>• Degree of possible momentum of artifacts<br>• Degree of flexibility of the process documentation<br>• Degree of possible diversity of results of the process | • Number of possible self-contained groups of artifacts that can be processed independently from decision points<br>• Degree of adaptivity of the process |
| Org. unit | • (not applicable) | • (not applicable) |
| Time | • Level of risk of different scenarios of a possible time to market | • Number of possible paths through the process<br>• Degree of adaptivity of the process |
| Event | • Extent of possible overall states the process can take up<br>• Degree of insecurity processed<br>• Degree of stability of a target-oriented process flow | • Degree of flexibility of the process to (possibly unexpected) events |
| Resource | • (not applicable) | • (not applicable) |

## 10.6 Computability of metrics

| Metrics | Deterministic computation | Heuristic computation | Complexity of algorithm[84] |
|---|---|---|---|
| **Size and density** | | | |
| Number of domains | 1 | | $O(n)$ |
| Number of nodes | 1 | | $O(m)$ |
| Number of edges | 1 | | $O(m)$, max. $O(n^2)$ |
| Number of classes | 1 | | $O(n)$ |
| Number of interfaces between domains | 1 | | $O(m)$ , max. $O(n_1 \cdot n_2)$ |
| Number of edges per node | 1 | | $O(m)$, max. $O(n^2)$ |
| Relational density | 1 | | $O(m)$, max. $O(n^2)$ |
| Number of unconnected nodes | 1 | | $O(m)$, max. $O(n^2)$ |
| **Adjacency** | | | |
| Activity / Passivity | 1 | | $O(m)$, max. $O(n^2)$ |
| Degree correlation (nodes) | 1 | | $O(m)$, max. $O(n^2)$ |
| Degree correlation (edges) | 1 | | $O(m)$, max. $O(n^2)$ |
| Degree distribution | 1 | | $O(m)$, max. $O(n^2)$ |
| Fan criticality | 1 | | $O(m)$, max. $O(n^2)$ |
| Synchronization points / distribution points | 1 | | $O(m)$, max. $O(n^2)$ |
| Number of independent sets | 1 | | $O(n + c \cdot m)$, max. $O(n + 2^n \cdot m)$ |
| **Attainability** | | | |
| Number of reachable nodes | 1 | | $O(n \cdot m)$, max. $(n^3)$ |
| Reachability of a node | 1 | | $O(n \cdot m)$, max. $(n^3)$ |
| **Closeness** | | | |
| Proximity | 1 | | $O(n \cdot m)$, max. $(n^3)$ |
| Relative centrality (based on between-ness) | 1 | | $O(n^2 \cdot m)$, max. $(n^4)$ |
| **Connectivity** | | | |
| Node connectivity | 1 | | $O(m)$ (breadth-first search) |
| Edge connectivity | 1 | | $O(m)$ (breadth-first search) |
| **Paths** | | | |
| Number of paths | 1 | | $O(m)$, max. $O(n^2)$ |
| Path length | 1 | | $O(m)$, max. $O(n^2)$ |
| Weight of an edge | | | $O(m)$, max. $O(n^2)$ |
| Centrality of path (based on centrality) | 1 | | $O(n^3 \cdot m)$, max. $(n^5)$ |
| Centrality of path (based on degree) | 1 | | $O(n \cdot m)$, max. $O(n^3)$ |
| Degree of progressive oscillation | 1 | | $O(n \cdot m)$, max. $O(n^3)$ |

| Metrics | Deterministic computation | Heuristic computation | Complexity of algorithm[84] |
|---|---|---|---|
| **Hierarchies** | | | |
| Height of hierarchy | 1 | | $O(m)$ |
| Width of hierarchy | 1 | | $O(m)$ |
| Tree criticality | 1 | | $O(m)$ |
| Snowball factor | 1 | | $O(n \cdot m)$, max. $O(n^3)$ |
| Forerun factor | 1 | | $O(n \cdot m)$, max. $O(n^3)$ |
| Tree-robustness | 1 | | $O(n^2 \cdot m)$, max. $O(n^4)$ |
| Maximum nesting depth | 1 | | $O(n)$ |
| **Clustering** | | | |
| Number of cliques | 1 | | $O(2^n)$ |
| Cluster-coefficient (local) | 1 | | $O(m^2)$ |
| Cluster-coefficient (global) | 1 | | $O(m)$, max. $O(n^2)$ |
| Module quality 1 | 1 | | $O(m)$, max. $O(n^2)$ |
| Module quality 2 | 1 | | $O(m)$, max. $O(n^2)$ |
| **Cycles** | | | |
| Number of cycles | 1 | | $O(n + c \cdot m)$, max. $O(n + 2^n \cdot m)$ |
| number of cycles per node | 1 | | $O(n + c \cdot m)$, max. $O(n + 2^n \cdot m)$ |
| Number of cycles per edge | 1 | | $O(n + c \cdot m)$, max. $O(n + 2^n \cdot m)$ |
| Number of feedbacks | | 1 | $O(n + c \cdot m)$, max. $O(n + 2^n \cdot m)$ |
| Activation of cycle | 1 | | $O(n + c \cdot m)$, max. $O(n + 2^n \cdot m)$ |
| Number of starting points for iterations | 1 | | $O(n + c \cdot m)$, max. $O(n + 2^n \cdot m)$ |
| Iterative oscillation | 1 | | $O(n + c \cdot m)$, max. $O(n + 2^n \cdot m)$ |
| **Several domains** | | | |
| Bipartite density | 1 | | $O(m)$, max. $O(n_1 \cdot n_2)$ |
| Number of organizational interfaces | 1 | | $O(m)$, max. $O(n_1 \cdot n_2)$ |
| **Cognition** | | | |
| Cognitive weight | 1 | | n.a. |
| Degree of non-planarity | | 1 | $O(n! \cdot n)$ |
| **Boolean Operators** | | | |
| McCabe Cyclomatic Number | 1 | | $O(m)$, max. $O(n^2)$ |
| Control-Flow Complexity | 1 | | $O(m)$, max. $O(n^2)$ |

[84] The complexity of the algorithm refers to the time complexity of computing the algorithm; *n* is the number of nodes, *m* the number of edges; if more than one domain is involved, the domains are indexed; *c* is the number of cycles in the domain; if a maximum complexity of the algorithm can be estimated, the estimation is given. For relevant algorithms, see [LÄUCHLI 1991].

## 10.7 Classification of metrics

On the following four pages, the classification of structural metrics is listed as a table that spreads over four pages. For layout reasons, the table is split unevenly.

| Metric | ID | Scope of analysis | | | | | Particularities of the model | | | Type of entity | | Direction of impact | | |
|---|---|---|---|---|---|---|---|---|---|---|---|---|---|---|
| | | Single entity | Module | Alignment | Overall as histogram | Overall as metric | Decision points | Comprehensibility of model | Quality of modeling | Nodes | Edges | Active | Passive | Related |
| **Size and density** | | | | | | | | | | | | | | |
| Number of domains | M01 | | | | 1 | | | 1 | | | | | | |
| Number of nodes | M02 | 1 | | | 1 | | | | | 1 | | | | |
| Number of edges | M03 | 1 | | | | | | | | | 1 | | | 1 |
| Number of classes | M04 | | | | | 1 | | | 1 | 1 | | | | |
| Number of interfaces between domains | M05 | | 1 | | | | | 1 | | | 1 | | | 1 |
| Number of edges per node | M06 | | | | 1 | | | 1 | | 1 | 1 | | | 1 |
| Relational density | M07 | | | | 1 | | | | | 1 | 1 | | | 1 |
| Number of unconnected nodes | M08 | 1 | | | | | | | | 1 | 1 | | | 1 |
| **Adjacency** | | | | | | | | | | | | | | |
| Activity / Passivity | M09 | 1 | | | | | 1 | | | 1 | | 1 | 1 | |
| Degree correlation (nodes) | M10 | 1 | | | | 1 | | | | 1 | | | | |
| Degree correlation (edges) | M11 | 1 | | | | 1 | | | | | 1 | 1 | 1 | 1 |
| Degree distribution | M12 | 1 | | 1 | | 1 | | | | 1 | | 1 | 1 | |
| Fan criticality | M13 | | 1 | | | | | | | | | 1 | 1 | |
| Synchronization / distribution points | M14 | 1 | | | | | 1 | | 1 | 1 | | 1 | 1 | |
| Number of independent sets | M15 | | | | | | | | | | | | | |
| **Attainability** | | | | | | | | | | | | | | |
| Number of reachable nodes | M16 | 1 | | | | | | | | 1 | | 1 | | |
| Reachability of a node | M17 | 1 | | | | | | | | 1 | | | 1 | |
| **Closeness** | | | | | | | | | | | | | | |
| Proximity | M18 | 1 | | | | | | | | 1 | | | | 1 |
| Relative centrality | M19 | 1 | | | | | | 1 | | 1 | | | | 1 |
| **Connectivity** | | | | | | | | | | | | | | |
| Node connectivity | M20 | 1 | | | 1 | | | | | 1 | | | | |
| Edge connectivity | M21 | 1 | | | 1 | | | | | | 1 | | | 1 |
| **Paths** | | | | | | | | | | | | | | |
| Number of paths | M22 | | 1 | | | 1 | | | 1 | 1 | 1 | | | 1 |
| Path length | M23 | | 1 | | | | | | | | 1 | | | 1 |
| Weight of an edge | M24 | 1 | | | | | | 1 | | | 1 | 1 | 1 | |
| Centrality of path (centrality) | M25 | | 1 | | | 1 | | | | | 1 | 1 | 1 | |
| Centrality of path (degree) | M26 | | 1 | | | 1. | | | | 1 | 1 | 1 | 1 | |
| Degree of progressive oscillation | M27 | | 1 | | | 1 | | | | | 1 | 1 | 1 | |

| Focus of analysis | | | | | | | | | | | Type of network | | |
|---|---|---|---|---|---|---|---|---|---|---|---|---|---|
| Robustness | | Grouping | | | Extent | | | Propagation | | | | | |
| Centrality | Redundancy | Cluster | Modules | Cycles | Size | Cross-linking | Perceivability | Adjacency | Paths | Hierarchies | Intra-domain | Inter-domain | Multiple-domain |
| | | | | | | | | | | | | | |
| | | | | | 1 | | | | | | | | 1 |
| | | | | | 1 | | | | | | 1 | 1 | |
| | | | | | 1 | | | | | | 1 | 1 | |
| | | | | | 1 | | | | | | 1 | 1 | |
| | | | | | | 1 | | | | | | 1 | |
| | | | | | | 1 | | | | | 1 | 1 | |
| | | | | | | 1 | | | | | 1 | 1 | |
| 1 | | | | | | | | | | | 1 | 1 | |
| | | | | | | | | | | | | | |
| | | | | | | | | 1 | | | 1 | | |
| | | | | | | | | 1 | | | 1 | | |
| | | | | | | | | 1 | | | 1 | | |
| | | | | | | | | 1 | | | 1 | | |
| | | | | | | | | 1 | | | 1 | | |
| | | | | | | | | 1 | | | 1 | | |
| | | | | | | | | | | | 1 | | |
| | | | | | | | | | | | | | |
| | | | | | | | | | 1 | | 1 | | |
| | | | | | | | | | 1 | | 1 | | |
| | | | | | | | | | | | | | |
| | | | | | | | | | 1 | | 1 | | |
| 1 | | | | | | | | | | | 1 | | |
| | | | | | | | | | | | | | |
| | 1 | | | | | | | | | | 1 | | |
| | 1 | | | | | | | | | | 1 | | |
| | | | | | | | | | | | | | |
| | | | | | | | | | 1 | | 1 | | |
| | | | | | 1 | | | | 1 | | 1 | | |
| 1 | | | | | | | | | | | 1 | | |
| 1 | | | | | | | | | | | 1 | | |
| 1 | | | | | | | | | | | 1 | | |
| | | | | | | | | | 1 | | 1 | | |

| Metric | ID | Scope of analysis | | | | | Particularities of the model | | | Type of entity | | Direction of impact | | |
|---|---|---|---|---|---|---|---|---|---|---|---|---|---|---|
| | | Single entity | Module | Alignment | Overall as histogram | Overall as metric | Decision points | Comprehensibility of model | Quality of modeling | Nodes | Edges | Active | Passive | Related |
| **Hierarchies** | | | | | | | | | | | | | | |
| Height of hierarchy | M28 | 1 | | | | 1 | | | | | | | | |
| Width of hierarchy | M29 | 1 | | | | 1 | | | | | | | | |
| Tree criticality | M30 | | | | | | | | | | | | | |
| Snowball-factor | M31 | 1 | | | | 1 | | | | | | 1 | | |
| Forerun-factor | M32 | 1 | | | | 1 | | | | | | | 1 | |
| Tree-robustness | M33 | | | | 1 | | | 1 | | | | 1 | | |
| Maximum nesting depht | M34 | | | | | 1 | 1 | | 1 | | | | 1 | |
| **Clustering** | | | | | | | | | | | | | | |
| Number of cliques | M35 | | 1 | | | 1 | | 1 | | 1 | 1 | 1 | 1 | |
| Cluster-coefficient (local) | M36 | 1 | 1 | | | | | | | 1 | 1 | | | |
| Cluster-coefficient (global) | M37 | | 1 | | 1 | | | | | 1 | 1 | | | |
| Module quality 1 | M38 | | 1 | | | | | | | 1 | 1 | | | |
| Module quality 2 | M39 | | 1 | | | | | | | 1 | 1 | | | |
| **Cycles** | | | | | | | | | | | | | | |
| Number of cycles | M40 | | 1 | | | 1 | | | | 1 | 1 | | | |
| Number of cycles per node | M41 | 1 | | | | 1 | | | | 1 | | | | |
| Number of cycles per edge | M42 | 1 | | | | 1 | | | | | 1 | | | |
| Number of feedbacks | M43 | 1 | | | 1 | | | | | 1 | | | | |
| Activation of cycle | M44 | | | | 1 | | | | | 1 | | | | |
| Number of starting points for iterations | M45 | | | | | 1 | | | | 1 | | | | |
| Iterative oscillation | M46 | | 1 | | | 1 | | | | 1 | | | | |
| **Several domains** | | | | | | | | | | | | | | |
| Bipartite density | M47 | | | 1 | | 1 | | 1 | | | 1 | | | |
| Number of organizational interfaces | M48 | | | 1 | | | | | | | 1 | | | |
| **Cognition** | | | | | | | | | | | | | | |
| Cognitive weight | M49 | | | | 1 | | | 1 | | | | | | |
| Degree of non-planarity | M50 | | | | 1 | | | 1 | | | | | | 1 |
| **Boolean Operators** | | | | | | | | | | | | | | |
| McCabe Cyclomatic Number | M51 | | | | 1 | | 1 | | | 1 | | | | |
| Control-Flow Complexity | M52 | 1 | | | 1 | | 1 | | | | | | | |

| **Focus of analysis** | | | | | | | | | | | **Type of network** | | |
|---|---|---|---|---|---|---|---|---|---|---|---|---|---|
| Robustness | | Grouping | | | Extent | | | Propagation | | | | | |
| Centrality | Redundancy | Cluster | Modules | Cycles | Size | Cross-linking | Perceivability | Adjacency | Paths | Hierarchies | Intra-domain | Inter-domain | Multiple-domain |
| | | | | | | | | | | | | | |
| | | | | | | | | | | 1 | 1 | | |
| | | | | | | | | | | 1 | 1 | | |
| | | | | | | | | | | | 1 | | |
| | | | | | | | | | | 1 | 1 | | |
| | | | | | | | | | | 1 | 1 | | |
| | | | | | | | | | | 1 | 1 | | |
| | | | | | | | | | | 1 | 1 | | |
| | | | | | | | | | | | | | |
| | | 1 | | | | | | | | | 1 | | |
| | | 1 | | | | | | | | | 1 | | |
| | | 1 | | | | | | | | | 1 | | |
| | | | 1 | | | | | | | | 1 | | |
| | | | 1 | | | | | | | | 1 | | |
| | | | | | | | | | | | | | |
| | | | | 1 | | | | | | | 1 | | |
| | | | | 1 | | | | | | | 1 | | |
| | | | | 1 | | | | | | | 1 | | |
| | | | | 1 | | | | | | | 1 | | |
| | | | | 1 | | | | | | | 1 | | |
| | | | | 1 | | | | | | | 1 | | |
| | | | | 1 | | | | | | | 1 | | |
| | | | | | | | | | | | | | |
| | | | | | | 1 | | | | | | | 1 |
| | | | | | | 1 | | | | | | | 1 |
| | | | | | | | | | | | | | |
| | | | | | | | 1 | | | | 1 | | |
| | | | | | | | 1 | | | | 1 | | |
| | | | | | | | | | | | | | |
| | | | | | | | | | 1 | | | | 1 |
| | | | | | | | | | 1 | | | | 1 |

## 10.8 GQM-Framework for metrics

| ID | Goal / Question | Metrics |
|---|---|---|
| G01 | Planning | |
| Q01 | To what extent is it possible to incorporate risks into the process planning? | Metrics: M06, M07, M20, M21, M40<br>For domains: Artifacts, tasks, points in time |
| Q02 | How can the focus be placed on important process steps? | Metrics: M09, M12, M18, M19, M20, M30, M31, M41, M43, M44, M45<br>For domain: Tasks |
| Q03 | What are bottlenecks in the schedule? | Metrics: M10, M20, M21, M43<br>For domains: Artifacts, tasks, points in time |
| Q04 | What parts of the process are substantially impacted by iterations? What level of uncertainty is handled by the process? | Metrics: M35, M36, M37, M40, M41, M42, M43, M44, M45, M46<br>For domains: Artifacts, tasks, points in time |
| Q05 | What is the stakeholder situation? | Metrics: M01, M02, M03, M04, M05, M06, M07, M08<br>For domains: Overall Network |
| G02 | Resource consumption | |
| Q06 | Is the process laid out in a homogeneous manner? | Metrics: M06, M07, M10, M11, M19, M25, M35, M36<br>For domains: Artifacts, tasks |
| Q07 | Where is it possible to remove redundancies to reduce waste? | Metrics: M05, M15, M18, M20, M21, M22, M47, M48<br>For domains: Organizational units, resources, tasks |
| Q08 | Are the resources easily accessible? | Metrics: M06, M08, M16, M17, M18, M19, M31, M32, M33<br>For domain: Resources |
| G03 | Quality | |
| Q09 | Does the process allow for the consistent transfer of information? | Metric(s): M06, M08, M16, M17, M18, M19, M31, M32, M33, M47<br>For domain: Artifacts |
| Q10 | Is the documentation in line with the process? | Metric(s): M02, M03, M04, M05, M12, M47<br>For domains: Artifacts, tasks |
| Q11 | What is the risk of error distribution across the process? | Metric(s): M06, M14, M15, M16, M17, M18, M31, M32, M33, M47<br>For domains: Artifacts, tasks |

| ID | Goal / Question | Metrics |
|---|---|---|
| G04 | Flexibility | |
| Q12 | What buffers are available in the process to absorb delays and errors? | Metrics: M14, M27, M31, M32 for domains: Artifacts, tasks, points in time |
| Q13 | How robust is the overall process against individual failures? | Metrics: M06, M12, M20, M21, M36 For domains: Overall Network |
| G05 | Organizational decomposition | |
| Q14 | Is the organization of workgroups and teams adequate? | Metrics: M05, M13, M35, M36, M37, M47, M48 For domains: Organizational units, tasks |
| Q15 | How well is the organizational structure suited to provide efficient communication? | Metrics: M06, M07, M08, M11, M12, M16, M17, M18, M19, M22, M23, M31, M32 For domain: Organizational units |
| Q16 | What is the internal structure of an organizational unit? | Metrics: M12, M14, M18, M19, M31, M32 For domain: Organizational units |
| G06 | Interfaces | |
| Q17 | Which entities of the process need to be synchronized? | Metrics: M08, M10, M13, M14, M31, M32 for domain: Organizational units, tasks |
| Q18 | How fast is communication in the process? | Metrics: M16, M17, M18, M19, M31, M32 For domain: Organizational units, tasks |
| Q19 | What are relevant communication channels? | Metrics: M12, M22, M23, M24, M25, M26, M42, M43 For domains: Artifacts, organizational units, tasks |
| G07 | Transparency | |
| Q20 | Are the organizational units aware of their impact on the overall process? | Metrics: M06, M16, M17, M50 For domains: Organizational units, tasks |
| Q21 | How transparent is the overall process organization? | Metrics: M01, M02, M04, M49, M50 For domains: Overall Network |
| G08 | Decision making | |
| Q22 | Which decision points have a high impact on the process? | Metrics: M09, M43, M44, M51, M52 For domains: Overall Network |

## 10.9 Complete results of case study 7.2

*Table 10-3: Key outliers (upper bound) for aggregate view on tasks (via artifacts)*

| | Activity | Passivity | No of reachable nodes | Reachability of a node | Relative centrality | Snowball-factor | Forerun-factor | Number of cycles per node | Number of cycles per edge | Number of feedbacks |
|---|---|---|---|---|---|---|---|---|---|---|
| **Max. value of scale per metric** | **17** | **32** | **131** | **120** | **3233** | **52.3** | **62.7** | **205, 467** | **101, 751** | **72** |
| Support dev. of body structure (AC 13) | | | | | 806 | | | | | |
| Simulate parts (AC 19) | | | | | | | | | | 67 |
| Coordinate setup of simulation model for mounted parts (AC 26) | 17 | | | | | | | | | |
| Release parts (AC 31) | | | | 120 | | | | | | |
| Release cockpit (AC 32) | | 32 | | | | | | | | |
| Release body structure (AC 34) | | | | 118 | | | | | | |
| Set up simulation model for passenger safety (AC 38) | | | | | | | | | 41,111 | |
| Set up sim. model for parts (AC 41) | | | | | | | | | | 67 |
| Set up sim. model for crash (AC 43) | | | | | | | 62.7 | 145, 754 | 101, 751 | |
| Set up simulation model for body-in-white properties (AC 46) | | | | | | | | | 68, 668 | |
| Set up simulation model for body structure properties (AC 49) | | | | | | | | | | 72 |
| Coordinate simulation of crash (AC 65) | 16 | 30 | | | 3233 | 51.9 | 62.4 | 205, 467 | | |
| Simulate crash (AC 66) | 17 | | | | | 52.3 | | | 101, 751 | |
| Pre-dev. concept / package (AC 75) | | | 128 | | | | | | | |
| Simulate passenger safety (AC 90) | | | | | | | | | 41, 111 | |
| Coord. sim. passenger safety (AC 91) | | | | | 1190 | | | | | |
| Coordinate body-in-white sim. (AC 92) | | | | | | | | 156, 927 | | |
| Simulate body-in-white (AC 93) | | | | | | | | | 68, 668 | |
| Develop body-in-white (AC 94) | 17 | | | | | | | | | |
| Develop vehicle strategy (AC 128) | | | 131 | | | | | | | |
| Develop strategy for variants and derivates (AC 129) | | | 131 | | | | | | | |
| Simulate body structure (AC 131) | | | | | | | | | | 72 |
| Coordinate development of body structure model (AC 135) | 17 | | | | | 51.7 | | | | |
| Support modification of body structure model (AC 138) | | | | 119 | | | | | | |
| Develop inter. lining conc. (AC 154) | | 19 | | | | | 54 | | | |

*Table 10-4: Key outliers (upper bound) for aggregate view on artifacts (via tasks)*

| | Activity | Passivity | No of reachable nodes | Reachability of a node | Relative centrality | Snowball-factor | Forerun-factor | Number of cycles per node | Number of cycles per edge | Number of feedbacks |
|---|---|---|---|---|---|---|---|---|---|---|
| **Max. value of scale per metric** | **21** | **21** | **96** | **84** | **976** | **44** | **43.3** | **331, 386** | **176, 546** | **51** |
| Sim. results aero-acoustics (AR 19) | | | | | | | | | | 51 |
| Sim. results aerodynamics (AR 20) | | | | | 468 | | | | | |
| Release approval cockpit (AR 35) | | | | 84 | | | | | | |
| Release approval seating (AR 36) | | | | 84 | | | | | | |
| Release approval int. lining (AR 39) | | | | 84 | | | | | | |
| Sim. model aero-acoustics (AR 45) | | | | | | | | | | 51 |
| Simulation model crash (AR 49) | | 13 | | | | | 32,7 | | | |
| Simulation model for body-in-white properties (AR 52) | | | | | | | | | 96, 486 | |
| Sim. model body structure (AR 55) | | | | | | | | | | 49 |
| Simulation results crash (AR 66) | 21 | 14 | | | 976 | 44 | | | | |
| Data from crash tests (AR 67) | | | | | | | | | 149, 867 | |
| Specifications for safety (AR 77) | | | | | | | | | 176, 546 | |
| Sim. res. passenger safety (AR 86) | 17 | | | | | 39,8 | | | | |
| Results from body-in-white properties simulation (AR 88) | | | | | | | | 331, 386 | 149, 867 | |
| Vehicle concept (AR 89) | | | | | | | | 294, 209 | 96, 486 | |
| Vehicle concept draft (AR 93) | | | 86 | | | | | | | |
| Specifications crash (AR 95) | 15 | | | | | | | | | |
| Package as CAD model (AR 99) | | | 86 | | | | | | | |
| Simulation results for components with deficits (AR 103) | | | | | | | 33,5 | | | |
| Strategic vehicle concept (incl. variants and derivates) (AR 115) | | | 96 | | | | | | | |
| Sim. results body structure (AR 117) | | | | | | | | | | 49 |
| Technology model (AR 123) | | 21 | | | 588, 3 | | 43,3 | | | |
| Technical specifications (AR 124) | | | | | | | | 295, 849 | 176, 546 | |
| Model of interior lining (AR 130) | | | 82 | 76 | | | | | | |

*Table 10-5: Key outliers (upper bound) for aggregate view on organizational units (via tasks and artifacts)*

| | **Relative centrality** | **Snowball-factor** | **Forerun-factor** |
|---|---|---|---|
| **Max. value of scale per metric** | **26.6** | **11.5** | **12** |
| Design Department (OU 02) | 12.9 | | |
| Body-in-white Design Department (OU 06) | 13.3 | 10.5 | 12 |
| Interior Design Department OU 07) | | 10.5 | 11.5 |
| Comp. Flow Analysis Department (OU 08) | | | 11 |
| Safety Applications Department (OU 09) | 26.6 | 11.5 | |

*Table 10-6: Key outliers (upper bound) for aggregate view on IT systems (via tasks and artifacts)*

| | **Relative centrality** | **Snowball-factor** | **Forerun-factor** |
|---|---|---|---|
| **Max. value of scale per metric** | **58.7** | **21** | **24.5** |
| Ansa (RE 4) | 57.1 | | 24.5 |
| Catia (RE 6) | 58.7 | 20.5 | |
| Medina (RE 16) | | | 23.5 |
| Nastran (RE 19) | | 20.5 | |
| Pam Crash (RE 20) | | 21 | |
| Text Editor (RE 27) | 43.5 | | 23.5 |

# 11. Keyword index

MIX
Papier aus verantwortungsvollen Quellen
Paper from responsible sources
FSC® C105338

If you have any concerns about our products,
you can contact us on
**ProductSafety@springernature.com**

In case Publisher is established outside the EU,
the EU authorized representative is:
**Springer Nature Customer Service Center GmbH**
**Europaplatz 3, 69115 Heidelberg, Germany**

Printed by Libri Plureos GmbH
in Hamburg, Germany